Offshore Wind Power

Edited by
John Twidell and Gaetano Gaudiosi

ISBN 978-0906522-639

5 Wates Way, Brentwood, Essex CM15 9TB, United Kingdom

Contents

Contents continues on the next page

Preface

Offshore wind power is already a major development in electricity generation for marine countries. However, the challenge is not simple and the eventual status is unknown. This book arose from international participants at the OWEMES[1] conferences, who have contributed invited chapters under our general editorship. We have considered principally the engineering, manufacturing, servicing and climatological aspects, with a bias to application in Western Europe; these are more than sufficient for one book. We are well aware that we have not included national and international law, financing, insurance, environmental and ecological impact, and many other key aspects.

Obviously most experience to date of utility-scale wind turbines comes from land-based windfarms. Likewise, most experience of designing, installing and operating offshore structures comes from the offshore oil and gas industries; knowledge of marine wind stems from the needs of shipping. Yet such established knowledge is insufficient for the needs of offshore windfarms, for which much research and development is in progress.

Fig 1 shows how Europe is expected to continue with its dominance of offshore windfarm installations, with most manufacturing being in Europe. Such an early lead should mean that European exports and expertise also become significant.

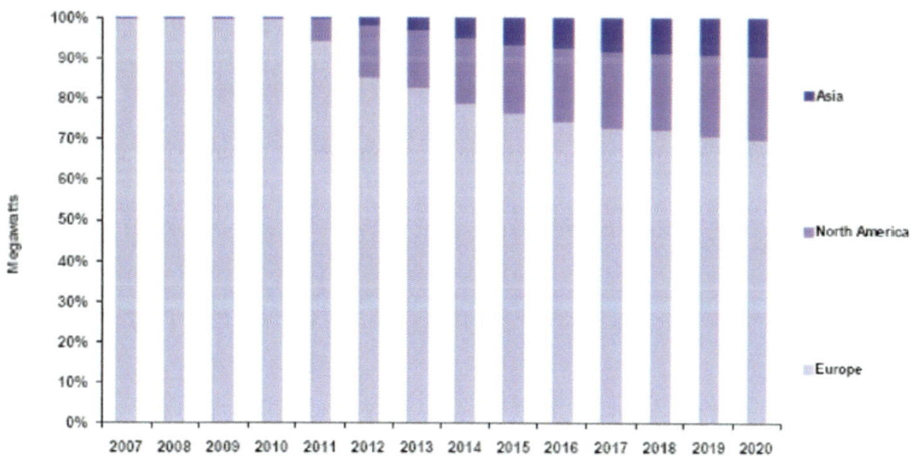

Figure 1. The predicted proportion of installed capacity of offshore windfarms by continent, 2007 to 2020.
(Taken, with permission, from 'Global Offshore Wind Energy Markets and Strategies 2008-2020', Emerging Energy Research, March 2008).

Fig 2 shows the installed offshore wind turbine capacity by country in Europe, from 2007 to that expected in 2020.

[1] Offshore Wind and other marine renewable Energy in Mediterranean andEuropean Seas, contact <gaudiosi@owemes.it>

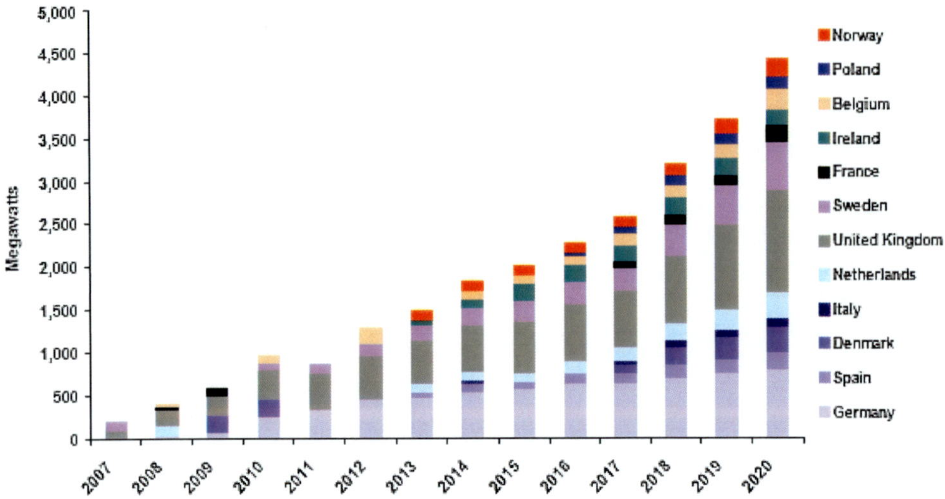

Figure 2. Annual rate of installation of offshore wind turbine capacity by European country, from 2007 to that expected in 2020. Ibid.

It is clear that offshore wind power is a major new development in human history. It is the first time that major engineering structures have been installed offshore at such replicated scale. Moreover, these are dynamic machines with automatic operation, expected to last at more that 97% availability for at least 20 years. The challenge is formidable.

The chapters of this book have been written by experts in the particular subjects. Our role has been to work with these authors for a coherent text of the book as a whole; we are grateful for their cooperation and acknowledge their dedicated effort. Brief biographies of the individual authors are given in the appendix to this Preface, so we give no further detail here. Nevertheless, we emphasize that without such an outstanding set of authors, this book could not have published. We are grateful to all of them.

John Twidell and Gaetano Gaudiosi

Authors' and Editors' Details

EDITORS

John Twidell works independently as an academic consultant and writer in renewable energy and sustainability, and as Director of the AMSET Centre Ltd. He previously held the Chair in Renewable Energy at De Montfort University and was Director of the Energy Studies Unit of Strathclyde University. He has also held appointments at Universities in Africa and the South Pacific. He has served on the Boards of the British Wind Energy Association and the UK Solar Energy Society, on committees of the Institute of Physics and as an adviser to the UK Parliamentary Select Committee on Energy. Practical experience in wind power has included management of a 3 MW (60 m) wind turbine and other smaller wind turbines on Orkney, northern Scotland. He is a Board member of Westmill Windfarm Co-operative. Presently, he is Editor Emeritus of the academic journal 'Wind Engineering', having been General Editor from 1998 to 2007. He is a visiting lecturer at the Environmental Change Institute, University of Oxford (MSc in Environmental Change, Energy Module), and the School of Aeronautics and Engineering, City University, London (MSc in Energy, Environment, Economics and Technology). His co-authored textbook with A.D. Weir, *'Renewable Energy Resources'* (second edition 2006, published by Taylor Francis) sells worldwide, mainly on postgraduate courses. He has written ~ 100 published papers on renewable and sustainable energy, and chapters in 9 other books.

Gaetano Gaudiosi graduated in Naval Engineering at Naples University in 1959, and now works independently as a consultant on offshore and onshore wind energy and other marine renewable energies. From its foundation in 2006, he has been president of the not-for-profit association OWEMES, emphasising the Mediterranean and Southern European seas. He is a Board member of ATENA (Associazione di Tecnica Navale) Rome Section and ANIV (Association Nazionale Ingegneria del Vento, 2001-2007). Previously he was an officer in the wind energy programs of ENEA (Project Manager 1980-1987 for MEDIT 300 KW and GAMMA 1.5 MW wind turbine; senior consultant and E.C. Italian project leader 1990-2001), Italian Member in the IEA Wind R& D Implement Agreement 1984-1990) and European Commission Expert for the evaluation of wind and ocean energy project proposals. From 1961 to 1980, he worked in nuclear energy, with experience in testing thermal critical phenomena in the reactor core (BWR and PWR) in Italy, USA and Sweden. From 1984, he has chaired the triennial European Seminar OWEMES in Italy (Rome 1994; La Maddalena 1997; Siracusa 2000; Napoli 2003; Civitavecchia 2006). He is author of papers on nuclear energy and wind energy, with particular attention to the offshore wind power applications presented in international conferences since 1992. He co-edited a special issue of the journal 'Wind Engineering', reviewing offshore wind energy. Teaching activity in Wind Energy has been carried out for many years in Italy and other countries, recently including the e-learning ENEA-UNESCO course on offshore wind energy.

AUTHORS

Chapter 1. Poul Erik Morthorst, Jørgen Lemming and Niels-Erik Clausen,

Poul Erik Morthorst is a Professor at Risø National Laboratory for Sustainable Energy, the Technical University of Denmark[1]. He has worked in the Department for Systems Analyses at Risø for more than

[1] http://www.risoe.dk/?sc_lang=en

30 years. His main activities cover the assessment of the economics of renewable energy technologies; especially wind power, evaluation of policy instruments for regulating energy and environment and development of long-term scenarios for energy, technology and environmental systems. He has participated in a large number of projects within these fields and has extensive experience in international collaboration. He has taken part in several national and international committees. He is member of the board of the Danish TSO, Energinet.dk, and appointed to the Danish Research Councils on Sustainable Energy and Environment and on Transport. Recently he was appointed member of the Danish Commission on Climate Change set up by the Danish Government. The main task of this commission is to analyse national and international proposals of how to reduce significantly emissions of greenhouse gases.

Jørgen Lemming is Chief Consultant at Risø National Laboratory for Sustainable Energy, the Technical University of Denmark. Dr Lemming joined the Department for Wind Energy at Risø in 2005, after many years in the Danish Energy Authority as an expert on wind power. He has extensive experience in national and international collaboration within the wind energy field and is much involved in the IEA annexes on wind energy.

Niels-Erik Clausen is Senior Adviser at Risø National Laboratory for Sustainable Energy, the Technical University of Denmark. Dr. Clausen joined Risø DTU in 2000, where he has worked with research as well as consultancy for private and institutional clients within application of wind energy. Research areas include Climate Change and renewable energy potential; design of wind turbines in areas with tropical cyclones; development of wind energy projects; environmental impact assessment of wind power projects. Presently he is responsible for a course on planning and development of windfarms at DTU's MSc degree programme in wind energy[2].

Chapter 2. John Twidell
see above.

Chapter 3 Luigi Cavaleri
Luigi Cavaleri received his Degree of Mechanical Engineer from the University of Padua, Italy, in 1965 and his Master of Aeronautics from the California Institute of Technology, Cal, USA, in 1969. Since then he worked for the Institute of Marine Science, CNR, In Venice Italy, firstly in 1969 as a researcher, then in 1992 as Director of Research, and finally in 2006 Director. Presently retired, but continuing as a Senior Scientist at the same institute. His main professional interest have always been wind waves, with connected activities in wind modelling and air-sea interaction processes. Author of almost 100 refereed papers and 3 books.

Chapter 4 Rebecca Barthelmie
Rebecca Barthelmie holds the Ewart Farvis Chair of Energy System at the University of Edinburgh and Professor of Atmospheric Science and Sustainability at Indiana University. Prior to this, she worked as a consultant and researcher in offshore wind energy at Risø National Laboratory, Denmark. She is a workpackage leader in two European Union funded projects; 'Flow' in UPwind and 'Wakes' in POWWOW. She is author/co-author of more than 80 international journal articles and more than 200 conference papers. She is an elected member of the European Union Technology Platform on Wind Energy and an editor of the journal Wind Energy.

Chapter 5 Zhe Chen and Frede Blaabjerg
Zhe Chen received the B.Eng. and M.Sc. degrees from Northeast China Institute of Electric Power

[2] http://www.dtu.dk/English/education/MSc_Programs/Wind%20Energy.aspx

Engineering, Jilin City, China, and the Ph.D. degree from University of Durham, U.K. He was a Lecturer and then a Senior Lecturer with De Montfort University, U.K. In 2002, he became a Research Professor and is now a Professor with the Institute of Energy Technology, Aalborg University, Denmark, where he is the coordinator of the Wind Power System Research program. His background areas are power systems, power electronics and electric machines; and his main current research areas are wind energy and modern power systems. He has more than 170 publications in his technical field. He is an Associate Editor (Renewable Energy) of the IEEE Transactions on Power Electronics, a Member of the Institution of Engineering and Technology (London, U.K.), and a Chartered Engineer in the U.K

Frede Blaabjerg received the M.Sc.EE. from Aalborg University, Denmark in 1987, and the PhD. degree from the Institute of Energy Technology, Aalborg University, in 1995. He was employed at ABB-Scandia, Randers, from 1987-1988. During 1988-1992 he was a PhD. student at Aalborg University. There he became an Assistant Professor in 1992, in 1996 Associate Professor and in 1998 Full Professor in power electronics and drives. He research expertise is especially in power electronics, static power converters, ac drives, switched reluctance drives, modelling, characterization of power semiconductor devices and simulation, wind turbines and green power inverter. He is the author or co-author of more than 300 publications in his research fields, including the book '*Control in Power Electronics*' (series editors. M.P. Kazmierkowski, R. Krishnan, F. Blaabjerg) 2002, Academic Press.

Chapter 6 Thomas Ackermann.

Thomas Ackermann has the degree of a *Diplom Wirtschaftsingenieur* (M.Sc. in Mechanical Engineering combined with an MBA) from the Technical University Berlin/ Germany, an M.Sc. in Physics from Dunedin University/ New Zealand and a Ph.D. from the Royal University of Technology in Stockholm/ Sweden. He is the editor of the book "Wind Power in Power Systems". In addition to wind power, his main interests are related to the concept of distributed power generation and the impact of market regulations on the development of distributed generation in deregulated markets. He has worked in the wind energy industry in Germany, Sweden, China, USA, New Zealand, Australia and India. He is CEO of 'energynautics', a research and consulting company in the area of sustainable energy supply and lecturer at the Royal University of Technology (KTH) in Stockholm, Sweden.

Chapter 7 Carlo Degli Esposti

Carlo Degli Esposti received the M.S. degree in electrical engineering in 1999 from the University of Bologna. He is presently a Project Manager and Technical & Economical Adviser for ETSO, the European Transmission Systems Organisation.

Chapter 8, Jan van der Tempel

Jan van der Tempel received his MSc in Civil Engineering, section Offshore, in 2000 at the Delft University of Technology. Worked at Royal Boskalis Westminster after graduating, focusing on installation methods for Horns Rev and sub-sea cable installation and protection. Went back to DUT for a PhD on design of support structures for offshore wind turbines, defining a frequency domain fatigue calculation method (April 2006). He designed the Ampelmann, a system for safe access to offshore structures, even in high wave conditions. Currently working as project manager and assistant professor at the DUT teaching offshore wind energy courses and CEO of the spin-out Ampelmann Company.

Chapter 9 Andrew Henderson

Andrew Henderson is an Offshore Wind Engineer at Garrad Hassan, responsible for Offshore Wind Technical Due Diligence. Previously Andrew worked for Spanish offshore wind farm developers, leading technical development in shallow and deep waters. He held the position of Assistant Professor in Offshore Wind Energy at the Technical University of Delft, where he was involved in major European and Dutch research projects including "Design Methods for Offshore Wind Turbines at Exposed Sites" and the "CA-Offshore Wind Energy" information dissemination project. As a Research Fellow at the University College of London, he undertook the research project on "The Development of Research Tools for Offshore Wind Farms Implementation" including detailed modelling of the performance of offshore wind turbines and the floating support structures.

Chapter 10, Göran Dalén

Göran Dalén received his M.Sc degree in Naval Architecture and Marine Engineering at Chalmers University of Technology in Gothenburg, Sweden in 1978. He has worked with Wind Energy for many years for the manufacturing industry as well as for developers and utilities such as Vattenfall and E.ON. He has also been involved in various EU projects including the coordination of the R&D part of the Downvind project. He has also been the Swedish head delegate in IEC TC 88 and representative in different steering committees for the Swedish Energy Board as well as for different working groups within the European commission such as the Technology Platform for Offshore Wind Energy. He is currently working for wpd Scandinavia as a project manager developing offshore projects in Germany and other European countries.

Chapter 11 Christian Nath, Axel Andraea, Kimon Argyriadis, Peter Dalhoff, Silke Schwartz,

Christian Nath received his degree as a Diplom-Ingenieur for Naval Architecture from the Technical University of Hannover in Germany in 1976. Upon his graduation, he received a research fellowship for a year at the University of California in Berkeley, where he worked as a Visiting Scholar in nonlinear Finite Element Analysis. After his return, he worked for Germanischer Lloyd in the structural analysis group. The time at GL was interrupted by a two-year research project with employment at the University of Hamburg. In the early eighties, he worked in the approval of wind energy as a structural engineer. From 1989-1992 he was project manager of a project of the Federal Research Ministry to install a test site for wind turbines in China. Since 1993 he was been responsible for wind energy at Germanischer Lloyd. He is now Global Head of Practice for Certification and Inspection within the Renewables Segment. Also from 1993, he has been chair of the German Wind Turbine Standardisation Committee and was member in several Working Groups of the IEC. He has contributed to several national and international research projects and is currently a member of the Executive Committee of the European Technology Platform TP Wind. His first offshore wind involvement was in the "Study on Offshore Wind Energy in the EG" carried out by Germanischer Lloyd and Garrad Hassan.

Axel Andreä received his degree as a Dipl.-Ing. from the University of Applied Sciences in Hamburg in 1989. He worked in the field of measurements at Windtest-Kaiser-Wilhelm-Koog GmbH for two years. After that he was engaged at aerodyn Energiesysteme GmbH and worked in load analysis, component and tower design. In 1995, he joined Germanischer Lloyd's Wind Energy Department where he stayed until 2008. The time at GL was interrupted by a one-year-leave for REpower Systems AG. At GL he held the position of the Head of the Department for Load Assumptions. He left GL in 2008 to work for support projects in developing countries in Southern America.

Kimon Argyriadis has a Naval Architect Degree (Dipl. Ing.) from Hamburg University. After a period in marine engineering consultancy, he joined Germanischer Lloyd (GL) in 1994. His field of activities covers load analysis for onshore and offshore wind turbines, and he is responsible for ocean-energy device certification. He is Expert-in-Charge for load assumptions and is involved in software and guideline development. He participates in national and international standardisation and is member of IEC Technical Committee 114 "marine energy", and Working Group 3, developing offshore wind turbine standard IEC61400-3. He has contributed to several national and international research projects and is currently a member of the European Technology Platform TP Wind.and the International Ship and Offshore Structures Congress (ISSC) specialist committee "Ocean wave and wind energy utilisation".

Peter Dalhoff is Head of the Projects Department, Business Segment Wind Energy, at Germanischer Lloyd. His involvement in wind energy began as a mechanical engineer in 1996 with GL. He performed design reviews and on-site inspections on wind turbines focusing on machinery and drive trains. As a project manager Peter Dalhoff was responsible for certification projects of different turbine manufacturers and wind farm developers. Since 2000, he has been engaged in Offshore Wind Energy and has led several offshore wind related projects, including project certifications, development and maintenance of offshore wind standards and participation in international research projects. He has been a team member in a number of technical due diligence processes as a technical advisor on behalf of banks and investors. He has organised wind related seminars and is a funding member of the Hamburg Offshore Wind conference (HOW), being responsible for its organisation.

Silke Schwartz has a degree in Naval Architecture from the technical University in Hamburg. Her engineering experience with Germanischer Lloyd had earned her a position as Expert-in-Charge for load assumptions for onshore and offshore wind turbines from January 2001, and Deputy Head of the load assumption department from 2004. She has worked with developing guidelines for load assumptions, research of ocean energy concepts and supervision of final thesis and performance of research concerning offshore loads. Since 2006, she has worked in Australia for Hydro Tasmania Consulting, a leading consultancy in the renewable energy sector.

Chapter 13 Lorenzo Battisti.

Lorenzo Battisti is currently Fluid-Machinery Associate Professor by the Mechanical and Structural Engineering Department, University of Trento, Italy and Head of the Turbomachinery Laboratory. He graduated in Mechanical Engineering in 1988 and post-graduated- at the von Karman Institute of Bruxelles- Belgium following the 40th (AGARD-NATO) Postgraduate Diploma – option Turbomachnery. The scientific activity is focussed on heat transfer in gas turbine and wind energy conversion, where a wide experimental and numerical activity on high temperature heat-transfer mechanism in gas turbine components and anti-icing de-icing systems for wind turbines has been carried out. He was member of the Technical Option Committee – Refrigeration, AC and Heat Pumps of United Nations Environmental Programme (UNEP) and co author of the 1998 Protocol, and member 2005-2007 of the IEA – XIX ANNEX "Wind Energy in Cold Climate". Former member of the ETN (European Turbine Network). Lorenzo Battisti is author of many papers and international patents on gas turbine cooling and anti-icing systems. (The kind assistance of Alessandra Brighenti is acknowledged in the final preparation of the text).

Chapter 1:

Development of Offshore Wind Power – Status and Perspectives

by Poul Erik Morthorst, Jørgen Lemming and Niels-Erik Clausen

1. INTRODUCTION

Offshore wind power capacity now only counts for a small amount of total installed wind power capacity worldwide, approximately 1%. Development has mainly taken place in North European countries, mostly around the North Sea and the Baltic Sea, where to date a few more than 20 projects have been implemented. At the end of 2007 almost 1100 MW of capacity was located offshore in five countries: Sweden, Denmark, Ireland, Netherlands and the United Kingdom (cf. Table 1). Most of the capacity has been installed on relatively shallow water dept (less than 20 m) and with a proximity to the coast of no more than 20 km, to keep the extra cost to foundations and sea cable as small as possible. In 2000, the largest installation was in Swedish waters, where the offshore windfarm Lillgrunden is installed with a rated capacity of 110 MW, so increasing the Swedish offshore capacity to 133 MW. UK experiences a constant expansion of offshore installations, Blyth Offshore of 4 MW in 2000, Barrow windfarm of 90 MW in 2006, Burbo Bank, also of 90 MW, in 2007, and Beatrice of 10 MW in 2007, with a further 629 MW under construction in 2008 (BWEA statistics).

Offshore wind is still some 50% more expensive than onshore wind, but due to expected benefits of more wind and lesser visual impact from larger turbines, several countries have very ambitious goals for offshore wind power.

2. DEVELOPMENT AND INVESTMENT COSTS OF OFFSHORE WIND POWER

The total offshore wind capacity is still limited, but growth rates are high. Offshore windfarms are installed in large units, often 100 to 200 MW total capacity, and only two units installed a year will results in future annual growth rates between 20-40%. Higher costs and temporally capacity problems in the manufacturing stages and in availability of installation vessels, causes some delays.

Table 1. Installed offshore capacity in offshore wind countries. Source BTM Consult

Country	MW installed in 2006	Accumulated MW, end 2006	MW Installed in 2007	Accumulated MW, end 2007
Denmark	0	398	0	398
Ireland	0	25	0	25
The Netherlands	108	127	0	127
Sweden	0	23	110	133
UK	90	304	90	394
Total in the world	198	877	200	1077

Offshore costs are largely dependent on weather and wave conditions, water depth, and distance to the coast. The most detailed cost information on recent offshore installations comes from the UK where 90 MW was added in both 2006 and 2007 and from Sweden with the installation of Lillgrunden in 2007. Table 2 gives information on some of the recently established offshore windfarms.

As shown in Table 2 the chosen turbine size for offshore windfarms ranges from 2 MW to 3.6 MW, the newer windfarms equipped with the larger turbines. Also the turbine farm sizes differ substantially from the fairly small Samsø windfarm of 23 MW to Robin Rigg with a rated capacity of 180 MW, which will be the worlds largest offshore windfarm. Investment costs per MW range from a low of 1.2 mill.€/MW (Middelgrunden) to almost the double of 2.7 mill.€/MW (Robin Rigg) cf. Figure 1.

Table 2. Key information on recent offshore windfarms.

Offshore windfarm	In operation	Number of turbines	Turbine size	Capacity MW	Investment cost/million €
Middelgrunden (DK)	2001	20	2	40	47
Horns Rev I (DK)	2002	80	2	160	272
Samsø (DK)	2003	10	2.3	23	30
North Hoyle (UK)	2003	30	2	60	121
Nysted (DK)	2004	72	2.3	165	248
Scroby Sands (UK)	2004	30	2	60	121
Kentich Flat (UK)	2005	30	3	90	159
Barrows (UK)	2006	30	3	90	–
Burbo Bank (UK)	2007	24	3.6	90	181
Lillgrunden (S)	2007	48	2.3	110	197
Robin Rigg (UK)	2008	60	3	180	492

The higher capital costs of offshore are due to the larger structures and complex logistics of installing the towers. The costs of offshore foundations, construction, installations, and grid connection are significantly higher than for onshore. For example, typically, offshore turbines are 20% more expensive, and towers and foundations cost more than 2.5 times the price for a project of similar size onshore.

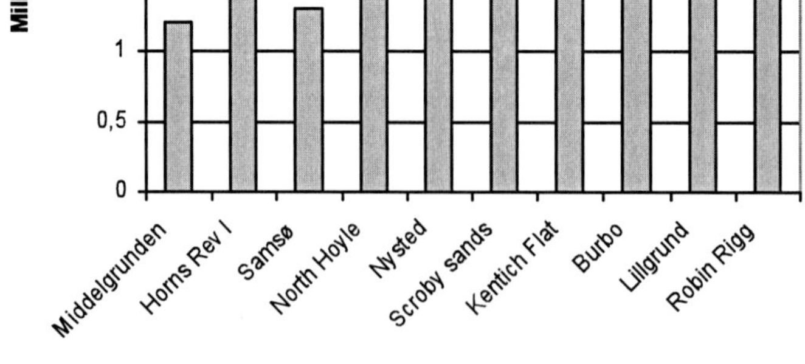

Figure 1: Investments in offshore windfarms, mill.€/MW (Current prices)

In general, the costs of offshore capacity have increased in recent years, as seen for onland turbines and these increases are only partly reflected in the costs shown above in Figure 1. Therefore, the average cost of future offshore windfarms is expected to increase. On average, investment costs for a new offshore windfarm today are expected be in the range of 2.0 to 2.2 million €/MW for a near-shore shallow depth facility

To illustrate more thoroughly the economics of offshore wind turbines, the two largest Danish offshore windfarms together with the Swedish Lillgrunden are chosen as examples. The Horns Rev project located approximately 15 km off the west coast of Jutland (west of Esbjerg) was finished in 2002. It is equipped with eighty 2 MW machines and thus have a total capacity of 160 MW. The Nysted offshore windfarm is located south of the isle of Lolland. It consists of 72 2.3 MW turbines and has a total capacity of 165 MW. Both windfarms have their own transformer station located at the sites, which through transmission cables are connected to the high voltage grid at the coast. The farms are operated from onshore control stations and no staff is required at the sites.

Lillgrunden is located south of the Øresund-bridge connecting Copenhagen and Malmö approximately 8-10 km off the Swedish coast. Lillgrunden is equipped with 48 2.3 MW turbines. The average investment costs related to the above-mentioned windfarms are shown in Table 3 split into main components.

Table 3. Average investment costs per MW related to offshore windfarms at Horns Rev, Nysted and Lillgrunden.

	Horns Rev and Nysted		Lillgrunden	
	Investments	Share	Investments	Share
	1000 €/MW	%	1000 €/MW	%
Turbines ex work, including transport and erection	872	49	1074	57
Transformer station and main cable to coast	289	16	244	13
Internal grid between turbines	91	5	–	–
Foundations	375	21	361	19
Design, project management	107	6	60	3
Environmental analysis etc.	54	3	–	–
Other Contractors	–	–	80	4
Miscellaneous	11	1	54	3
Total	1798	~100	1873	~100

Note: 2007-prices.
Horns Rev/Nysted recalculated to 2007-prices.Exchange rate 1 Ä = 7.45 DKK=9.31 SEK

In Denmark all of the above costs components have to be born by the investors except the costs of the transformer station and the main transmission cable to the coast, which is born by the TSO's in the respective areas. The total cost of each of the two Danish offshore farms is close to 260 mill. €, while the Swedish cost around 215 mill. €.

Compared to land-based turbines the main differences in the cost structure are related to two issues:

- Foundations are considerably more costly for offshore turbines. The costs depend on both the sea depth, and the chosen principle of construction[1]. For a conventional turbine sited on land, the share of the total cost for the foundation normally is approx. 5-9%. As an average of the three

[1] At Horns Rev monopiles have been used, while the turbines at Nysted are erected on concrete foundations.

above mentioned projects this percentage is 20% (cf. Table 3), and thus considerably more expensive than for on-land sites. But it should be kept in mind that considerable experiences are gained in establishing these windfarms and therefore a further optimization of foundation can be expected in future projects.

- Transformer station and sea transmission cables. Connections between the turbines and to the centrally located transformer station and from thereon to the coast generate additional costs compared with on-land sites. For Horns Rev, Nysted and Lillgrunden windfarms the average cost share for the transformer station and sea transmission cables is between 13 and 21% (cf. Table 3). For Horns Rev and Nysted a minor share of this amounting to 5% is related to the internal grid between turbines.

Finally, in relation to Horns Rev and Nysted a number of environmental analyses, including an environmental impact investigation and visualizing the windfarms, and also additional research and development were carried out. The average cost share for these analyses for the two windfarms account for approximately 6% of total costs, but part of these costs are related to the pilot character of these projects and is not expected to be repeated next time an offshore windfarm will be established.

3. THE COST OF ENERGY GENERATED BY OFFSHORE WIND POWER

Though the costs are considerable more for offshore windfarms, they are somewhat moderated by greater total electricity production from the turbines due to the stronger offshore wind speeds. For an on-land installation utilization time (equivalent full load hours) is normally around 2000-2300 hours per year, while a typical offshore installation has an utilization time of 3000 hours per year or more. The investment and production assumptions used to calculate the costs per kWh are stated in Table 4.

Table 4. Assumptions used for economic calculations.

Offshore windfarm	In operation	Capacity/MW	Cost/Million €/MW	Utilization time/Full load hours per year
Middelgrunden	2001	40	1.2	2500
Horns Rev I	2002	160	1.7	4200
Samsø	2003	23	1.3	3100
North Hoyle	2003	60	2.0	3600
Nysted	2004	165	1.5	3700
Scroby sands	2004	60	2.0	3500
Kentish Flat	2005	90	1.8	3100
Burbo	2007	90	2.0	3550
Lillgrunden	2007	110	1.8	3000
Robin Rigg	2008	180	2.7	3600

In addition the following economic assumptions are used to calculate the cost of energy produced

- Over the lifetime of the windfarm annual operation and maintenance costs are assumed to 16 €/MWh, except for Middelgrunden where O&M costs based on existing accounts are assumed to be 12 €/MWh for the entire lifetime. These assumptions on O&M-costs are subject to high uncertainty.
- The number of full load hours is assumed for a normal wind year, corrected for shadow effects in the farm and for unavailability and losses in transmission to the coast.
- The balancing of the power production from the turbines is normally the responsibility of the farm owners. According to previous Danish experiences balancing requires an equivalent cost of

approx. 3 €/MWh. In addition, balancing costs are subject to high uncertainty and might differ substantially between countries.

- The economic analyses are carried out as simple national economic ones, using a discount rate of 7.5% p.a. over the assumed lifetime of 20 years. No taxes, depreciation, risk premium etc are taken into account.

Figure 2 shows the total calculated costs per MWh for the windfarms stated in

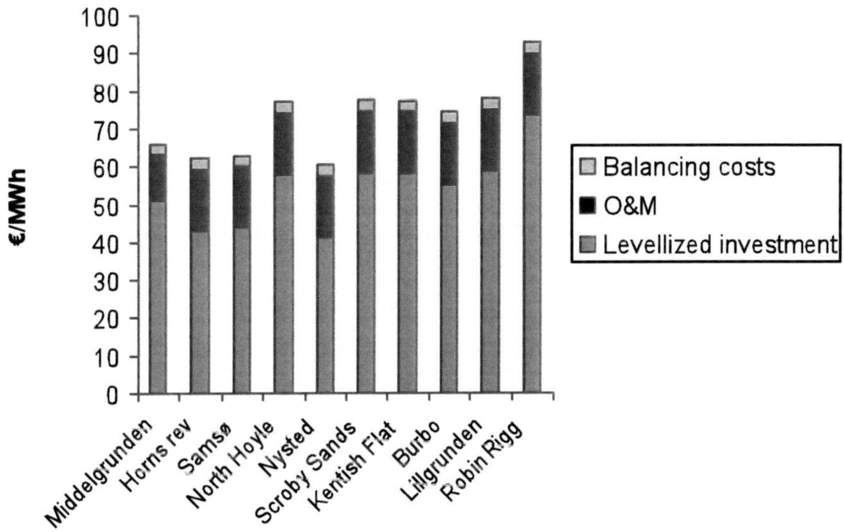

Figure 2. Calculated production cost for selected offshore windfarms, including balancing costs (2006-prices).

As shown in Figure 2 total production costs differ significantly between the illustrated windfarms, Horns Rev, Samsø and Nysted being among the cheapest, while especially Robin Rigg in UK appears to be expensive. Lillgrunden is at approximately the same level as the recent British offshore windfarms, Scroby Sands, Kentish Flat and Burbo. Partly differences can be related to depth of sea and distance to shore, partly to increased investment costs. Observe that O&M-costs are assumed to be at the same level for all windfarms (except Middelgrunden) and are subject to considerable uncertainty.

Costs are calculated as simple national economic ones, thus these costs will not be those of a private investor, which will have higher financial costs, require a risk premium and a profit. How much a private investor will add on top of the simple costs will among other things depend on the perceived technological and political risk of establishing the offshore farm and, on the competition between manufacturers and developers.

4. OFFSHORE WINDFARMS UNDER CONSTRUCTION AND IN PLANNING STAGE

At present several offshore windfarms are under construction in UK waters (Robin Rigg, Rhyl Flats, Inner Dowsing and Lynn) and in Dutch waters the second offshore farm Q7-WP consisting of 60 2 MW turbines. Much more offshore capacity is in the planning stages. In the United Kingdom, for example, London Array Limited received consent in December 2006 for the world's largest offshore windfarm to be built in the London Array. At 1,000 MW of capacity, it will be capable of powering one-quarter of the homes in London.

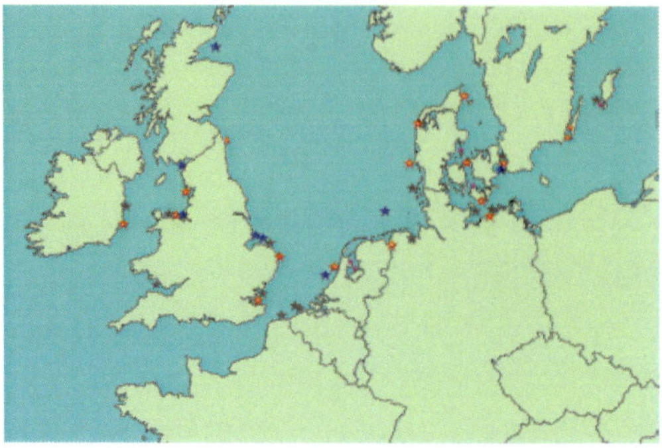

Figure 3. Map of existing and planned offshore windfarms in North-West Europe © 2002 www.offshorewindenergy.org. "Stars": red = (built MW wind turbines), purple = (built small wind turbines), blue = (under construction), grey = (planned)

In Denmark a number of new offshore windfarms are expected to be developed in the future. The Committee for Future Offshore Wind Turbine Locations published its final report in April 2007 (ref. 13). The report charts a number of possible offshore areas where offshore turbines could be built to an overall capacity of some 4,600 MW. Offshore wind turbines with a capacity of 4600 MW could generate approximately 18 TWh, or just over 8% of total energy consumption in Denmark. This corresponds to approximately 50% of Danish electricity consumption. The committee has examined in detail 23 specific possible locations each of 44 square kilometres to an overall area of 1012 square kilometres divided between 7 offshore areas.

The committee has assessed society's interests in relation to grid transmission conditions, navigation, the marine environment, the landscape, raw material exploitation etc. The committee has also assessed options for connecting major offshore windfarms to the national grid, including examining the engineering, economic and planning options for landing power and the consequences for the underlying grid of the various potential areas for construction. At the same time, the committee

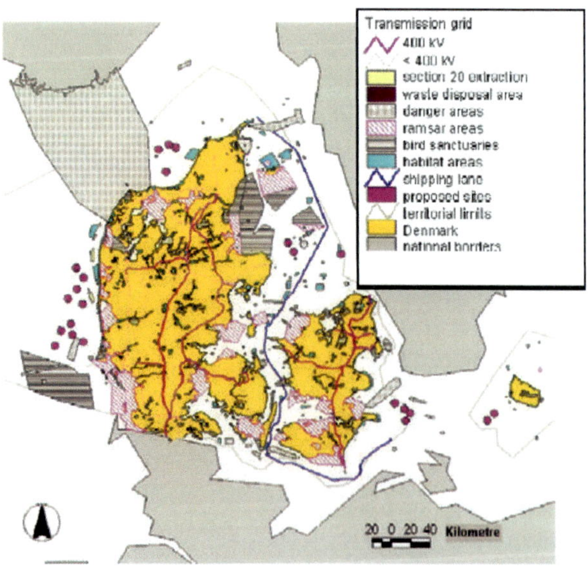

Figure 4. Proposal for siting of future Danish Offshore windfarms. Source: Danish Energy agency.

described scenarios for technological development of wind turbines capable of installation at greater sea depths. The committee attached importance to a planned and coordinated expansion of wind power and the transmission network with a view to obtaining the greatest possible economic benefits.

In 2008 the Danish Parliament decided to tender two additional offshore windfarms of 200 MW each to be established before 2012.

5. FUTURE TECHNOLOGICAL DEVELOPMENT

In the long-term perspective the offshore technology development has to be been seen in relation to areas as aerodynamics, structural dynamics, structural design, machine elements, electrical design and grid integration. The development can be structured in:

- Incremental developments
- New main component concepts
- New Wind turbine concepts

Right from the start back in the 1970s the wind turbine industry has been characterized by *incremental development*. In the future this development is especially to be seen in the following areas:

- Development of more efficient methods to determine wind resources
- Development of more efficient methods do determine the external design conditions e.g. normal and extreme wind conditions, wave conditions, ice conditions etc.
- Development of more efficient methods to design and construct the wind turbines blades, transmission and conversion system, load carrying structure, control system and grid interconnection system. Condition monitoring can trough the introduction of new and more advanced sensor systems open up for the development of important improvements of the reliability of offshore wind turbines which can be crucial for the development of more cost efficient and competitive technology
- Innovations with more efficient designs, introduction of new control elements e.g. new sensor systems, more intelligent communication between wind turbines, and introduction of new more advanced materials.
- Innovations in the wind turbine production, transportation and installation methods.

The incremental development of the technology is where the main research and development priority is in the industry and in the research community. The learning (cost reduction) in the industry comes from a combination of incremental development in design and construction of wind turbines and cost reduction due to increased production volume.

Development of *new main components concepts* has also been seen from the mid 1970s and new component concepts competes with existing concepts and thereby is continuously a challenge for the existing main component concepts. The main areas for the competition today are:

- New wind turbine blade concepts with new materials, new structural designs and new aerodynamic features
- New transmission and conversion systems e.g. wind turbines with gearboxes versus wind turbines without gearboxes with multipole generators
- New electrical generator concepts
- New power electronic concepts
- New grid integration concepts
- New foundation concepts e.g. gravitation foundations, monopole foundations, tripod foundations and floating wind turbines

The development of new main component concepts is a very dynamic part of the technology development of the wind energy field and opens often up for new innovative components. This development is extremely dependent of a very reliable verification of the performance of the new component concepts through research and experimental verification.

The competition between *new wind turbine concepts* was intense from the late 1970s to mid 1990s. The most important concepts were:

- 3-bladed upwind wind turbines with ridged (stiff) rotors, connected to the electrical grid through a gearbox and an induction generator
- 2-bladed downwind wind turbines with teetered blades, connected to the electrical grid through a gearbox and an induction generator
- 2- or 3-bladed Darrieus wind turbine (vertical axis wind turbine), connected to the electrical grid through a gearbox and an induction generator.

Also other concepts were on the market but in the end the 3-bladed upwind wind turbine until now has been the winner in the competition. But in general the technological development combined with a rapid offshore development might open op for new concepts. On the other hand is the experience with existing concepts so valuable that it is a big challenge for new concepts to compete with the existing ones.

In general the future technological development of offshore wind energy is expected to be mainly incremental and more fundamental research is very important to continue the innovation in the industry. In the future development it might prove to be important to distinguish offshore wind turbines from onshore ones. The onshore development is more mature than offshore wind technology and new innovative concepts are more likely for offshore applications.

The development offshore goes from shallow water to very deep water. The development until now is mainly seen in areas with shallow water. Technologies used offshore are expected to differ depending on the water depth and can be divided into:

- Shallow water
- Intermediate depth (50 m > depth > 20 m) bottom mounted
- Floating concepts

It should be mentioned that availability and reliability is crucial for the development of a competitive offshore wind energy technology and will in coming years be the dominating factor for the development.

6. SCENARIOS FOR THE FUTURE OFFSHORE DEVELOPMENT OF WIND POWER

In the last two decades wind power has developed rapidly. For 15 years annual growth rates in total accumulated capacity has ranged within 20 to 35%. At present the wind power market is characterized by a strong demand implying supply constraints within the turbine manufacturing industry and no signs indicate that demand will diminish within the coming years (BTM market world market updates ref. 1). Turbine manufacturers and sub suppliers are expanding their manufacturing capacity and new companies are entering the arena. In the light of a growing concern for climate change and security of energy supply – increasing prices for fossil fuels, high crude oil prices – expectations are that the wind power industry also in coming years will witness a rapid growth.

Based on existing studies on future wind power development (ref. 3 and ref. 12) Risø has evaluated future opportunities in offshore wind power and calculated a future scenario for offshore wind power deployment. It should be strictly underlined that results shown in this section mainly have character of a scenario for offshore wind power deployment and are subject to large uncertainties.

The following assumptions are made for offshore wind power development:

- Mainly based on existing plans offshore wind power development is assumed to grow by approx. 34% p.a. until 2015. Growth rates are expected to fall after 2015 to approx. 27% in the period 2015-20, to 20% in 2020-30 and, to a little more than 5% in 2030-50.
- The capacity factor will on average be about 38% for offshore turbines (3300 full load hours) for the whole period until 2050, covering that new wind turbines will have a higher production being moderated by a lower availability of sites with high wind speeds.
- The global expected final electricity consumption will follow existing forecasts (ref. 12 and ref. 3) that is approx. 2.8% annual growth until 2030, followed by an assumed slower growth of 1.5% p.a. in the period 2030-50.

On these assumptions, a future scenario for offshore wind power development is given in Table 5.

Table 5: Scenario for global offshore wind power development.

Year	Offshore wind GW	Yearly growth offshore wind /%	Offshore of total wind power/ %	Production from offshore wind/ TWh/y	Expected global electricity consumption /TWh/y	Penetration of offshore wind, %
2006	0.9		1.2	3	15500	0.0
2015	12.8	34	2.6	42	21300	0.2
2020	42.4	27	4.0	140	23800	0.6
2030	251.1	19.5	9.5	829	29750	2.8
2050	773.8	5.5	18.4	2559	40100	6.4

As shown in Table 5 total offshore wind power production is calculated to 2,559 TWh in 2050, offshore wind power supplying a little more than 6% of global final electricity consumption and constituting approx. 18.4% of total wind power capacity. The assumed growth implies that the accumulated global offshore wind power capacity will double each 2nd to 3rd year until 2015, each 3th year from 2015-20, and, finally, each 5th year from 2020-30.

7. LONG-TERM COST PERSPECTIVES FOR OFFSHORE TURBINES

Until 2004 the cost of wind turbines in general followed the development of a medium-term cost reduction curve (learning curve) showing a learning rate of approximately 10% that is each time wind power capacity was doubled the cost was reduced by approx. 10% per MW-installed. This decreasing cost-trend was interrupted in 2004-6 where the price of wind power in general increased by approx. 20%, mainly caused by increasing material costs and a strong demand for wind capacity implying bottlenecks in wind power manufacturing and sup-supplier capacity.

A similar increase in price is witnessed for offshore wind power, although a small number of realized projects in combination with a large spread in investment costs make it difficult exactly to identify the price level for offshore turbines. On average expected investment costs for a new offshore windfarm will today be in the range of 2.0 to 2.2 mill.€/MW.

In the following the long term cost development of offshore wind power will be estimated using the learning curve methodology. However, learning curves are not developed to be applied for that long a time period and for this reason the estimated figures are mainly to be seen as the results of a long term scenario development.

The long term cost perspectives for offshore wind power is shown in Table 6, given the following conditions:

- The total capacity development of wind power (including both on-land and offshore capacity) is assumed to be the main driving factor for all turbines, irrespective of onshore or offshore location, because the major part of turbine costs are related to the general wind power industry development. However, faster development of offshore capacity is expected and also a number of cost issues (foundation, transmission cables etc.) are specific for offshore, which by now are expected to have considerable cost reduction potentials. For this reason, the total wind power capacity development is used in combination with higher learning rates for offshore development than seen for onshore.

- The present rapid growth within the overall wind power industry of approx. 25% increase in total accumulated capacity will continue until 2015, including both on- and offshore capacity. Up and downs will exist, but on average the growth rate will be 23% annually. The high demand will be driven by increasing demand for energy in developing regions (as China), increasing environmental concerns and by increasing fossil fuel prices.
- As the wind industry grows more mature capacity growth rates for on- and offshore wind power will decline, to 17% on average in the period 2015-20, to approx. 10% in 2020-30 and, to 2.4% in 2030-50
- The existing manufacturing capacity constraints for the wind turbines will persist until 2010. Although we gradually will see an expanding industrial capacity for wind power, a continued increasing demand will also continue to strain the manufacturing capacity and not before 2011 increasing competition among wind turbine manufacturers and sub suppliers will again imply unit reduction costs in the industry.
- For the period 1985 to 2004 a learning rate of approx. 10% was estimated (ref. 6). With the return of competition in the wind industry again in 2011 this learning rate is again expected to be realized by the industry. Because offshore wind power is a relatively young and immature area, this learning rate is assumed to persist until 2030, when the learning rate is assumed to fall to 5% until 2050.

Given these assumptions minimum, average and maximum cost-scenarios are reported in Table 6.

Table 6. Scenarios for cost development of offshore wind turbines, constant 2006-€.

Year	Investment costs/ million €/MW			O&M	Capacity factor
	Min	Average	Max	€/MWh	%
2006	1.8	2.1	2.4	16	37.5
2015	1.55	1.81	2.06	13	37.5
2020	1.37	1.60	1.83	12	37.5
2030	1.20	1.40	1.60	12	37.5
2050	1.16	1.35	1.54	12	37.5

As shown in Table 6 average cost of offshore wind capacity is calculated to decrease from 2.1 million €/MW in 2006 to 1.35 million €/MW in 2050 or by approx. 35%. A considerable spread of costs will still exist, from 1.16 million €/MW to 1.54 million €/MW. A capacity factor of constant 37.5% (corresponding to a number of full load hours of approx. 3300) is assumed for the whole period, covering an increasing production from newer and larger turbines moderated by sites with lower wind regimes and increasing distance to shore and thus increasing losses in transmission of power.

8. NEW OFFSHORE CONCEPTS

Although the offshore market is only approximately 1% of the world market, many new technology developments are first seen offshore. There are many reasons for this. The development offshore started much later than on-shore development and is not as mature. At the same time offshore is the most challenging environment for application of wind power with a harsh environment and difficult access, which calls for autonomous designs with very high reliability. Furthermore, the relatively high cost for foundations and grid connection drives the size of the wind turbines towards larger units in order to reduce generation costs. In some countries, e.g. USA and Norway, locations outside the visibility zone are considered in order to eliminate possible conflicts with people living near the coast. The water depth outside the visibility zone (> 25 km) in these waters is significant and leads to new challenges.

As offshore oil and gas runs low the production facilities are likely to be transformed from pure fossil fuel based towards hybrid/renewable energy facilities adding wind, wave and solar devices for generation of electricity and later also fuels for the transport sector. A first sign of this development is seen in the North Sea at the Beatrice oil field off the cost of Scotland. The prototype installation consist of two 5 MW wind turbines at water depths of 42 m and the power from the wind turbines can cover approximately one third of the needs of the nearby oil production platform (Figure 5).

Figure 5. The Beatrice windfarm consisting of two 5 MW wind turbines in 42 m deep water. The windfarm is connected to an existing oil production facility. Source www.beatricewind.co.uk

In the US in 2004, the Offshore Wind Energy Consortium financed by the US Department of Energy, General Electric and the Massachusetts Technology Collaborative announced a project to consider technology for water depths from 50 ft to 100 ft (20-35 m). The same year a company, Atlantis Power LLC, launched a financing scheme for $2 million for 3 x 2 MW wind turbines to be operated in 120 m water depth. In March 2006, GE announced a $27 million partnership with the U.S. Department of Energy to develop 5 to 7 MW turbines by 2009, supplanting the company's current 3.6-megawatt turbines.

Japan has also been investigating offshore wind development (ref.8). Japan has a national target of 3000 MW by 2010 (current status is 1500 MW total on and offshore) which will be the equivalent of 0.5% of national electricity consumption. There are several areas within wind speeds above 8 or 9 m/s at 60 m height but the contour of 20 m water depth is only about 2 km from the coastline. Ryukyu University has developed the 'hexa-float' system made of concrete with 10 m sides and a 10 kW prototype is planned. Also under consideration is a stable floating platform for two turbines in a diamond shape which has been tested in a water tank as has the spar type floating structure.

In Norway two competing projects, Hywind (Norwegian Hydro, Statoil) and Sway (Statoil, Statkraft, Lyse Energi, Shell) both are developing floating offshore windfarm concepts for deep water (200-300m). Both are based on wind turbines rated at 3-5 MW or larger and the sub-sea structure is made of concrete. The main difference between the two concepts is the mooring principle. Recently Hywind has received financial support of 59 million NOK from the Norwegian Government for their prototype off the coast of Norway to be installed in 2009, while the companies behind Sway have managed to raise the necessary funds for the prototype from private investors.

Figure 6. The floating wind turbine concept Hywind. Source: Norwegian Hydro and Solberg production

9. CONCLUSIONS

Today offshore wind power only counts for approx. 1% of total installed wind power capacity and development has mainly taken place in North European countries round the North Sea and the Baltic Sea, where by now a little more than 20 projects have been implemented. At the end of 2007 almost 1100 MW of capacity was located offshore in five countries: Sweden, Denmark, Ireland, Netherlands and the United Kingdom.

A rapid growth is taking place and many new projects are in the planning process, especially in UK and Denmark. Based on existing studies on future wind power development Risø DTU has evaluated future opportunities in offshore wind power and calculated a future scenario for offshore wind power deployment. Thus, offshore wind power is calculated to increase its share of total wind capacity from approximately 1% today to a little more than 18% by 2050, supplying more than 6% of global electricity consumption. However, this development requires cost-reduction of future offshore turbines mainly achieved by incremental improvements in design and production. The Risø DTU scenario calculates a cost-reduction of approximately 35% by 2050 compared to today (in real terms), thus average cost of offshore wind capacity is calculated to decrease from 2.1 mill.€/MW in 2006 to 1.35 mill€/MW in 2050

In spite of the relatively low volume in the offshore market many new technology developments are first seen offshore. Offshore wind power is still not a mature technology and new concepts will undoubtedly emerge. However, these concepts will only be relevant and interesting if they can ensure higher reliability and/or lower costs of energy of future offshore wind plants.

REFERENCES

1. BTM-consult: *World market update 2007*, March 2008.
2. *IEA Annual report 2006*
3. Wind Force 12
4. Long Term Research and Development Needs for Wind Energy for the time 2000 – 2020. IEA Wind Energy Implementing Agreement, October 2001
5. Redlinger, R.Y.; Dannemand Andersen, P.; Morthorst, P.E. (2002): *Wind energy in the 21st century*, Palgrave 2002.
6. Neij, L. Morthorst et al (2003): *Final report of EXTOOL: Experience curves, a tool for energy policy programs assessment*. Lund, Sweden, 2003.
7. Jensen, P.H.; Brøndsted P.; Mortensen, N.G.; Øster, F.; Sørensen, J.N.(2007):Risø Energy Report 6, *Future options for energy technology* (pages 31-36; 7.1 Wind).
8. Kogaki, T. (2004) *Feasibility of Offshore Wind Development in Japan*. National Institute of Advanced Industrial Science and Technology.
9. Frandsen, S.; Morthorst, P.E.; Bonefeld, J.; Noppenau, H., *Offshore wind power: Easing a renewable technology out of adolescence*. 19. World energy congress 2004, Sydney (AU), 5-9 Sep 2004.
10. *Danish Offshore Wind – Key Environmental Issues*, DONG Energy et al, November 2006, The book can be ordered from www.ens.dk.
11. *Changes in bird habitat utilization around the Horns Rev 1 offshore windfarm, with particular emphasis on Common Scoter*, Report Commissioned by Vattenfall A/S 2007, National Environmental Research Institute, University of Aarhus, Denmark
12. *IEA World Energy Outlook 2007.*
13. *Danish Energy Agency, Future Offshore Wind Turbine Locations – 2025*, April 2007.

Chapter 2:

Fundamentals of Wind Turbines

John Twidell

1. INTRODUCTION

This chapter is written as a basic explanation of the technology, scaling and impacts of windpower. Those already acquainted with the windpower industry will probably wish to turn directly to later chapters. However, if those chapter subjects are completely new, then this chapter may help.

We begin by briefly reviewing the growth of the industry generally, and then proceed to the fundamentals of the technology generally, before concluding with aspects of offshore wind power and institutional factors.

2. WIND POWER GROWTH

After 35 years of modern development, the generation of electricity from wind is now accepted as 'mainstream' grid power generation by most utilities worldwide. This is because the technology is reliable, cost-effective, and avoids the environmental disadvantages of fossil- and nuclear-fuelled generation. *Large machines* (usually now of about 3 to 5 MW in capacity, with rotor diameters of between 80–110 metres) are used to supply power grid networks. *Small machines* (say 0.1 to 20 kW) are used for grid-connected microgeneration and for autonomous and remote power. Between these scale ranges, most machines are of older manufacture, with few made today.

Wind energy technology has become a most sophisticated and advanced multidisciplinary engineering enterprise, with designs sensitive to both environmental obligations and economic benefits. World wind generating capacity reached 100,000 MWc in early 2008, with annual global manufacturing worth €20 billion, increasing exponentially at about 33% per year The energy produced each year equals the essential and sustainable electricity requirements of about 30 million European-style houses, abating about one tonne of fossil fuel per house per year.

Wind power generation for utility supply through grid networks is predominantly from *windfarms*, defined as 'arrays of about 4 or more turbines connected as a group to export electricity to a grid network'. The economics of wind power for company investment is favoured by such a scaling, with most windfarms having around 10 to 50 turbines. Smaller numbers of turbines may be called a *cluster*, which is more likely to be owned by individuals or small cooperatives.

Each turbine on a windfarm needs to be sited about 7 or more rotor diameters from its nearest neighbour turbine, so that one does not steal the wind from the other. With very large machines, this means neighbouring turbines may be ½ km apart, and a windfarm perhaps 3 km x 3 km in ground area (this area can be used as before, e.g. for farming or for uninterrupted ecology). In practice, this means that several landowners and planning authorities may have to be involved with just one windfarm, so the development may become very complicated. It is therefore attractive to have very large turbines in *offshore windfarms* in the sea, where, in effect, the state is the only owner of the seabed. Moreover, wind is generally stronger and less turbulent over the sea than over land, so more power can be generated per unit of rotor area.

Appreciating these machines requires an understanding of their fundamental concepts, their environmental impact, and of where they are best located for cost-effectiveness. It is also important to be aware of the materials used for construction and of how turbines are sized. Since the vast majority of wind turbines have a horizontal axis with two or more blades, this is the type covered here.

3. HOW WIND TURBINES FUNCTION

3.1 Blade rotation

To understand why a wind turbine blade rotates, we first consider at an aeroplane wing (Figure 1). The air meeting the wing has relative wind speed v_r, at an angle of attack _ of about 5°. The wing profile makes the air pass (in a smooth, laminar flow) over the top of the wing faster than it passes beneath. In accordance with Bernoulli's theory, this causes a lift force on the wing. Since the wing is joined at its root to the fuselage, the aeroplane stays up. The drag force, perpendicular to the lift force and made as small as possible, is overcome by the airplane engines.

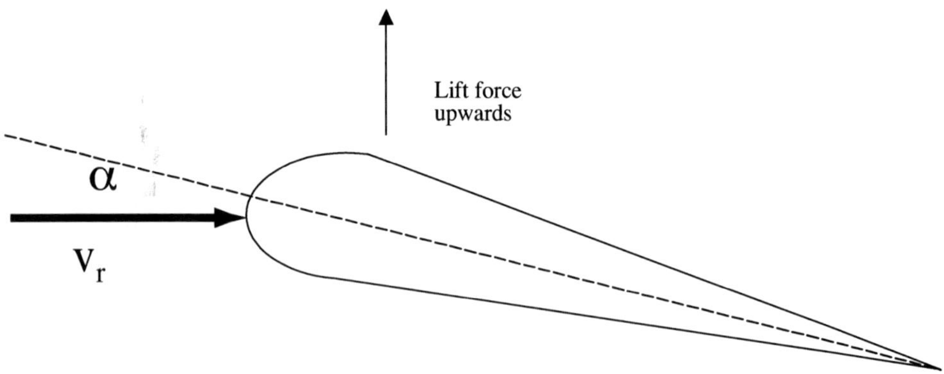

Figure 1. Airplane flying horizontally: relative air speed v_r at angle of attack α to the chord line

Now, turn the wing so it becomes a wind turbine blade. Imagine looking down onto the section of a vertical blade, as in Figure 2 (this figure also applies to all positions of the rotating blade). Because the blade is turning, the wind direction relative to the blade is similar to that relative to an aeroplane wing. The lift force has a component in the plane of the rotor that causes the rotor to turn and so transmit power to the turbine. The other component of the lift force is along the axis of the turbine and so does not affect the rotation. As with an airplane, there is also a drag force, which is made as small as possible. The component of the drag force in the rotor plane is small compared with that of the lift force, so it is not indicated on the figure.

The setting of the blade, angle γ, is such that the lift force causes the blade to move in the direction of rotation, and so power is transmitted via the root of the blade to the turbine shaft. Note that the direction of the relative wind, as experienced by the blade because of its own motion, is in a quite different direction from that of the unperturbed wind. In Figure 2, this direction is at angle ø to the plane of rotation, where ø is the flow angle, equal to the sum of the angle of attack, α, and the blade setting angle γ. The design of the blade profile allows optimum values of angle of attack and blade setting, both of which should remain constant for maximum power extraction, whatever the speed of the unperturbed wind. Conversely, to decrease and control the power, these angles may be changed by active pitching of the blades.

Because the relative wind speed increases in amplitude from the blade root to the blade tip, blade section design has improved from standard aeroplane shapes to specialist wind turbine designs. It is also necessary to limit the energy capture from the wind when the machinery has reached its rated (maximum) limits; this also affects blade design.

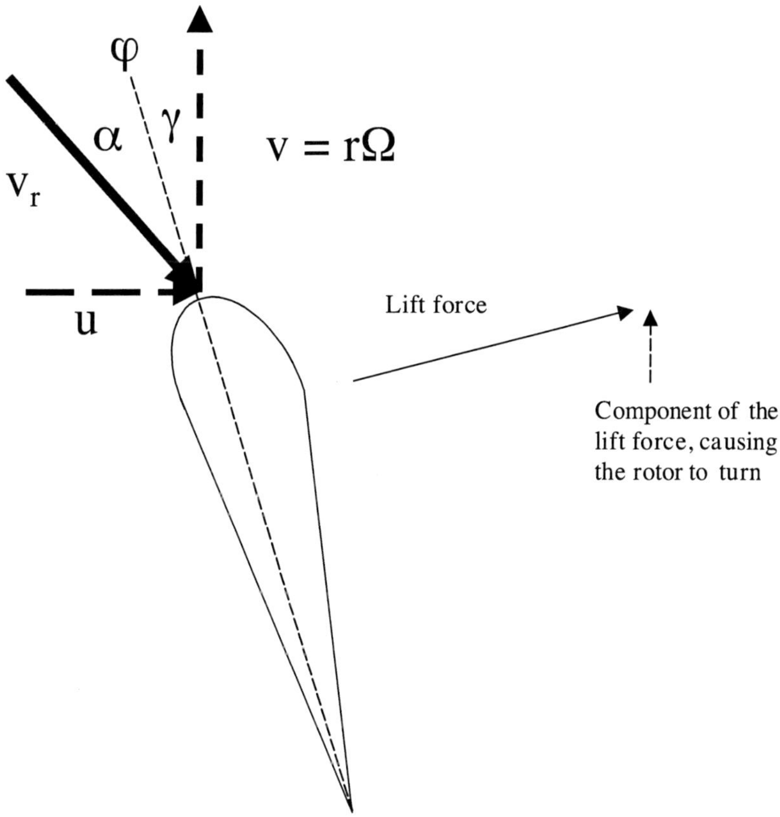

Figure 2. Looking down on a vertical wind turbine blade. The blade is turning perpendicular to the oncoming free wind, with tip speed v, but is set at angle γ. The relative air speed is v_r at angle of attack α to the chord line. The unperturbed free wind speed is u.

3.2 Speed of Rotation: Tip-Speed Ratio λ

Therefore, if you understand how aeroplanes fly, or how yachts sail into the wind, you should now understand that Bernoulli's theory of lift also explains why wind turbine blades rotate. Next, we need to know the optimum rotational speed of the blades about the hub for a particular wind speed.

A very slow rotational speed will allow wind to pass unperturbed through the gaps between the blades, and so the rotor will be inefficient. A very fast rotational speed will make the rotor appear solid to the wind, the air will become grossly turbulent, and again the wind will not turn the blades efficiently. At some value between very slow and very fast is the optimum rotational speed for transmitting power to the rotor. It is easy to determine this value for a wind of unperturbed speed u arriving at a blade of constant blade setting angle γ, with constant angle of attack α.

Variable speed and fixed speed

Figure 2 shows the relative wind impinging on the rotating blade at an angle of attack α, blade setting angle γ, and relative speed v_r. The tip of the blade moves at a speed $R\Omega$, where R is the blade rotor radius (blade length) and Ω is the angular speed of rotation (radians/second). If γ and α are constant, then their sum, φ, is constant. Thus, the cotangent of φ (= $R\Omega/u$) is also constant. This dimensionless ratio, $R\Omega/u$, is the 'tip-speed ratio', symbol λ, a most important parameter for wind turbines.

For each particular shape of wind turbine blade, whatever the size, the tip-speed ratio λ should remain constant, regardless of wind speed. Greatest efficiency is therefore achieved by allowing the rotor to change its rate of rotation as the wind speed changes, so maintaining a constant value for λ; this is called **variable speed** operation.

The latest wind turbines have generators and power electronics that allow variable speed, yet still produce alternating current (AC) electricity at 50 Hz to be produced for export to the utility grid (60 Hz in the US and other parts of the Americas). If traditional generators are used, then the rotor has to remain at, effectively, a constant rotational rate to export constant frequency (e.g. 50 Hz or 60 Hz) power; called **fixed speed** operation. The components for fixed speed generation are cheaper than for variable speed, so this method was, and still is, used, despite the loss of approximately 20% of energy production by not having variable speed with constant λ. However, variable speed wind turbines are becoming standard for modern large machines, as on offshore windfarms, especially since the blade-pitch control for variable speed also allows remote control by grid operators.

Tip-speed ratio and turbine design

Both variable and fixed speed turbines reach their maximum, rated, power in strong winds. Therefore, the blade has to be designed and adjusted to become inefficient in wind speeds greater than the rated wind speed (usually about 12 metres/second). This value, however, depends on the 'windiness' of the site. Therefore, each turbine has to be designed and tuned to the expected wind conditions of its site. The order of design procedure is as follows:

1. The maximum power of the wind turbine is decided
2. From meteorological information, the most likely wind speed, u_l, and the distribution of wind speeds, is deduced for the site
3. The length of the blades, R, is calculated to optimize annual energy production at the least cost, remembering that rated power cannot be exceeded
4. The most probable rate of rotation is calculated, so the blade tips have a speed λu, where λ usually has a value of about 7–10
5. An appropriate combination of gearbox (if needed), generator and power electronics is chosen, to produce electricity at 50 Hz or 60 Hz.

For a given wind speed, it follows that Ω is proportional to $1/R$, i.e. the larger the wind turbine, the slower it should rotate to maintain optimum efficiency. The requirement for a blade tip-speed ratio of 7–10, whatever the size of the rotor, fixes the period of rotation at $T = 2\pi/\Omega$. Therefore, in a moderate wind speed of about 7 m/s, a small turbine of rotor diameter 10 m should have a rotational period of about 0.4 seconds; however, a very large turbine of 100 m diameter should have a period of about 4 seconds. Thus, the larger the wind turbine, the more slowly it rotates.

3.3 SPEED OF ROTATION AS WIND SPEED VARIES

Fixed speed turbines

Some manufacturers of wind turbines use traditional electricity generators for AC at grid frequency (50 Hz or 60 Hz), and these generators are usually wired directly to the grid. The generators only produce electricity at the grid frequency if rotating at an exactly constant (synchronous generator) speed, or nearly constant (induction generator) 'fixed speed', of say 1500 rpm. Matching the rotor to this requires a gearbox; for instance, with a generator speed of 1500 rpm, and a rotor shaft at 30 rpm, a gearbox with a ratio of 50:1 is required. If the gearbox has only one speed ratio, then the designer has to design for one wind speed (usually, the 'most probable wind speed'); consequently the rotor always turns at the one 'fixed speed' when it is generating, whatever the wind speed. At wind speeds other than the design optimum, the capture efficiency is decreased.

Some 'fixed speed' turbines have two values of 'fixed speed', with the slower rotation rate in light winds. There are three ways to accomplish such two-speed operation, (i) to have two gearbox ratios with the same generator, (ii) to have one gearbox ratio, but two generators, and (iii) to have one gearbox ratio and a generator with switchable windings, so that fewer windings are used in light winds and all the windings in stronger winds.

For fixed blade setting angle, when the wind is at speeds other than the design wind speed(s), generation can still occur, but the efficiency of energy capture is less. At much larger wind speeds, this inefficiency becomes a benefit, as the generator is then less likely to be overloaded. Such design may be called 'stall regulated'.

Variable speed turbines

Optimum energy capture by the rotor requires having a constant tip-speed ratio whatever the speed of the wind. Therefore when the wind speed changes, the rotor speed should change accordingly. This is possible, in two main ways:

1. The wind turbine generator is entirely *'decoupled'* from the grid by the following process:
   - *generate AC, with a synchronous generator,*
   - *rectify all the power to direct current (DC)*
   - *invert (convert) the DC to standard 50 HZ or 60 Hz AC for grid connection.*

There are added costs of power-electronic equipment, and, in the past, some inverters produced unwanted electrical harmonics. However, such AC/DC/AC systems are increasingly common, especially because of the improved energy capture, improved low-speed wind efficiency, and better and cheaper power electronics.

2. *A 'doubly-fed' induction generator* (also called 'double-fed) is used (see Figure 3), as increasingly developed, especially for wind power. Such generators allow variable speed rotation by controlling and changing the currents in the rotor using power electronics. Power is extracted from the stator (as with a conventional induction generator), and via an AC/DC/AC converter, from the rotor (not conventional). Many companies now use this method, which provides an example of the many improvements in technology that wind turbines have promoted.

Elimination of the gearbox

Moving a wire through a magnetic field generates electricity. Standard commercial electricity generators have the equivalent of four or six magnetic fields from windings (pole-pairs). In all but the smallest wind turbines, a gearbox is needed so the requirement for

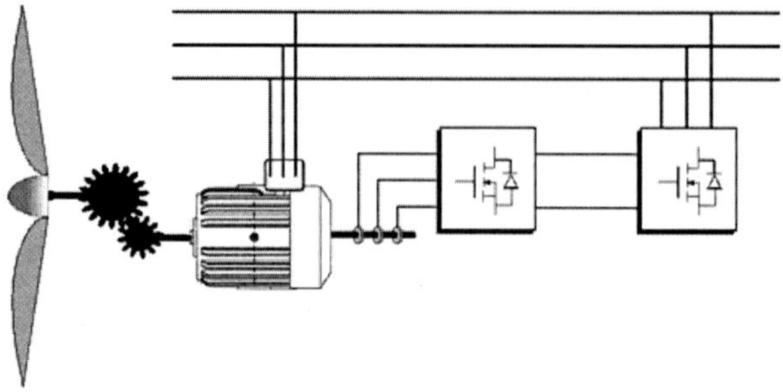

Figure 3. Doubly-fed wound-rotor induction generator with cascaded converter for slip energy recovery from the rotor. 25% to 30% of power through the converters and the remainder from the directly connected stator.

[figure credit, MultiScience Publishers, Wind Engineering vol 30 p 201-224, H. A. Sobumah & N. C. Kar]

optimum tip-speed ratio is met. However, if the generator has many more pole-pairs (a multi-pole generator from windings or actual magnets), then the gearbox can be eliminated. This has potential benefits for cost and noise reduction. The necessary 50–100 pole-pair generators have a large diameter, which become a defining feature for nacelle shape (e.g. of the German Enercon turbines). This is an example of how sophisticated electrical engineering and electronic control is reducing the amount, and hence cost, of the mechanical components.

3.4 Number of Blades

As a blade rotates, it moves into the space occupied by a previous blade. The limit to speed of rotation is that this space should not contain air strongly perturbed by that previous blade. Therefore, fast-turning rotors should have few blades. If large torque at slow speed is required, then having more blades is beneficial.

The general rule for the optimum number of blades on a rotor depends on its function; for instance:

- *electricity generation requires large speed at small torque, so the rotor has few blades*
- *water-pumping (and historic milling) requires large torque at slow speed, so this rotor has many blades.*

The minimum number of blades is one, which is possible with a dense counterweight; however, in practice the rotor motion becomes very uneven because of the vertical gradient of wind speed. The single blade passes through oncoming wind which is faster with the blade up than with the blade down. Having two blades gives smoother rotation and is relatively common, but motion is still not steady, and the visual impact can be slightly disturbing. With a three-bladed rotor, the rotation is smooth and visually the most acceptable. Moreover, the slightly slower speed than the equivalent one or two-bladed rotors means that aerodynamic noise is reduced. However, blades are expensive, so the fewer of them there are, the cheaper the turbine. Three-bladed turbines have become the norm on land and for the initial offshore windfarms that use designs developed for onshore sites. However, two-bladed machines may become more accepted for offshore wind farms, since acoustic noise is not objectionable.

3.5 Power Extracted From the Wind

Interaction of the wind with the blades transmits power to those blades. However, this wind must not stop at the rotor; otherwise, more wind cannot 'arrive'. Therefore, the departing wind takes away kinetic energy, so the extraction of power from the wind by the turbine is less than 100%. The Lanchester-Betz's linear momentum theory concludes that the maximum fractional energy extraction is 16/27, i.e. 60%, which is the benchmark criterion for designers. The dimensionless fraction is called the 'power coefficient', C_p, which is a function of the tip-speed ratio, λ. In practice, $C_{p'max}$ is less than 60%, and about 40% for a rotor with clean blades at optimum tip-speed ratio. Rotors with fewest blades have largest $C_{p'max}$.

3.6 Energy in the Wind

Let the unperturbed wind speed be u. Every second, this wind imparts energy to the rotor, of cross-sectional area A, *from* a cylinder of air of length u and cross-section A. This cylinder, of mass m, has kinetic energy $P_o = mu^2/2$. However, $m = \rho Au$, where ρ is the air density. So:

$$P_o = (\rho Au^3)/2.$$

Therefore, the power in the wind varies as the **cube** of the wind speed. This rapid and highly non-linear relationship explains much about wind turbine design, siting and performance. For example, a doubling of wind speed allows eight times more power capture. The consequences are as follows:

- *great care is taken to site machines for strong and steady winds*
- *since wind speed varies, the design (rated) wind speed is about twice the average wind speed*
- *by designing for strong winds, turbines may perhaps not turn in low-speed wind*
- *a rapid increase in wind speed can quickly bring generation to the maximum rated value of the generator, so rapid and efficient control action is required.*

3.7 Power Curve

The main practical characteristic is the change in power production with wind speed. Figure 5 is an example of power versus wind speed for a single speed machine in increasing wind. As wind speed increases,

- *generation starts at the cut-in wind speed (about 4 m/s);*
- *power increases in proportion to the wind speed cubed until the generator's rated power is reached;*
- *power then flattens, either because the blades or blade tips are turned for control (increase the angle of attack for 'active stall', decrease the angle of attack for 'pitch control') or the blade is designed to become very inefficient ('passive stall') at the rated condition;*
- *the rotor may be stopped and braked ('parked') in gales when the extreme wind turbulence would cause damage.*

3.8 Design Size

At least four factors dominate the design of blades and components:

- *the forces from the wind vary as the* ***square*** *of the wind speed (as with all moving fluids)*
- *the rotational frequency and its harmonics may cause unwanted, but predictable, vibrations*

- *unpredictable turbulence in the wind may cause persistent, rapid and often violent oscillations*
- *the rise and fall of the blades (which weigh many tonnes) during rotation means they are subject to gravity stress cycles numbering 100 times more than the equivalent fatigue cycles on aircraft wings at landing and take-off.*

Therefore, the designer has to work within two different sets of criteria, one time-dependent and the other frequency-dependent. The limiting factors on machine size relate to:

- *the enormous torque on the main shaft at the hub of a slowly rotating rotor with large cross-sectional area*
- *the gravity- and turbulence-induced fatigue stresses on long blades*
- *operational experience.*

Cost-effectiveness relates to engineering costs being, very roughly, proportional to mass. Starting from a small machine, increasing size has a linear relationship with mass as the blades become longer. However, power capture is proportional to the rotor area and hence the square of linear dimension, so there is considerable benefit in choosing a larger size. Then, as machines are increased in size beyond a 'certain range' of rotor diameters, the power-train torque dominates, and components have to be increased in volume, i.e. as linear dimension cubed. Thus machine cost, per unit of rated electrical capacity, changes from being inversely proportional to linear dimension, to being proportional to the linear dimension squared. The 'certain range' of rotor diameter has increased steadily for commercial machines as manufacturing and operational experience is gained. Fifteen years ago, this optimum had *rotor diameters* of about 30 metres (~250 kW_e-rated turbines); now, this figure is now about 120 metres ~(5 MW_e), with designs being prepared for 200 m diameters (7 MW_e).

Blades today always incorporate composite material structure, and of these, wood laminates are the toughest and lightest. Carbon fibre composites, being lightweight and tough, may be included at key locations. Usually, blade angle (pitch) is adjustable at the blade/hub fixing for rotational control. The manufacture of blades is an associated speciality, undertaken by a relatively small number of companies.

Avoiding problems

The dangers of enforced resonance in structures require the designer to avoid vibrational frequencies near to the fundamental rotor frequency and its harmonics. Designers of bridges and aircraft have similar challenges; structures are either stiff (i.e. rigid and heavy) or flexible (i.e. dynamic and lightweight). To date, the great majority of commercial wind turbine towers and structural components are 'stiff' and, hence, more heavy and expensive than otherwise. Cost-effectiveness of future wind turbines may depend more on having dynamic and compliant design than on increased size. Control of rotation is now accurate and reliable, so conditions leading to resonant vibration in blades and other structural components can be avoided, as is particularly required for variable speed turbines.

Remote monitoring and control

The use of information technology and sophisticated communications for real-time monitoring, condition monitoring and control of wind turbines will increase. For instance, this will include the use of meteorological information and subsequent action to increase the accuracy of predicting electricity sales to the grid. It is already standard for companies to monitor and, if necessary, control machines worldwide, e.g. via satellite communication. No other commercial plant is operated in this fashion to such an extent.

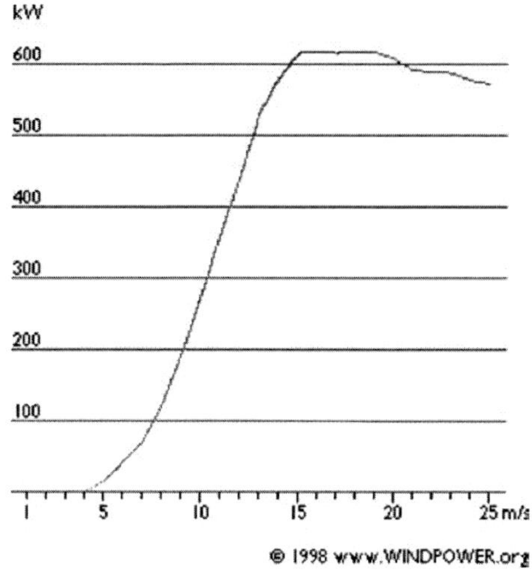

Figure 5. Power vs wind speed characteristic for a wind turbine rated at 600 kW. [credit to the Danish Wind Energy Association]

4. OFFSHORE OR ON LAND?

Section 3 has explained that the unit cost of electricity from wind turbines tends to decrease with increase in rotor blade length, i.e. large wind turbines produce cheaper electricity. This trend continues until other technical factors limit the size, e.g. strength of materials. Having once decided to use very large wind turbines, say greater than 60 m rotor diameter (~ 3 MW capacity), for a multi-turbine windfarm, the problem becomes where to locate such an array. An obvious location is the sea, especially if the sea-depth is not great.

Table 1.1 is an elementary and generalised comparison of marine-based (offshore) and land-based (onshore) wind power installations for very large wind turbines. The factors include both institutional processes and technical characteristics.

Factor	Onshore, land-based, trends.	Offshore, marine-based, trends.
Large Turbine size	Delivery by road becomes difficult.	Delivery by barge or ship is relatively easy.
Wind strength	Reduced over land, unless high sites; land and structure-induced turbulence.	Stronger and less turbulent over sea. Hence significantly increased turbine capacity factor (full-load hours).
Array size of windfarm	More than ~ 1 km 'length' leads to land ownership complexity and local planning, objections.	Single 'state ownership' of seabed and national planning allows large arrays.
Windfarm capacity	‹ 50 MW likely, so limited.	› 100 MW, so unlimited, conceptually › 1000 MW
Experience	Considerable.	Limited.
Logistics	Straightforward.	Complex and specialised.
Foundations	Straightforward, relatively cheap.	Complex and expensive.
Erection	Easy, but requires large land cranes.	Straightforward with techniques and cranes from offshore oil & gas rigs.
Electrical connection	Straightforward. Grid connection may be near.	Complex; plus need for strengthened land line to grid transmission.
Maintenance	Straightforward.	Complex and very weather dependent.

Large structures, for example, 60 metre-long wind turbine blades, are difficult to transport on land. In addition, in the UK (excepting Scotland) and elsewhere in Europe, there are potentially only a few permitted land-based (as opposed to offshore) sites for wind farms comprising such large machines. Therefore, there is increasing experience of offshore wind farms. The benefits include large-scale implementation with component delivery by ship, steadier and stronger winds than on the adjacent land, and the opportunity for the windfarm to generate at a scale of several hundred MW and be connected directly to national transmission grids. The difficulties include the extremes and dangers of the sea, and the increased cost of foundations, interconnections, installation and maintenance. Acoustic noise is not so important, so the larger tip-speed ratio of two-bladed rotors is permissible, with a consequent gain in efficiency and potential reduction in capital costs. Synergistic development with, amongst others, fisheries, fish protection and leisure facilities is important. We can expect the ship-borne delivery of such large machines to be of great benefit in coastal regions worldwide, including developing counties.

Designs for offshore wind turbines obviously arise from land-based experience. A major addition to the design process is to include the effects of wave action and wave-induced oscillations on the foundation structures and tower. The loss of generated energy due to faults is potentially much larger offshore than onshore (because of the difficulties of access for repairs, including the delays caused by bad weather), so reliability is crucial. Damage and corrosion related to the salt-water conditions is potentially much more serious than on land, requiring specialist design from established marine industries. As offshore experience increases, so the turbine designs become more specialised and different from onshore designs.

The impacts of offshore windfarms are distinct from those on land. There is less concern about visual impact and acoustic noise, but related concerns about low-flying aircraft, radar and birds. Other concerns are distinct, such as interference with seaways and commercial fishing.

5. ENVIRONMENTAL IMPACT

In many parts of the world, there is such a dearth of electricity generation that wind turbines are welcomed by the public. Where there are alternative choices, however, environmental impact is of major significance for development – though it should be noted that 'impacts' may be judged as either beneficial or harmful. The impacts, both positive and negative, of wind turbines and the factors influencing them include:

- *visual impact*
- *acoustic noise*
- *bird strike*
- *other environmental interaction at sites (e.g. access roads)*
- *electromagnetic interference, TV, microwave communication and radar*
- *very-low-flying aircraft*
- *sustainability, e.g. fossil-carbon abatement, local employment*
- *fisheries and ship-lanes for offshore windfarms*

Windfarm area and use

Turbines should be separated by at least five to ten tower heights; this allows the wind strength to re-form, and for air turbulence created by one rotor not to adversely affect another downwind. Consequently, only about 1% of land area is actually taken out of use by the towers and the access tracks of onshore windfarms. The taller and larger the turbines, the greater the separation. Megawatt-sized machines should be spaced between 0.5 km and 1 km apart. Neither buildings nor commercial forestry can be established between, so the land is thereafter safeguarded against such onshore development, and can remain for agricultural use, leisure activities or natural ecology. Nevertheless, a windfarm of, say, 50 or more multimegawatt turbines requires an area of about 3 km x 3 km which brings difficulties in many land areas, due to one or more factors, such as multiple land-ownership, agricultural variety, complex terrain, visual impact, roads and communication networks. Lack of such complexity is a great benefit for windfarms.

Visual impact

Wind turbines are always visible from places in clear line of sight; however, as noted before, the larger the machines, the greater the distance between them and the slower the rotation rate. The need for a long 'fetch' of undisturbed wind, and the economic bias towards large machines, means that even onshore turbines will be potentially visible from distances of tens of kilometres. However, at such distances on land, the majority of the public will have their sight obscured by features such as hills or trees, or by buildings. The people most likely to notice the machines on land are walkers. For the former, it is an aesthetic matter – 'beauty is in the eye of the beholder'. For offshore machines, visual impact is important, but very different from that of land-based machines.

Acoustic noise

Noise is generated and modulated mostly from blade tips (high-frequencies), from blades passing towers and perturbing the wind (low-frequencies), and from machinery, especially gearboxes. Since noise is essentially a sign of inefficiency, and because of complaints, manufacturers have reduced noise generation intensities greatly over the last five years. The critical noise intensity at the nearest building is usually considered to be 40 dBA (acoustic decibels), or less, as judged necessary for sleeping. This level of acceptance is usually attained at distances of about 250 metres or less. However, attitudes to noise are strongly psychological; the owner of a machine probably welcomes the noise as a sign of prosperity, whilst neighbours may be irritated by intrusion of 'their space'.

Bird strike

There have been many independent studies of birds killed by rotating blades. This undoubtedly happens, but perhaps at a similar or smaller frequency than strikes by a car, against the windows of a building or by grid transmission cables; every death is regretted. The counter-argument, again attested by experts, is that land around wind turbines may provide excellent breeding conditions. The exception to this argument is the possibility of strikes by large migratory birds flying in the dark, or by raptors intent on their prey.

Electromagnetic interference

TV, FM and radar waves are perturbed in line of sight by electrically conducting materials. Therefore, the metallic parts of rotating blades can produce dynamic interference in signals. It is easy, but not necessarily cheap; to install TV and FM repeater stations to provide another direction of signal for receivers. Radar interference occurs of an intermittent nature and is therefore potentially dangerous, especially for flight controllers. However, careful study has enabled 'windfarm removal' technology for flight control. Low flying aircraft always depend on good maps for avoiding obstructions. There are many sites of wind turbines close to airfields, and no significant difficulties occur.

Sustainability

There is a scientific need for sustainable and zero-emission technology. Wind turbines generate power with no chemical emissions, and therefore abate unacceptable pollution from fossil and nuclear generation. Wind power has become an essential aspect of fossil-carbon abatement for global environmental benefit. As humans, we are part of ecology, and so we should preserve and improve conditions for all species. Wind turbines are key components of such a sustainable lifestyle.

BIBLIOGRAPHY

Burton, T, Sharpe, D, Jenkins, N and Bossanyi, E., (2001) *Wind Energy Handbook*, Wiley. {This is the wind turbine 'bible', with advanced fundamental theory and professional experience of designing, manufacturing and implementing wind power}.

Gipe, P. (1995) *Wind Power*, James and James, London. {An update of other similar publications and editions, with thorough and personal analysis of wind power development, especially in the USA; a bias to the independent owner}.

Golding, E. W. (1976) *The Generation of Electricity by Wind Power*, reprinted with additional material by E. and F. N. Spon, London. {The classic text that became a guide for much modern work}.

Hansen, M O (2000), *Aerodynamics of wind turbines*, James & James, London. {Clearly presented but advanced text from an experienced lecturer; moves from fundamental aeronautics to blade element theory, with physical explanations of further intricacies and applications}

Manwell J.F., McGowan J and Rogers A.. (2nd edition 2008) *Wind Energy Explained,* Wiley. {Excellent textbook at engineering degree standard. Covers all technical aspects from first principles to early postgraduate level. Touches environmental and economic aspects. Much recommended}.

Pasqualetti M.J., Gipe P. and Righter R.W. (2002) *Wind Power in View* Academic Press with Reed Elsevier, San Diego; Academic Press, London. { An edited set of chapters, mostly by experts other than engineers, concerning the visual and other non-engineering impacts of wind power installations. Important insights into personal aesthetics and cultural heritage}.

Twidell J. W. and Weir A.D., *Renewable Energy Resources,* (2nd ed. 2006) Taylor & Francis {Basics of all renewables technology, explanation and analysis}.

Chapter 3:

Wind and waves in the Mediterranean Sea

Luigi Cavaleri and Corrado F. Ratto

Wind, wave and current conditions are the main inputs for the generation of electricity from offshore windfarms. At each potential site, these factors have to be characterised and assessed. As a major example, in this chapter we characterise the wind and wave conditions in the Mediterranean Sea. Similar analysis would apply to other seas and oceans. After a short description of the meteo-oceanographic characteristics of the Mediterranean Sea, we discuss the various sources of data, in particular distinguishing two main data sources: measurements and modelling. Then, we describe how data can be combined to obtain reliable statistics. Finally, we report the main results of interest. However, we note that more comprehensive statistics than can be reported in this book can be found in the 'Wind and Wave Atlas of the Mediterranean Sea' (The MedatlasGroup, 2004) and in the papers cited as references.

1. THE METEO-OCEANOGRAPHIC CHARACTERISTICS OF THE MEDITERRANEAN SEA

The Mediterranean Sea is one of the most extensive enclosed basins. Open only to the Atlantic Ocean through the Gibraltar strait, it extends from 6° West to 36° East, and from 30° to 46° North. It spans more than 3,000 km in longitude and about 1,600 km in latitude. This remarkable extent is fractured into a number of sub-basins because of the complicated orography that characterises the Mediterranean Sea. Although the southern, often flat, African coast is relatively uniform with the only indentation being the Sirte gulfs, its northern side is bordered by high mountain ranges that extend from Spain to Turkey, and combine with the Italian and Greek peninsulas to complicate the meteorology of the area. This leads to substantially different meteo-oceanographic climates in the various sub-basins. Since a detailed knowledge of their characteristics is required, a detailed study is the only solution.

Nevertheless, some general characteristics can be identified. The Mediterranean divides the northerly cold climate of Europe from the hot, tropical climate of Africa. At the border between these two opposing regimes, quite different climates are experienced in Summer and Winter. Summer is usually hot and dry, with a limited number of storms. In contrast, Winter can be cold, rainy, and stormy. The intermediate seasons, Spring and Autumn, are characterised by changeable conditions.

A dominant feature, of especially the western part, is the Azores anticyclone. Its position affects the weather conditions substantially. If withdrawn to the Atlantic Ocean, the storms moving along its northern border enter the Mediterranean Sea through lower France and Spain. This often leads to a cyclogenesis in the western part of the basin, with dominant winds from the southern quadrants. For instance, these are the conditions favourable for South-East winds in the Adriatic Sea and flooding in Venice. When the anticyclone extends over Europe, the storms enter the Mediterranean Sea from its northern border, leading to northerly winds and, as a particular case, to the phenomenon of bora, the cold, gusty, violent wind that blows from north-east on the Northern Adriatic Sea. There are indications (Gimeno et al., 2003) that, in very general terms, the dominant position of the anticyclone follows an 11-year cycle that stimulates a similar cycle of weather in the various sub-basins.

In general, the Mediterranean Sea does not experience the extensive and extreme storms found, for example, in the Norwegian and the North Seas. However, in some specific areas, violent winds and, depending on the marine space, large wave heights are a substantial possibility. The typical example is the Gulf of Lion and its protrusion to the coasts of Sardinia. Here the largest measured wave heights in the Mediterranean Sea have been reported, with significant wave height close to 10 metres (De Boni et al., 1993). Also the bora can reach very high-speeds, reportedly above 100 km/h (30 m/s). The coast of Egypt is a third area where large waves may occur, because of the winds blowing down from the Aegean Sea.

2. THE SOURCES OF WIND AND WAVE DATA IN THE MEDITERRANEAN SEA

There are basically two sources for marine wind and wave data:

1. *observations*, either on site or from remote sensing, of which:
   - visual observations from ships of opportunity;
   - data measured from buoys and platforms;
   - data derived from remote sensing, both measuring satellites and radar;

2. *numerical models* operational at meteo-oceanographic centres or created and extensively used for some specific project, of which:
   - general circulation models (GCM);
   - limited area model (LAM);
   - diagnostic models;
   - wave models (WAM).

All these data have different characteristics, accuracy and reliability. However, none satisfies all the requirements for a reliable atlas. The logical approach is to combine them in an optimal way for maximum information. This requires a critical analysis of their basic characteristics.

2.1 Observations

Measurements of atmosphere and ocean parameters are of the greatest importance in meteorology and oceanography, as advanced d by the invention of suitable instruments for observation and by the organization of networks of observation stations to gather atmospheric and oceanic data. In particular, networks of stations are very important for large scale studies because they provide spatial correlation among different physical variables; examples are (i) the interaction of wind and sea waves, and (ii) the direct comparison between numerically simulated data and observations. A brief description of the most important and reliable measuring techniques for wind and waves follows.

Visual observations

Taken from ships as opportunities occur, visual observations have historically been the only sources for wind and wave data in the open sea. Many decades of data exist, and they have been used to produce a whole range of products and atlases. However, when compared with measured data, the data derived from visual observations are clearly limited. Their accuracy is debatable, particularly in stormy conditions when the data are most important. Moreover, visually recorded data are not uniformly distributed geographically, but are mostly concentrated along specific routes and ships tend to avoid stormy areas, so biasing the derived statistics towards low values.

Therefore, given the large availability of alternative data, which, (see later sections) are accurate and complete in space and time, it is best not to use visual data.

Buoy and platform data

In general, this is the most reliable source of data. In particular, for many years buoys have been the standard method for the collection of wave data in the open sea. The data are measured from a floating buoy, moored at a given location. The local depth can reach a few hundreds metres, with the exclusion of very shallow water (a few metres). The most common supplier is Datawell (Waverider, Wavec, Directional Waverider), although other companies are in the market.

Buoys are by definition surface followers. However, different principles can be used and exploited. The physical quantities measured are the surface elevation and the two orthogonal components of the surface slope, or alternatively the three-dimensional motion of the buoy. Proper mathematical handling of these quantities gives the elevation frequency spectrum and the dependence on frequency of mean wave direction, skewness and kurtosis ('peakedness'). These latter two statistical quantities are defined respectively as μ_3/σ^3 and $(\mu_4/\sigma^4\text{-}3)$, where μ_i and σ^2 are the i-th moment and the variance of the elevation distribution. Further integration leads to the integral parameters: significant wave height $H_{s,}$ mean direction q_m, peak and mean period T_p and T_m.

Platforms fixed to the seabed are used for instruments and larger equipment, for instance radar installations associated with buoys moored close by are the most common arrangements. However, instrumented platforms are not common in the Mediterranean Sea, as they are, for instance, in the Norwegian and North Seas. Therefore, for locally measured data for wind and waves, buoys are by far the most common sources. However, a major limitation, at least for the buoys used in the Mediterranean Sea, is the lack of wind data.

A substantial number of these buoys are distributed along the coasts of the Mediterranean Sea. Yet it is obvious that the resulting data are far from sufficient for completely characterising the climate of the Mediterranean Sea. The main reason is that, as we have already pointed out, the Mediterranean has very strong spatial gradients of wave height, period and direction, because of its complicated geometry and its subdivision into different sub-basins. Besides, the buoys are mostly located very close to the coasts, so being unrepresentative of the conditions in the open sea. Nevertheless the buoys are the most accurate source of information now available for wave data. The error is estimated in the order of a few percent. The error increases substantially for the highest wave heights, because of the tendency of the buoys to slip around the highest crests, in so doing underestimating the overall wave height.

Satellite data

Satellites are by far the most abundant source of marine wind and wave data. Two basic kinds of instruments are used: altimeters and scatterometers.

Altimeters measure wind speed and wave height along the nadir ground track of the satellite. Typically, data are provided at 7 km intervals. The data are sufficiently accurate , particularly after correcting them for any systematic error (bias), with such corrections being derived from devoted sea campaigns and opportunistic local measurements (buoys and platforms). However, some uncertainty exists in all inner seas, as the Mediterranean Sea, because the calibration campaigns have been carried out mainly in oceans characterised mainly by swell and stronger storms. Therefore the calibrations are not necessarily correct in an environment that is dominated by wind sea, and this introduces an element of uncertainty, particularly in the derived wind speed. Further limitations come from the orbit followed by the satellites, whose passes on the ground are rather far from each other. This leaves large areas completely uncovered, or alternatively with too few data to derive reliable statistics. Nevertheless, altimeters have been flying continuously since 1991, so that a large amount of data are presently available.

The situation is different for the wind data derived from scatterometers. This instrument measures wind speed and direction on the sea surface on a wide swath, typically 500 km, on one or both sides of its ground track. The data are provided at 25 or 50 km intervals, depending on the instrument, and are representative of the average conditions in the corresponding area. As with altimeters, some uncertainty exists concerning the accuracy of the measurements in the inner seas because of the different conditions with respect to the oceans where the instruments have been calibrated. Besides, the scatterometer data are extremely sensitive to the environmental conditions, e.g. rain. This limits their availability, particularly in stormy areas.

2.2 Model simulations

In general, model data are exactly what they are: model data, and therefore only an approximate representation of the truth. As such, they have intrinsic errors, which must be taken into account when using the data. The errors are associated with the equations used in the models and their methods of resolution, but also with the data assimilated within these equations. Data assimilation, i.e. the optimal combination of the previous model forecast and the data collected from all the possible sources and made readily available to obtain the best possible estimate of the situation at a given time, is a very delicate process. Therefore, if errors are present in the data, these are reflected in the derived analysis. In particular, since surface wind is a variable of primary importance in forcing marine waves, circulations and surface fluxes, any errors in modelling the wind induce further inaccuracies in the modelled waves, circulations, and surface fluxes. We will discuss these points in detail.

There are basically three kinds of available meteorological models suitable for wind simulation over the sea: general circulation models, limited area models and diagnostic models. Moreover, there are a specific class of wave model for the simulation of gravity waves over the sea. In the following we will briefly describe all of these categories.

General Circulation Models

For the present analysis, we have made use of the daily results produced at the European Centre for Medium-Range Weather Forecasts (ECMWF, Reading, U.K.). ECMWF runs operationally a spectral model, i.e. the horizontal fields are described by a two-dimensional expansion of spherical harmonics truncated at, for example, '511' (T511). The truncation identifies the resolution, here defined as half the smallest resolved wave length, used to describe the fields. For T511 it is 40,000/(2 x 511) ≅ 39 km. Advection is calculated with a semi-Lagrangian scheme, while the physics is carried out on a reduced Gaussian grid in physical space. The vertical structure of the atmosphere is described by a multi-level hybrid s coordinate system. The physical parameterisations describe the basic physical processes connected to radiative transfer, turbulent mixing, subgrid-scale orographic drag, moist convection, clouds and surface soil processes. The prognostic variables include wind components, temperature, specific humidity, liquid/ice water content and cloud fraction. Parameterisation schemes are necessary in order to describe properly the impact of subgrid-scale mechanisms on the large-scale flow of the atmosphere. Forecast weather parameters, such as the two-metre temperature, precipitation and cloud cover, are computed by the physical parameterisation of the model. The ten-metre wind is derived with a boundary layer model from the lowest s level, 0.997, corresponding to about 30 metre height. A compact description of the model can be found in Simmons (1991) and Simmons et al. (1995).

The horizontal resolution and the number of levels with which the atmosphere is described in the model, has varied with time. T213 (95 km resolution) and 31 levels were used from 1991 until 1998, when ECMWF passed to T319. This change had a limited effect, because the Gaussian grid, used to model the processes in physical space, was not changed. The big step came in November 2000, when the Centre passed to T511 (39 km resolution), with 60 levels on the vertical and a 40 km resolution Gaussian grid. Cavaleri and Bertotti (2003a) have clearly shown how a different resolution implies a different quality of the results. In the Mediterranean Sea this corresponded to an appreciable increase of the derived wind speeds, a fact to be considered in the evaluation of the calibrated data and of the final statistics.

Concerning the results directly produced by the model, there are already a number of obvious limitations. The model outputs above the sea surface are not reliable at very small values of wind speeds because the situation is dominated by sub-grid processes. This is particularly true close to land, an obvious example being the land-sea breezes. In the coastal areas, the model winds are unreliable in all conditions because of the dominant influence of orography, not properly represented in the meteorological model because of limited resolution (95 km for T213, 39 km for T511). Also the limited accuracy with which the actual coastline is represented in the model introduces errors in the coastal wind fields.

In coastal areas the winds blowing from land to sea are strongly underestimated, much more than the ones blowing parallel to or towards the coast. This problem is not yet completely understood, but dominant roles are likely to be played by the orography and by the modelling of the marine boundary layer (see Cavaleri and Bertotti, 1997; Cavaleri et al., 1997; Cavaleri and Bertotti, 2003b).

There is a tendency to underestimate the peak wind speeds more than the average and small values. This is probably connected to the resolution of the model, but the proper parameterisation of the physical processes active in heavy storm conditions is likely to play a role. Indeed studies concerning other areas of the world suggest that forecast-analysis fields may be affected by relevant errors at the smallest grid scales (Wilke et al., 1999; Tournadre, 1999).

Chèruy et al. (2004) show that the sub-synoptic scale structure of wind can be very important in both forecasting sea-state and determining climatological balances. They have examined the 10-m wind above the region covering latitudes 25°N to 65°N and longitudes 105°W to 40°E produced by three different models The considered databases correspond to the NCEP-NCAR global reanalysis, with a grid spacing of 2.5 deg, the general circulation model of the CNRS – Laboratoire de Meteorologie Dynamique zoomed (LMDZ), at a spatial resolution of approximately 50 km near the centre of the zoom, and the Limited Area Model BOLAM (BOlogna Limited Area Model), with a grid spacing of 0.1 deg. They concluded that surface wind at smallest resolved horizontal scales of considered numerical models (about 900 km, 250 km, and 40 km, respectively), is affected by systematic power deficits associated with the smoothing action of operators introduced in the numerical codes for numerical stabilization purposes. This power deficit is due to greatly affect the sea-state forecast operations in the Mediterranean area, where local, intermittent, strong winds (Mistral., Libeccio, etc.) have a strong influence on the seasonal average.

Limited area models

The European Centre for Medium-Range Weather Forecast predicts the behaviour of the atmosphere in the medium-range up to ten days ahead, providing numerical weather prediction maps on a synoptic scale every 12 hours. National meteorological institutions usually receive these products through a dedicated telecommunications network and can further process the ECMWF output, providing a higher horizontal resolution in a short range prediction. In this case, a limited area model (LAM) may be used, which takes into account more detailed information, mostly local sampled data and a higher resolution orography. The output of a LAM is "limited", both in time and space, being valid in the range from several hours to two or three days and within a limited spatial area, but can offer a forecast map having grid points separated by 20 km or less. The former resolution is generally used for an operational LAM running on a rough topography in the Mediterranean area, while much higher-resolutions, to a few kilometres, are required for local regions, such as restricted basins.

Since the horizontal domain of a limited area model is restricted, numerical errors may appear due to possibly incorrect lateral boundaries as derived from the father model. The boundary values of limited area models are interpolated from the parallel computations of a larger area model, i.e. a GCM or a lower-resolution LAM. In the area near the boundaries, a weighted mean of the two model solutions is sometimes used to smooth out any inconsistencies between model fields.

Moreover, features such as fronts could be as narrow as 1 grid length; fully 3-dimensional features such as frontal-waves and polar-lows would need to be at least 3 grid lengths in extent before they can be resolved satisfactorily, even at the analysis stage. This effectively limits the accuracy to which these features may be forecast. Verifying the position and nature of these features, especially over areas with rather sparse data, can be done using satellite imagery.

Diagnostic models

Diagnostic models follow two main modelling approaches: (i) the steady-state solution of the linearised Navier-Stokes equations (Jackson and Hunt, 1975), and (ii) the reconstruction of three-dimensional wind fields from available meteorological data and coherent with some physical constraints, like mass-conservation (Ratto et al., 1994). None of these wind models forecast forward in time, so they do not solve explicitly the primitive momentum and energy equations, and do not provide any direct simulation of turbulence. Nevertheless, linearised wind models, e.g. WASP (Mortensen et al., 1993), and mass-consistent models, e.g. WINDS (Ratto et al., 1990), represent a valid complement to prognostic models. The mass-consistent code WINDS (*Wind–field Interpolation by Non–Divergent Schemes*) has been developed at the Department of Physics of the University of Genoa to perform three-dimensional wind flow simulations over complex terrain. For a recent review of both kinds of model, see Homicz (2002).

In particular, the class of mass-consistent diagnostic models only solves numerically the three-dimensional continuity equation and generates an Eulerian wind field by satisfying the mass-conservation constraint in every node of the computational domain. Consequently, they are particularly economical to operate and can easily deal with different initial and boundary conditions to simulate a multitude of different meteorological situations. Finally, they can be nested into numerical weather prediction models through downscaling procedures in order to increase the resolution and the detail of the simulation domain.

Wave Models

Concerning the wave data, since July 1992 ECMWF runs a wave model in parallel with the meteorological model. Similarly to the weather forecast, the aim is to produce a forecast of the wave conditions. The wave model used at ECMWF is WAM, an advanced third generation model developed with the co-operative effort of most of the experts available at the time. The two master references are WAM-DI (1988) and Komen et al. (1994). Only a brief description, sufficient for the purposes of the report, is given here. The interested reader is referred to the above references for a full description and explanation.

The model is based on the spectral description of the sea surface, i.e. at all the points of the grid covering the area of interest the wave conditions are represented as the superposition of a finite, but large, number of sinusoidal components, each characterised by frequency f (Hz), direction θ_m (flow, degrees clock-wise with respect to geographic north), and height h, hence energy F is proportional to h^2. The evolution in time and space of the whole field is governed by the so-called energy balance equation

$$\frac{\partial F}{\partial t} + c_g \cdot \nabla F = S_{in} + S_{nl} + S_{dis} \tag{1}$$

where the left-hand terms represent the time derivative and the kinematics of the field, and the right-hand terms the physical processes at work for its evolution. More specifically:

- $\partial/\partial t$ is the derivative with respect to time,
- c_g is the group speed,
- ∇F represents the spatial gradient of energy in the field,
- S_{in} is the power input by wind,
- S_{nl} are the nonlinear wave-wave power interactions,
- S_{dis} is dissipated power, e.g. by white-capping in deep water conditions.

Given the area covered by the model, the input information is provided by the driving wind fields, i.e. at each point by the modulus and direction of U_{10} (wind speed). More specifically, U_{10} is used, together with the wave conditions existing at that time at that point, to evaluate the friction velocity, hence the surface stress that expresses the actual transfer of momentum from wind to waves. The term S_{dis} summarises the rate of energy loss by the various dissipation processes at work. For all practical terms the only dissipative term of significance in deep-water is white-capping, i.e. the breaking that appears at the crest of the waves under the action of wind. More processes appear when the waves enter a shallow water area, the two most prominent ones being bottom friction and depth induced breaking. It is worthwhile to point out that the shallow water area is the least accurate part of the model. Together with other factors discussed later, this hampers the use of model data in the very shallow coastal zones. However, this is not relevant for the present project, that, because of its purposes, focuses on the deep water zone.

The fourth-order nonlinear wave-wave interactions S_{nl} represent the conservative exchange of energy between the different wave components that takes place continuously during the evolution of the field.

All the above processes are correctly represented in the WAM model via their proper equations. The model is numerically integrated through a semi-implicit scheme, while advection is done with a first order upwind scheme. Because of requirements for numerical stability, the integration time step depends on the grid resolution, being 20 minutes for resolution 0.5 degree, and 15 minutes for 0.25 degree resolution.

Two versions of the WAM model have been operational at ECMWF, one for the global ocean and one for the Mediterranean Sea. The reason for doing this is the maximum resolution allowed for the global version because of computer power limitations, and the contemporary requirements to go to a higher resolution to describe properly the geometrical characteristics of the Mediterranean basin.

The wave model for the Mediterranean Sea became operational in July 1992. A 0.5 degree resolution was used in both latitude and longitude, for an overall number of about 950 points (WAM considers only sea points in its description of the basin). The resolution was later increased to 0.25 degree, for an overall almost 4000 points.

The original wave model for the Mediterranean Sea included the area between 6° West and 36° East in longitude, and 30° and 46° North in latitude. The area was later extended to include the Baltic Sea. In 1998 a much more extended version was made operative. It includes the North Atlantic Ocean, the Barents Sea, the Baltic Sea, the Mediterranean Sea, and the Black Sea. The resolution is still 0.25 degree, but only in the latitude direction. For each latitude a different number of points have been used in the longitude direction, uniformly distributed at 27.5 km distance, more exactly at a distance corresponding to 0.25 degree in the latitude direction. This implies that the grid points are staggered and some further manipulation is required during the advection phase.

The number of frequencies has been kept constant at $N_f = 25$, with $f_1 = 0.04$ Hz and $f_{n+1} = 1.1 \diamond f_n$. N_d = 12 directions have been used for most of the time, increased to 24 when the area of interest was expanded in 1998.

The information available at any moment of the integration process is represented, at any grid point, by the energy available for each wave component, i.e. by the two-dimensional spectrum $F(f,\theta)$. From this spectrum a number of quantities can be evaluated. An integration over direction provides the one-dimensional spectrum $E(f)$, i.e. a description of the energy distribution with frequency. A further integration with respect to frequency provides the overall energy E, from which the significant wave height H_s is derived. Different integrations provide estimates of the mean period T_m, sometimes called the energy period, and of the mean direction θ_m.

The integral wave parameters available from the ECMWF archive have been retrieved for the Mediterranean area starting from 1 July 1992. The fields are available at 00, 06, 12, 18 UT of each day. The data have been taken with 0.5 degree resolution between the geographical limits present in the first version of the model at ECMWF, i.e. between 6° West and 36° East for longitude, and 30° and 46° North for latitude. This corresponds to a 85x33 point grid, out of which about 950 are sea points. Values on land are given as –1.0, an impossible value for any wave parameter.

At the Department of Physics of the University of Genoa, a high-resolution version of the WAM model, with a resolution of 0.25 degree, uses the limited area meteorological model BOLAM as input and is available for forecasting the wave field all over the Mediterranean basin.

3. COMBINING THE DIFFERENT DATA INTO A CONSISTENT DATA SET

The previous section explains the data required for reliable statistics of wind speed and direction, and of wave height. Recently there have been some attempts to derive also the wave period from the altimeter signal (Gommenginger et al., 2003). However, the results are not yet reliable enough for routine applications. In any case, as we will soon see, wave period information can be derived from properly calibrated model data.

From the previous section, it is clear that model data represent the most abundant, widely and uniformly distributed, and continuous source of both wind and wave data. On the other hand, they present definite problems, such as a steady underestimate in the Mediterranean Sea, varying from place to place. An extensive comparison with buoy data shows that the underestimation is constant enough to consider a calibration of the model data using satellite data. While the latter data are sparse in space and time, we can always evaluate the corresponding model data at the satellite measurement position. In so doing we can construct, for every point where the satellite has provided data, a sufficiently long series of co-located data to derive, by a direct comparison, the calibration coefficient of the model data. This will vary from point to point, according to the instrument used (whether altimeter or scatterometer), and to which model we are considering. In the previous section we have mentioned that the resolution of the ECMWF model has been changing in time (not to mention the continuous upgrading of the physics and numerics of the model). A different resolution means a different quality of the wind, hence wave, fields, and therefore different calibration coefficients.

If the same model parameter is compared with different sources, we can expect slightly different calibration coefficients to be derived from the different comparisons. Incidentally, this is a good proof that also measured data, especially the remote sensing ones, are far from perfect, although less so than the model data, at least in the enclosed seas (see Cavaleri and Bertotti, 2003a). In any case, by comparing satellite with buoy data, it is possible to estimate the relative reliability of the single satellite instruments. This leads to a series of weighting coefficients to be used when we derive the final calibration factor of the model data. The interested reader is referred to the final report of the Medatlas project (The Medatlas Group, 2004) for a detailed discussion of the subject.

By a direct comparison with measured data, we can obtain calibration coefficients for wind speed and wave height. A direct inspection of the scatterometer and model wind directions shows that the model is successful in this respect , with limited errors, at least for not too small values of the wind speed and wave height and period, when the processes are mostly governed by very local factors, not discernable in large scale models. As for wave period, in a wind sea dominated basin, as it is mostly the case for the Mediterranean Sea, this is strictly related to the wave height. In practice, the underestimate by the model does not change the average steepness of the waves. This implies that any correction of the wave height implies a similar correction of the wave length. As this depends on the square of the wave period (deep water are assumed throughout the basin), a square root provides the correction coefficient for the wave period.

4. RESULTS

In the following we present the state-of-the-art of the research in the knowledge of wind and waves climatology in the Mediterranean. In particular, we will show three different approaches developed with different methodologies: the Wind and Waves Atlas of the Mediterranean Sea (The Medatlas Group, 2004), the Offshore Wind Climatology over the Mediterranean Basin (Lavagnini et al., 2006), and the Italian Offshore Wind Atlas (Cassola et al., 2006).

4.1 The Medatlas project

The model data can be extracted at 0.25 or 0.5 degree resolution. In practice, because of the relatively poor resolution used in the first years of the considered period, and because of the mostly smooth in space variations of the average wind and wave characteristics, the 0.5 resolution is more consistent with the actual accuracy of the data. Even more, a careful analysis of the spatial distribution of the calibration coefficients reveals variability not consistent with the physics of the processes. This suggests smoothing the results of the data at larger intervals, the final data representing a careful average for the surrounding area. It is clear that, in so doing, we miss many coastal details. However, the models at global or Mediterranean scale cannot properly represent the details of area fields with complicated orography or coastal features. Typical examples are the East coast of the Adriatic Sea and the Aegean Sea. In these area, the large number of islands and indentations of the coast strongly limit the reliability and accuracy of the results. In any case, should a wind or wave farm be built in a specific area, a devoted high-resolution study is obviously required. Any result provided at large scale, like the ones in this book, can provide only general indications for a specific location, the background to start from for a careful detailed study of the area.

Once the calibration coefficients have been derived at the chosen points, combining them with the model time series allows reliable information to be obtained, from which a large number of statistics can be derived, both locally and synoptically.

We show here some maps providing a synoptic view of the main wind characteristics in the Mediterranean Sea. For waves we provide a map showing the probability of relatively large waves. Given the focus of this book, the information on waves is mainly for engineering purposes. As previously mentioned, for extensive statistics the reader is referred to the atlas produced during the Medatlas project (The Medatlas Group, 2004), available from the authors, including one of the authors of the present chapter (L.C.). The atlas provides multiple statistical tables, similar to those reproduced here, on both an annual and a seasonal basis. Moreover, extensive tables of two-dimensional distributions (wave height and wind speed, wind speed and direction, wave height and period, etc.) are reported for a large number of points.

We provide here three maps showing the main characteristics of the basin:

1. Figure 1, in particular, shows the annual-mean values of wind speed throughout the Mediterranean Sea. The distribution is exemplified by isotachs at 0.5 m/s intervals. The area of high values in the gulf of Lion is evident, with a maximum speed between 7.5 and 8.0 m/s. Two secondary maxima with speeds between 6.5 and 7.0 m/s can be seen between the coasts of Greece and Libya, east and west of Crete island. Another secondary maximum with speeds between 6.0 and 6.5 m/s is present in the Ionian Sea, between the Italian and the Greek coasts. The minima near the coastal areas, at this scale, appear to correspond to speeds between 4.5 and 5.0 m/s.
2. Figure 2 provides the primary and secondary directions of wind speed. The information is given at points at 1 degree intervals. The arrows indicate flow directions toward the reference point. Some general characteristics are clear, such as the dominant north-west (incoming) winds west and south of Sardinia, the sirocco and bora in the Adriatic Sea, East of Italy, and the prevailing counter clockwise circulation in the Eastern part of the basin.
3. Figure 3a shows the probability, in percent, of having wind speeds larger than 11 m/s while Figure 3b shows the probability, in percent, of having significant wave heights of more than 4 metres. Expectably, this reflects the distribution of the wind fields, but with due consideration to the relevance of fetch. The dominance of the area west of Sardinia, where the mistral is stronger, is evident.

4.2 The Offshore Wind Climatology over the Mediterranean Basin

The offshore wind climatology over the whole Mediterranean region has been obtained (Lavagnini et al., 2006) using ECMWF 6 hourly wind data, for a period of 24 years. More precisely, these authors considered the horizontal wind components from the ECMWF re-analysis for the period January 1979 – February 1994 and the operational analysis for the period 1994-2002 at a grid resolution of 0.5x0.5 degrees, at 850 and 700 hPa pressure levels, and at 10 m above the sea surface. Probability Density function parameters have been calculated for 12 directions of 30-degrees each for each grid point, mean wind speed U, and Weibull Distribution factors A, and k.

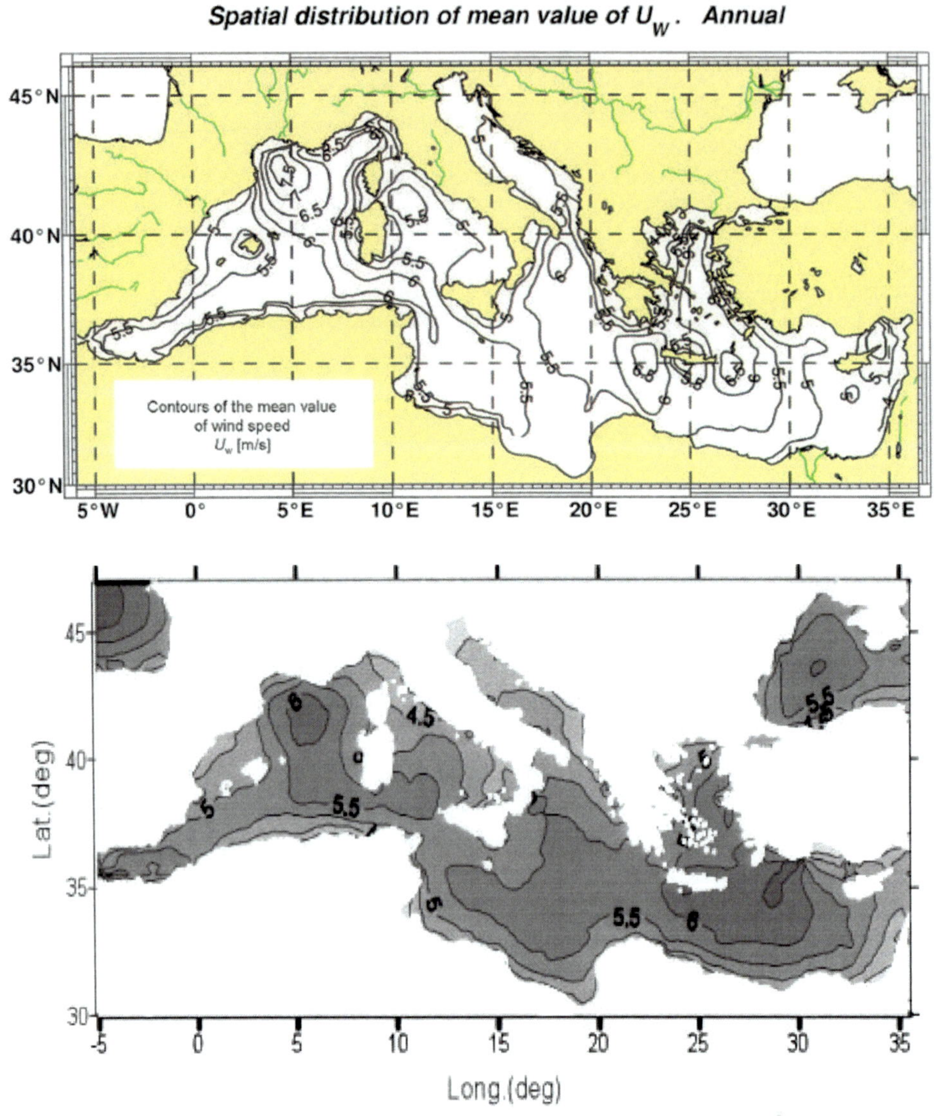

Figure 1. The distribution of annual mean wind speeds in the Mediterranean Sea. Values are in m/s. Isotachs are at 0.5 m/s intervals (after The Medatlas Group, 2004 (above), and after Lavagnini et al., 2006 (below)).

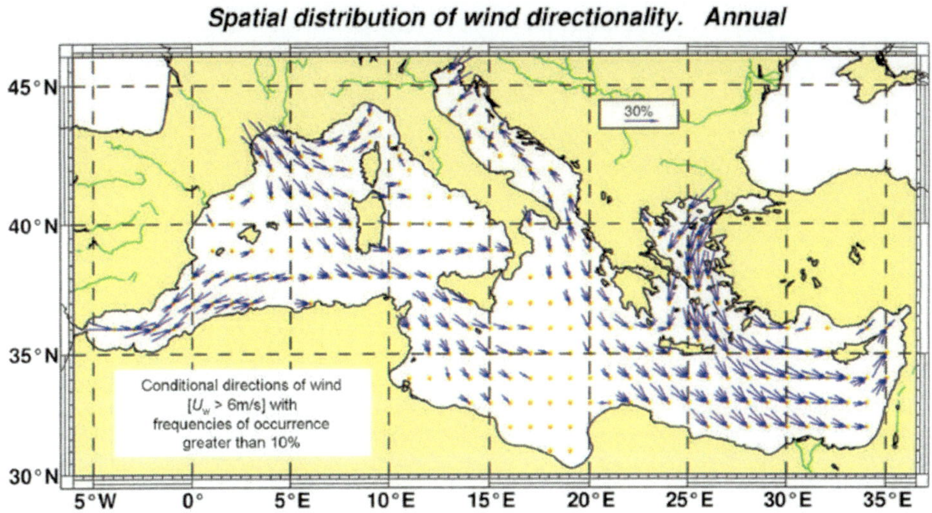

Figure 2. The annual primary and secondary wind flow directions throughout the Mediterranean Sea (after The Medatlas Group, 2004).

As the wind field of ECMWF at 10 meters is less accurate near the coast and in narrow basins, these statistics have been corrected, for each grid point, with the statistics produced using 2-year runs (from 01.10.2000 to 30.09.2002) of the Limited Area Model QBOLAM, that is a parallel version of the finite-difference, primitive-equation, hydrostatic model BOLAM. The model domain covers the whole Mediterranean Sea with horizontal grid step of 10 km. To compare the climatology from the two models, Lavagnini et al. (2006) have considered the outputs of QBOLAM and ECMWF for the two-year overlapping period. For each ECMWF grid point, they have compared the values of mean wind speed U, and A and k Weibull parameters with the values estimated at the same grid point using the ECMWF data. This procedure has allowed modification of the Weibull parameters of the long-term data series of ECMWF wind fields, in coastal areas.

The annual mean values of wind speed throughout the Mediterranean Sea are shown in Figure 1b. The distribution is again exemplified by isotachs at 0.5 m/s intervals. Now the principal maximum with speeds between 6.5 and 7.0 m/s appears in the south-western part of the Turkish coast, between Crete and Cyprus, while in the Gulf of Lion there is only a secondary maximum with speeds from 6.0 to 6.5 m/s. The minima near the coastal areas, at this scale, appear to correspond to speeds between 3.5 and 4.0 m/s.

Figure 4 shows the scale, A, and shape, k, parameters of the Weibull probability density functions characterising the whole Mediterranean Sea.

4.3 Evaluation of the offshore wind potential along the Italian coasts

At the end of 2002, the Italian Wind Atlas (IWA) was produced through a collaboration between the Department of Physics of the University of Genoa and the Italian Experimental Electrotechnical Centre (CESI) (Burlando et al., 2002). This Atlas covers the whole Italian land territory. Thus, the realisation of an Italian Offshore Wind Atlas (IOWA) has been considered as a natural extension and completion of the information already reported in the IWA. In particular, the IOWA adopts the same methodology followed for the realisation of the IWA. This methodology has been applied along the shorelines of the whole Italian Peninsula, in order to guarantee the coherence and comparability between these two atlases (Cassola et al., 2006).

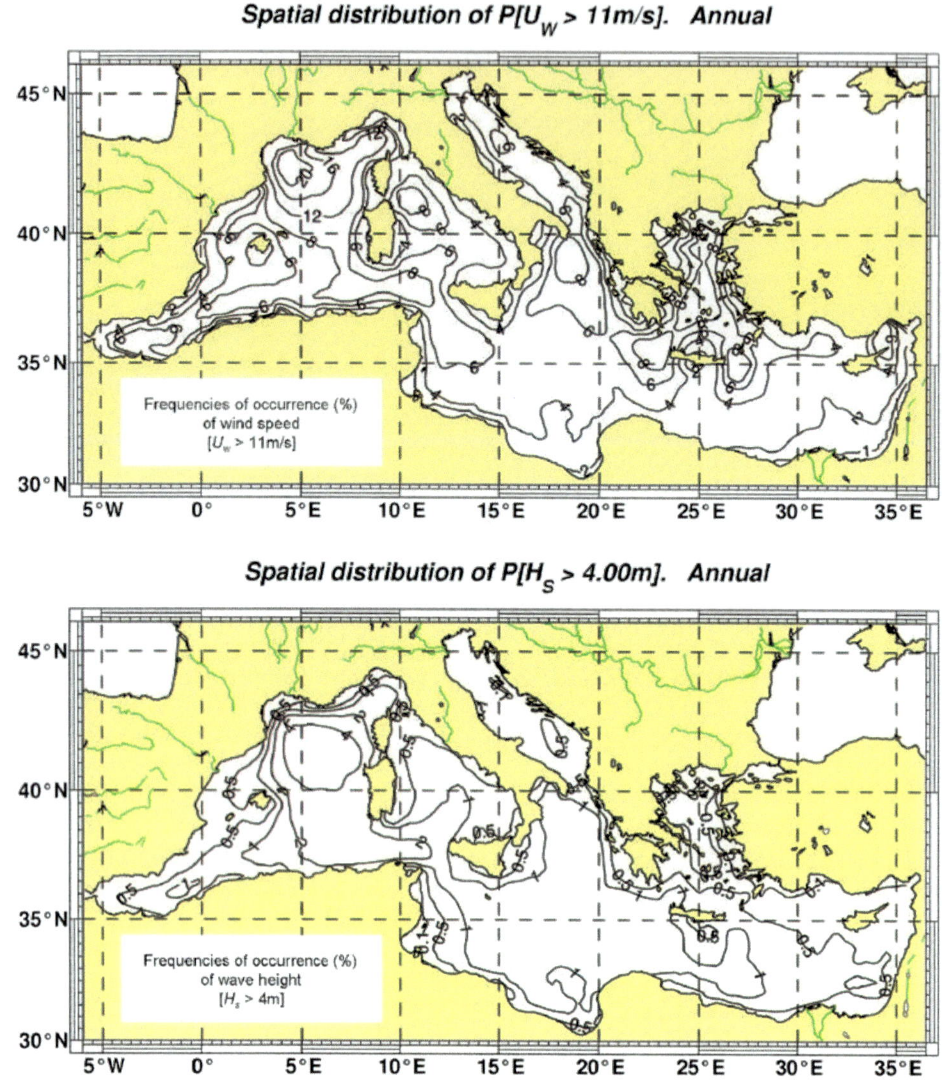

Figure 3. The distribution of the annual probability, in percent, of exceeding 11 m/s wind speed (above) and of exceeding 4 metre significant wave height (below) (after The Medatlas Group, 2004).

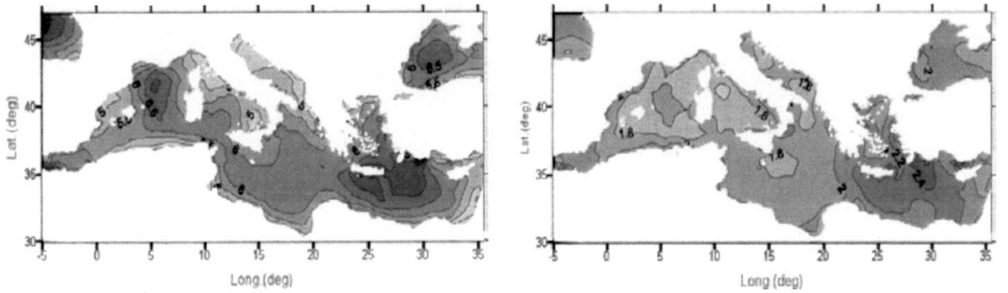

Figure 4. Scale parameter A [m/s] (above) and shape parameter k (below), at 10 m over Mediterranean Sea; (after Lavagnini et al., 2006).

The wind fields at different heights above sea level all along the Italian coast of the Mediterranean Sea have been computed through a procedure that combines statistical analyses of wind speed aloft with numerical modelling of wind flows off-shore. The first step of this methodology consists on the statistical analysis of 10 years of wind speed and direction data at 5000 m above sea level (a.s.l.) from the "re-analyses" of the GCM of the ECMWF. The Italian offshore territory is then divided in a number of "geographic areas" of the order of 200_200 km^2 which cover the sea as well as part of the inland in order to take into account the local effects induced by the coastal topography. Each area is discretised with a grid step of nearly 1 km, and is partially overlapped with its neighbouring areas in order to guarantee the continuity and smoothness between adjacent areas.

The simulations have been performed for each area by means of the diagnostic mass-consistent code WINDS, initialised by altitudinal wind-velocity values. The other inputs to the code have been: (i) stability conditions of the lower atmospheric layers; (ii) terrain elevation and (iii) land cover data, parameterised through the corresponding roughness length. All the simulations have been performed under neutral stability conditions in order to reduce computing times and take into account mainly the most intense winds.

Finally, the simulated wind fields have been corrected taking into account the mean wind values measured in about 150 anemometric stations all over the Italian coastal areas and compared with the Medatlas maps. As final results, maps of the mean annual wind speed at different elevations above sea level have been obtained.

Maps of mean cubic wind speed, Weibull parameters and wind power can be computed as well. The calculation of the corresponding maps of annual mean cubic wind speed, scale and shape parameters of the Weibull distribution and annual mean energy production of a whatever commercial wind turbine is straightforward.

For instance, Figure 5 shows the annual mean wind speed at 10 m height above sea level all along the Italian coasts. The wind speeds are once more shown by isotachs at 0.5 m/s intervals.

The largest wind speeds, between 6.5 and 7.5 m/s, appear in the south-western parts of Sardinia (Sardinia sea) and Sicily coasts (Sicily Strait). The absolute minimum (between 3.5 and 4.0 m/s) appears at the corner facing the border between Italy and Slovenia (Northern Adriatic). Other secondary minima can be found near the coastal areas of Liguria (between 3.5 and 5.0 m/s), Lazio (between 5.0 and 5.5 m/s), and Marche (between 5.0 and 5.5 m/s).

The maps presented in Figure 1a, Figure 1b, and Figure 5 can be compared. If considering a narrow region (say, 25 km wide) surrounding the Italian coasts, such comparison gives the result reported in Table 1. The classification of sea basins is that of Meteomar (see, e.g. http://www.eurometeo.com/english/meteomar/id_ml).

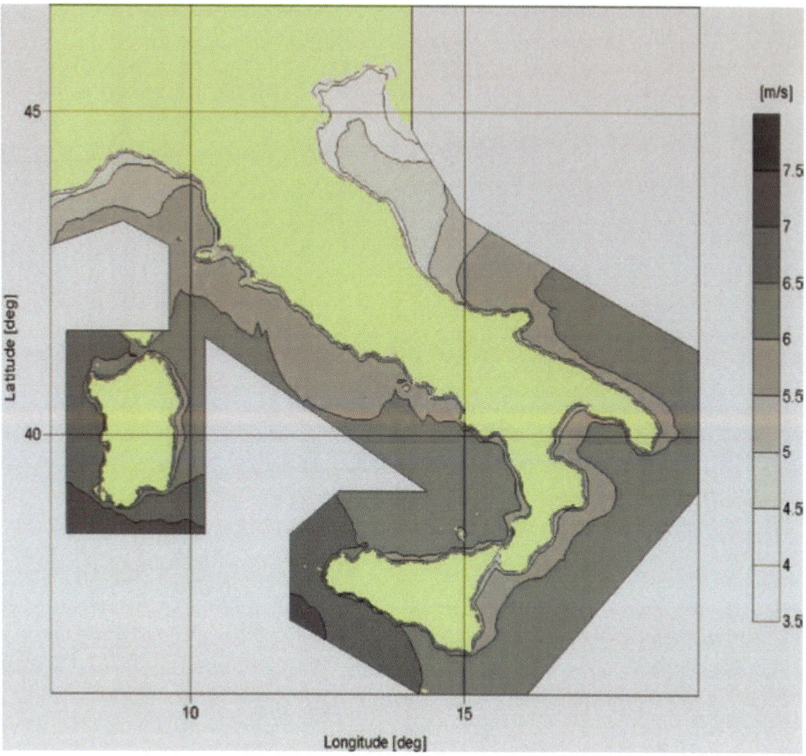

Figure 5. The distribution of annual mean wind speeds at 10 m above sea level along the Italian coasts. Values are in m/s. Isotachs are at 0.5 m/s intervals (after Cassola et al., 2006).

Table 1. Comparison among the winds speeds within a band, about 20 to25 km wide, surrounding the Italian coasts, as evaluated by the Wind and Waves Atlas of the Mediterranean Sea (The Medatlas Group, 2004), the Offshore Wind Climatology over the Mediterranean Basin (Lavagnini et al., 2006), and the Italian Offshore Wind Atlas (Cassola et al., 2006).

Sea	Medatlas	Lavagnini et al.	Cassola et al.
Ligurian	4.5-5.5	3.5-5.0	3.5-5.0
Northern Tyrrhenian	4.5-5.0	3.5-4.5	4.5-5.5
Sardinia	4.5-5.5	5.0-5.5	5.5-7.0
Central Tyrrhenian – W side	4.5-5.5	4.5-5.0	5.0-6.5
Central Tyrrhenian – E side	4.5-5.0	3.5-5.5	5.0-6.0
Southern Tyrrhenian – W side	4.5-5.5	5.0-6.0	5.0-6.5
Southern Tyrrhenian – E side	4.5-5.0	4.5-5.5	5.0-7.0
Sardinia Channel	5.0-6.0	5.0-6.0	5.0-7.5
Sicily Strait	4.5-6.0	4.5-5.5	5.0-7.0
Southern Ionian	4.5-5.0	4.5-5.5	4.5-6.0
Northern Ionian	4.5-6.0	4.0-5.5	4.5-6.0
Southern Adriatic	4.5-5.5	4.0-5.5	5.0-6.0
Central Adriatic	4.5-5.0	4.0-5.0	4.0-6.0
Northern Adriatic	4.5-5.0	3.5-4.5	4.0-4.5

In the considered band, the three maps are similar, because of the systematic decrease from the southwestern side of the Italian coasts to the northeastern one.

The results by Lavagnini et al. are a little different from those of the Medatlas Group in Ligurian, Northern and Central Tyrrhenian, Sicily Strait, Northern Ionian, Southern, Central., and Northern Adriatic seas (in Southern Tyrrhenian sea – west side and Southern Ionian seas). The results by Cassola et al. are always larger (by about 1.0 m/s) than those by Lavagnini et al. and larger (by about 0.5 m/s) than those of the Medatlas Group, except for the Ligurian sea and the Northern Adriatic sea.

Note that recently, thanks to a renewed collaboration between the Department of Physics of the University of Genoa and the Italian Experimental Electrotechnical Centre (CESI), a new version of the Italian Wind Atlas, including offshore areas, was produced and made available in http://atlanteeolico.cesiricerca.it/viewer.htm.

4.4 Comparison of different evaluations

The main difference between the annual mean values of wind speed of the Medatlas project (Figure 1a) and of Lavagnini et al. (Figure 1b) is that the former has largest values in the Gulf of Lion (between 7.5 and 8.0 m/s), while the latter has the principal maximum off south-western Turkey, between Crete and Cyprus (between 6.5 and 7.0 m/s).

Indeed, the main differences between the two evaluations can be found in the western part of the Mediterranean Sea. There is a systematic difference between the Medatlas map and that by Lavagnini et al. of about 1.0 m/s (in the Gulf of Lion the difference reaches 1.5 m/s), while in the Eastern part of the Mediterranean Sea the discrepancies are less (here, in some small areas, the Lavagnini et al. map has larger values than those of the Medatlas).

Finally, we have compared the winds speeds within a band about 20 to25 km wide, surrounding the Italian coasts, as evaluated by the Medatlas Group (2004), by Lavagnini et al. (2006), and by Cassola et al. (2006). The trend is fairly similar, because of the systematic decrease from the south-western to the north-eastern Italian coasts. The results by Lavagnini et al. are usually less than those of the Medatlas Group. The results by Cassola et al. are always larger (by about 1.0 m/s) than those by Lavagnini et al. and usually larger (by about 0.5 m/s) than those of the Medatlas Group.

ACKNOWLEDGEMENTS

Figures 1a), 2, and 3 and part of the text have been extracted from the Atlas of Wind and Waves in the Mediterranean Sea produced by The Medatlas Group. Dr. Alfredo Lavagnini has provided us with the text of the paper by Lavagnini et al. (2006) prior to publication. We warmly thank Dr. Massimiliano Burlando for his valuable work in many aspects of the organisation of this chapter. We are also grateful to the many colleagues with whom we had valuable and constructive exchanges of ideas and results; among others, Mrs. Paola Latona, Mr. Federico Cassola, and Mr. Luca Villa.

REFERENCES

Burlando M., A. Podestà, F. Castino, and C.F. Ratto, 2002, "The wind map of Italy", The World Wind Energy Conference and Exhibition, 2-6 July, Berlin, Germany. Available on CD.

Cassola, F., M. Burlando, L. Villa, Latona, P., and C.F. Ratto, 2006, "Evaluation of the offshore wind potential along the Italian coasts", to appear in proceedings of the European Seminar "Offshore Wind and other marine renewable Energy in Mediterranean and European Seas", Civitavecchia (Rome), Italy, April 20-22, 2006.

Cavaleri, L., and L. Bertotti, 1997, "In search of the correct wind and wave fields in a minor basin", *Mon. Wea. Rev.*, **125**, 1964-1975.

Cavaleri, L., L. Bertotti, H. Hortal., and M. Miller, 1997, "Effect of reduced diffusion on surface wind fields", *Mon. Wea. Rev.*, **125**, 3024-3029.

Cavaleri, L., and L. Bertotti, 2003a, "The characteristics of wind and wave fields modelled with different resolutions", *Q. J. R. Meteorol. Soc.*, **129**, pp.1647-1662.

Cavaleri, L., and L. Bertotti, 2003b, "The accuracy of modelled wind and wave fields in enclosed seas", *Tellus*, **56A**, 167-175.

Chèruy, F., A. Speranza, A. Sutera, and N. Tartaglione, 2004, "Surface winds in the Euro-Mediterranean area: the real resolution of numerical grids" *Ann. Geophys.*, **22**, pp. 4043-4048.

De Boni, M., L. Cavaleri, and A. Rusconi, 1993, "The Italian wave measurement network", *23rd Int. Conf .Coast. Eng.*, pp.116-128, 4-9 October 1992, Venice, 3520-3528.

Gimeno, L., L. de la Torre, R. Nieto, R. García, E. Hernández, and P. Ribera, 2003, "Changes in the relationship NAO–Northern hemisphere temperature due to solar activity", *Earth and Planetary Science Letters*, **206**, 15-20.

Gommenginger, C.P., M.A. Srokosz, and P.G. Challenor, 2003, "Measuring ocean wave period with satellite altimeters: A simple empirical model", *Geophys. Res. Lett.*, **30**, 2150-2161, doi:10.1029/2003GL017743.

Homicz, G.F., 2002, "*Three-Dimensional Wind Field Modeling: A Review*" SAND REPORT SAND2002-2597, Sandia National Laboratories, Albuquerque – New Mexico, and Livermore – California, USA, 68 pp.

Jackson, P. S. and J.C.R. Hunt, 1975, "Turbulent wind flow over a low hill", *Q J. R. Meteorol. Soc.*, **101**, pp.929-955.

Komen, G.J., L.Cavaleri, M.Donelan, K.Hasselmann, S.Hasselmann, and P.A.E.M.Janssen, 1994, *Dynamics and Modelling of Ocean Waves*, Cambridge University Press, 532 pp.

Lavagnini A., A.M. Sempreviva, C. Transerici, C. Accadia, M. Casaioli, S. Mariani, A. Speranza , 2006, "Offshore Wind Climatology over the Mediterranean Basin". *Wind Energy*, **9** (under press).

Mortensen, N. G., L. Landberg, I. Troen, and E. L. Petersen,1993, *Wind Analysis and Application Program (WASP)*, Risø National Laboratory, Roskilde, Denmark.

Ratto, C. F., R. Festa, O. Nicora, R. Mosiello, A. Ricci, D. P. Lalas, and O. A. Frumento, 1990. "Wind field numerical simulation: a new user-friendly code". In W. Palz (Ed.), *European Community Wind Energy Conference*, Madrid, Spain, pp. 130–134. H. S. Stephens & Associates.

Ratto, C. F., R. Festa, C. Romeo, O. A. Frumento, and M. Galluzzi, 1994, "Mass-consistent models for wind fields over complex terrain: The state of the art". *Environ. Soft.*, **9**, 247–268.

Simmons, A., 1991, Development of the operational 31-level T213 version of the ECMWF forecast model, *ECMWF Newsletter*, **56**, 3-13.

Simmons, A., R.Mureau, and T.Petroliagus, 1995, Error growth and predictability estimates for the ECMWF forecasting system, *Q.J.Roy.Meteor.Soc.*, **121**, 1739-1771.

The Medatlas Group: Gaillard., P., P. Ravazzola, Ch. Kontolios, L. Arrivet, G.A. Athanassoulis, Ch.N. Stefanakos, Th.P. Gerostathis, L. Cavaleri, L. Bertotti, M. Sclavo, E. Ramieri, L. Dentone, C. Noel, C. Viala, and J.-M. Lefevre, 2004 "*Wind and Wave Atlas of the Mediterranean Sea*", 420 pp.

Tournadre, J.,1999, "Analysis of mesoscale variability of oceanic surface wind fields: contribution of active microwave sensors". Procs of IGARSS 99 IEEE Press, **4**, pp. 1975-1977.

WAM-DI Group, 1988, The WAM model – a third generation ocean wave prediction model, *J.Phys.Oceanogr.*, **18**, 1775-1810.

Wilke, C. k., R.F. Milliff, and W.G. Large, 1999, "Surface wind variability on spatial scales from 1 to 1000 km observed during TOGA COARE", *J. Atmos. Sci.*, **56**, 2222-2231.

Chapter 4:

Climatological and Meteorological Aspects of Predicting Offshore Wind Energy

R.J. Barthelmie, S.C. Pryor and S.T. Frandsen

1. INTRODUCTION

1.1 Objectives and chapter structure

This chapter reviews climatological and meteorological issues of power output from offshore wind farms. In section 1, we introduce the atmospheric structure and variability of wind resource estimation, the short-term fluctuations in the wind energy density and the behaviour of wind turbine wakes. We also describe the existing offshore wind farms from which examples are presented. Section 2 concerns the prediction of power output over the lifetime of a wind farm. This assessment of the resource normally occurs at the planning stage and includes lifetime variations in wind speed at hub-height over several decades, which is influenced by climate change. Since climatologically representative measurements of wind speed at hub-height at the location of a proposed wind farm are usually not available, we demonstrate how short-term site measurements may be related to longer-term historical data from a nearby site and how data measured close to the surface may be extrapolated vertically to hub-heights.

Offshore wind farms have regular multiple rows and columns of turbines, so wake effects (i.e. the effects on down-stream rotors of the perturbed wind and turbulence from upstream turbines) are significant and have to be incorporated in the generated power estimate of the entire wind farm. Hence, in section 3, we describe tools for evaluating wake behaviour and we illustrate the potential importance of atmospheric boundary-layer phenomena not currently incorporated in standard models used for prediction of wind speeds and wake losses. In section 4, we review short-term prediction of power production by offshore wind farms, before providing, in section 5, a summary of key points raised in this chapter.

1.2 Scales of atmospheric variability

The atmosphere is constantly in motion. It may be characterised by (i) temporal periodicities and (ii) inherent horizontal spatial scales. These range:

(a) from turbulence (period £ seconds, spatial scale £ metre), traditionally referred to as *micro-scale*,
(b) to local, *meso-scale*, phenomena, such as thunderstorms (period ~ hours, scale hundreds of metres to a few kilometres),
(c) to synoptic, *macro-scale*), transient weather systems (period ~ days to months, scale continental to global).

The periods and scales are linked by energy exchange and each play a role in dictating the temporal variability of energy density in the wind at a given location.

1.3 Atmospheric stability

Atmospheric stability relates to the vertical exchange of momentum and of sensible and latent heat. The stability of the atmosphere indicates whether the atmosphere will develop turbulence or waves of growing amplitude. With stable stratification, the atmospheric flow is laminar (i.e. non-turbulent) because air flow at higher altitude (aloft) is decoupled from surface dissipation of momentum. Unstable conditions are characterised by turbulent mixing between air at different altitudes, with a significant contribution to this mixing from buoyancy forces (i.e. thermally generated eddies). At near-neutral stability conditions, the ambient turbulence is generated principally or solely from mechanical forcing (i.e. vertical wind shear due to surface drag).

So, why should wind turbine designers take atmospheric stability into account?

Under near-neutral atmospheric stability, the average change of wind speed with height has a logarithmic profile:

$$U_z = \frac{u_*}{\kappa} \ln\left[\frac{z}{z_0}\right] \tag{1}$$

where

- U_z is the wind speed at height z
- κ is the dimensionless von Karman constant (= 0.4)
- z_0 is surface roughness
- u_* is the friction velocity.

u_* is related to the momentum flux (t) at the surface by:

$u_*^2 = \frac{\tau}{\rho}$, where r is the air density,

and by the dimensionless drag coefficient (C_D) from:

$u_*^2 = C_D U_{10}^2$, where U_{10} is the wind speed at 10 m above the surface.

Thus the change of wind speed with height is dictated by two time/space variables,

- surface roughness length z_0, as dictated by the height of roughness elements,
- the wind speed, U_z.

The z_0 of land surfaces is documented in the Davenport-Wieringa roughness-length classification [1], while water surfaces have a dynamic roughness dictated by the presence of wave-surface features [2]. Generally, values of z_0 for water surfaces are orders of magnitude less than for land surfaces.

When the atmospheric stability is not near-neutral, the change of wind speed with height is also dependent on the prevailing atmospheric stability. Stability corrections to the logarithmic profile of wind speed are small near the surface, but increase with height [3, 4]. Hence, as hub-heights of new wind turbines are increased, the role of stability in modifying the wind profile becomes more important, because the correction to the logarithmic profile is height dependent:

$$U_z = \frac{u_*}{\kappa}\left[\ln\frac{z}{z_0} - \Psi_m\left(\frac{z}{L}\right)\right] \tag{2}$$

where the function $\Psi_m\left(\frac{z}{L}\right)$ is dependent on height (z) and the Monin-Obukhov length (L).

The Monin-Obukhov length is a metric of the atmospheric stability and is given by the ratio of mechanically generated to buoyancy generated turbulence [5].

$$L = \frac{-\left(\overline{u'w'}^2 + \overline{v'w'}^2\right)^{3/4}}{\left(\kappa(g/\overline{\theta_v})(\overline{w'\theta_v'})\right)} \tag{3}$$

unit = [($m^4 s^{-4}$) –] / [($m s^{-2}$)/ K) ($m s^{-1}$) K = ($m s^{-1}$)^3 / ($m^2 s^{-3}$) = m
where the over-bar indicates a time average. Below S.I. units are given in brackets.

- g ($m s^{-2}$)is acceleration due to gravity
- θ_v (K) is the virtual potential temperature
- where T (K) is temperature, q (kg/m^3) is the specific humidity, a is a constant (= 0.61 kg/m^3), P is the atmospheric pressure (mBar), P_0 is a standard pressure (1000 mbar) R is the universal gas constant (8.31 (J mol^{-1} K^{-1})) and c_p is the specific molar heat capacity of air (J mol^{-1} K^{-1}))
- $w'\theta'$ is the kinematic heat flux (K m s^{-1}). Heat fluxes which are normally expressed as W m^{-2} can be put into kinematic form by dividing by the air density and the specific heat capacity for air [6].
- u' , v, $'w'$ (m s^{-1}) are respectively the deviations of the wind speed components from their temporal averages (where u represents the west-east component, v the south-north component and w the vertical component). These components can be used to compute u_*:

and

$$u_*^2 = \sqrt{\overline{u'w'}^2 + \overline{v'w'}^2} \tag{4}$$

As L goes towards infinity, mechanically driven turbulence dominates and conditions are increasingly near-neutral. Table 1 summarizes the stability classification used herein [7]

Table 1. Stability classes defined by Monin-Obukhov length

Monin-Obukhov length (L)	Stability class
L= 0 to 200m	Very stable
L= 200m to 1000m	Stable
1000 m > L > -1000m	Near-neutral
L= -200m to -1000m	Unstable
L= 0 to -200m	Very unstable

Note that stability corrections are not equal in magnitude under stable and unstable conditions (i.e. for the same absolute value of the Monin-Obukhov length [5]). They are computed as follows:

For 0 < L < 1000 m (i.e. stable conditions):

$$\Psi_m\left(\frac{z}{L}\right) = \frac{4.7z}{L} \tag{5}$$

For -1000 m < L < 0 (i.e. unstable conditions):

$$\Psi_m\left(\frac{z}{L}\right) = -2\ln\left[\frac{(1+x)}{2}\right] - \ln\left[\frac{(1+x^2)}{2}\right] + 2\tan^{-1}(x) - \frac{\pi}{2} \tag{6}$$

Where $$x = \left[1 - \left(15z / L\right)\right]^{1/4} \tag{7}$$

The turbulent kinetic energy per unit mass ($\overline{e}$) may be derived from the velocity vector variances:

$$\overline{e} = \frac{1}{2}\left(\overline{u'^2} + \overline{v'^2} + \overline{w'^2}\right) \tag{8}$$

In wind energy applications, the direction of the wind and the vertical components are ignored, so the dimensionless turbulence intensity (I) is computed by aligning the co-ordinates such that $v \rightarrow 0$ and hence defined as:

$$I = \frac{\sigma_U}{U} \tag{9}$$

where:

- U is the mean horizontal wind speed
- σ_U is the standard deviation of U ($\sigma_U = \sqrt{\overline{U'^2}}$).

Atmospheric turbulence intensity offshore is strongly linked to the dynamic nature of the sea surface, which itself is related to the low-level wind speeds [8]. These wind speeds are dependent on the stability of the atmosphere. At small wind speeds, turbulence intensity is large on average due to the dominance of thermally generated turbulence. The turbulence intensity decreases to a minimum at wind speeds between 8 and 12 m s^{-1}. At larger wind speeds, the sea surface waves increase in amplitude, and the rougher sea surface increases mechanically generated turbulence intensity of the wind.

The surface roughness of water is generally less than land, so, the offshore value of u_* is small compared with that for most land surfaces. Consequently mechanically generated turbulence over sea waves generally has less impact on stability and on total turbulence intensity than mechanically generated turbulence over land. Conversely stated, the influence of the heat flux (i.e. buoyancy generated turbulence) is generally larger over water surfaces than over land. Hence, the impact of deviations of atmospheric stability from near-neutral is generally more significant for offshore windfarms than for onshore windfarms. Turbine blade stress depends on the wind shear and turbulence, which in turn depend on the atmospheric stability. Consequently, the nature of the vertical wind profile, through which the turbine blades sweep and to which the turbine tower is subject, affects turbine design and performance. It is therefore essential that measurements are made and extrapolated for a resource and performance estimate prior to windfarm design and installation. Moreover, as discussed below, atmospheric stability also substantially impacts the propagation of wakes downstream of the wind turbines.

In addition to its influence on the wind speed profile, atmospheric stability also affects the distance for wind speeds to equilibrate with changed surface conditions and fluxes. This applies as winds pass from land to sea or vice versa. For instance, stable conditions, with reduced vertical momentum transfer, can persist for more than 200 km from the coast-line [9], so producing very large wind-shear in the atmospheric surface-layer. However, this is only important for windfarms if conditions deviate from near-neutral at approximately $U > 4$ m s^{-1}, relating to turbine cut-in wind speeds (see below). For wind speed > 25 m s^{-1} (typically turbine cut-out wind speed) conditions can be assumed to be neutral. Since L is proportional to u_*^3, conditions tend to become more neutral as wind speed increases. However, stable conditions can persist to $U \sim 15$ m s^{-1} while unstable conditions appear to be less frequent offshore and not to persist at wind speeds more than about 10 m s^{-1} [4] (Figure 1).

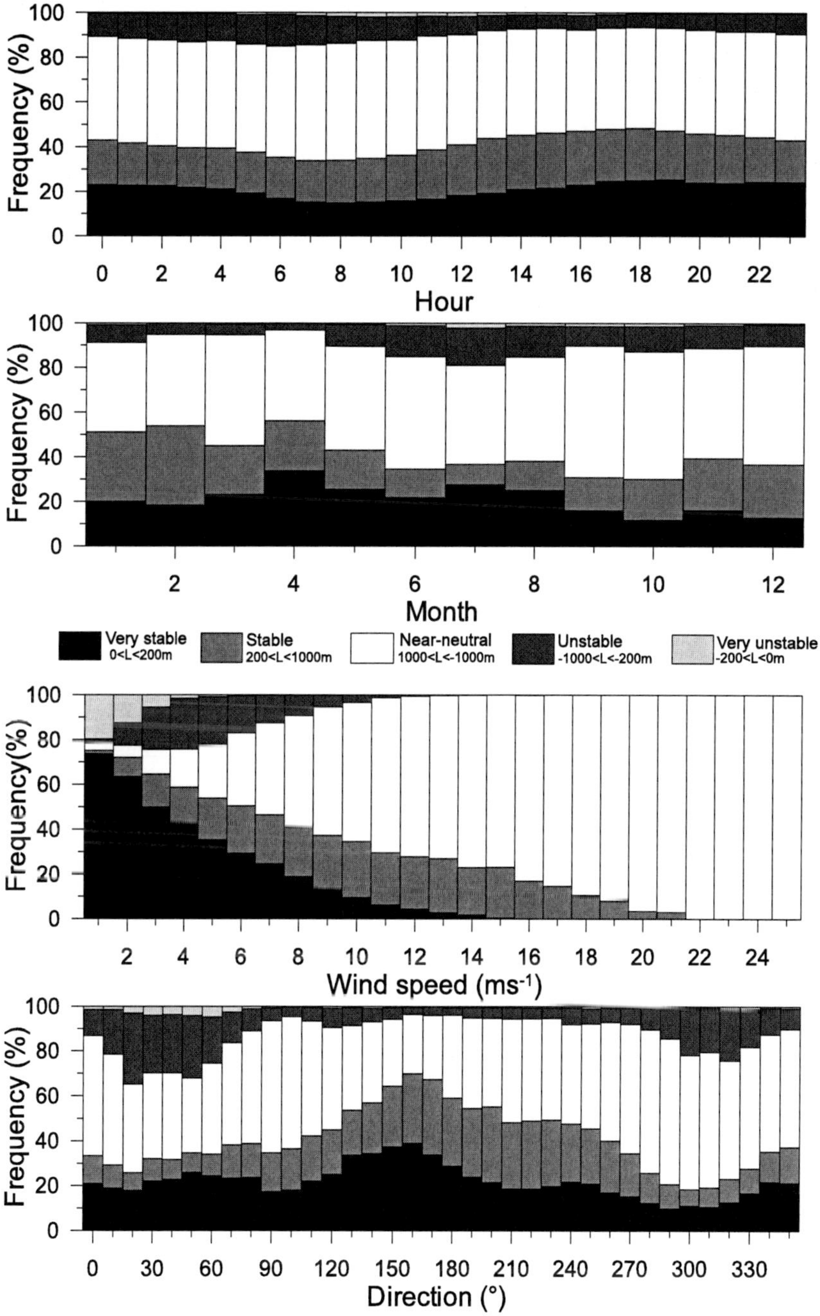

Figure 1. Typical stability climate of an offshore site in Denmark (Vindeby). The four frames show the percent frequency of different stability classes (see Table 1) in terms of the diurnal and seasonal variability (above) and their distribution with wind speed and direction. See section 1.5 for a description of this site.

The division of stability conditions at coastal or sheltered sites between stable/ neutral/unstable conditions is dictated principally by; (i) position of the site with respect to the coast and (ii) the prevailing synoptic meteorology. Offshore sites studied in the north of Europe usually show a large number of stable conditions in Spring, which is related to the lag in sea temperature compared to air temperature (of the order one month) (Figure 1). Similarly unstable conditions are more common in Autumn (Fall) and Winter. Since sea surface temperatures change little over the course of the day, stability offshore does not usually have the pronounced diurnal cycles that are experienced at land sites [10].

In addition to the processes presented above, atmospheric stability also influences a number of other coastal phenomena, such as low-level jets which affect wind and turbulence in the Baltic Sea [11] (and possibly other locations where they have been less intensively studied) and sea breezes [12]. Large scale effects, such as channelling, are also related to temperature differences between land and sea [13].

1.4 Describing wind speed distributions and relating wind speeds to energy density and power production

Although several different probability density functions have been matched to observed wind speed distributions, the two-parameter Weibull probability density function is most frequently applied. This generally gives a good fit to observed wind speed distributions, particularly over water surfaces [14]. The probability distribution, p(U), of wind speed, U, has the following form [15, 16]:

$$P(U) \equiv \frac{k}{A}\left(\frac{U}{A}\right)^{k-1} \exp\left[-\left(\frac{U}{A}\right)^{k}\right] for U \geq 0, A > 0, k > 0 \tag{10}$$

The parameters are the dimensionless shape parameter (k), which describes the 'peakedness', of the distribution and the scale parameter (A), which is a measure of the central tendency.

The cumulative probability distribution is given by:

$$P(U) \equiv 1 - \exp\left(\frac{U}{A}\right)^{k} \tag{11}$$

where P(U) describes the probability that the wind speed has a value less than or equal to U.

The Weibull A and k parameters can be used to compute (or can be computed from) the mean wind speed ($\bar{U}$):

$$\bar{U} = A\Gamma\left(1 + \frac{1}{k}\right) \tag{12}$$

where Γ is the gamma function, defined as

$$\Gamma(1 + \frac{1}{k}) = \int_{v=0}^{\infty} v^{1/k} e^{-v} dv$$

for which standard functions are published in mathematical texts.

Also, for percentiles (X*100) of the wind speed distribution:

$$U_x = A\left(-1 \cdot \ln\left(1 - X\right)\right)^{1/k} \tag{13}$$

The instantaneous wind energy density (i.e. power per unit area of wind front), E is :

$$E = 0.5\rho U^3 \tag{14}$$

Or an 'expected' average energy density may be derived from the Weibull distribution parameters:

$$E = \frac{1}{2}\rho A^3 \Gamma\left(1 + \frac{3}{k}\right) \tag{15}$$

The power production of a wind turbine is non-linear with wind speed (Figure 2). Electricity generation begins at the cut-in wind speed (U) of ≈ 4 m s^{-1}, and the power generation increases rapidly to the rated (name plate) power at typical wind speeds of approximately 12 to 15 m s^{-1}. The power production is constant for wind speeds greater than this threshold until cut-out wind speed is reached, the turbine rotation is stopped for safety reasons. The accurate relationship between ambient wind speed and electricity generation is specific to each turbine. Standard power curves are supplied by the manufacturer for each type of turbine and these can be used to predict energy generation from the expected wind speed distribution at the site. Such predictions are in practice approximations, since the exact wind speed distribution at the site is unknown, individual turbines may perform slightly differently from the standard and the structure of the wind (e.g. turbulence) may differ from the standardisation site. For example, at the Middelgrunden offshore wind farm in Copenhagen harbour, measured annual energy output has been 5.7% more than predicted using the manufacturer's power curve [17].

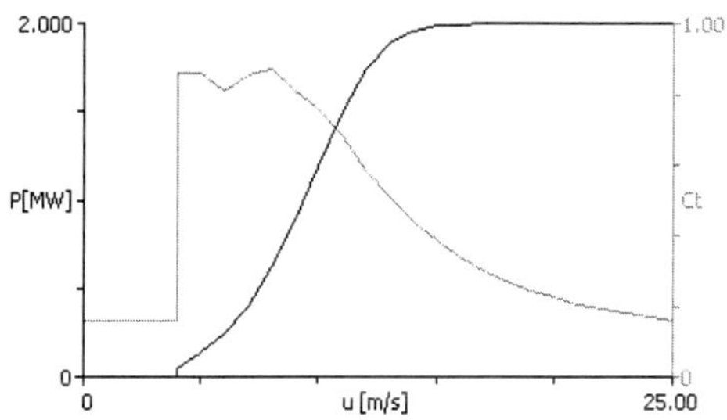

Figure 2. Bonus 2 MW power curve (i.e. power produced by the wind turbine, left ordinate) and thrust coefficient (Ct, right ordinate) as a function of wind speed (u).

1.5 Existing offshore wind farms in Denmark

Global installed wind capacity by the end of 2006 was about 73 GW and the annual average growth rate for installed global capacity is almost 25%/y, including increasing capacity from offshore installations [18]. In 2004 wind energy supplied 18.5% of Danish electricity. By 2006 Danish offshore wind farms had a capacity of over 400 MW, relative to a total national wind capacity of over 3000 MW. In this chapter we present specific examples drawn from Danish offshore wind farms, particularly those located at; Vindeby, Horns Rev and Middelgrunden (Figure 3). The wind farm at Vindeby [19] was the first offshore wind farm in the world, and this and the other two sites have extensive associated research into resource, wakes and loads (e.g. [20], [21, 22]). Details of the wind farms are given in Table 2.

Table 2. Offshore wind farms in Denmark operating in 2006.

Name	Year of installation	#	Turbine type	Spacing	Hub-height (m)	Rotor diameter (m)
Vindeby	1991	11	Bonus 450 kW	8.6 D	38	35
Tunø Knob	1995	10	Vestas 500 kW	5.1 D (in row), 10.2 D between row	43	39
Horns Rev	2002	80	Vestas 2MW	7 D	60	80
Nysted	2003	72	Bonus 2.3 MW	10.5D (E-W), 5.8D (N-S)	69	82.4
Middel-grunden	2002	20	Bonus 2 MW	2.4 D (bow)	64	76
Samsø	2003	23	Bonus 2.3 MW	60	80	
Frederiks-havn	2003	5	2 Vestas 3 MW 1 Bonus 2.3 MW 1 Nordex 2.3 MW			

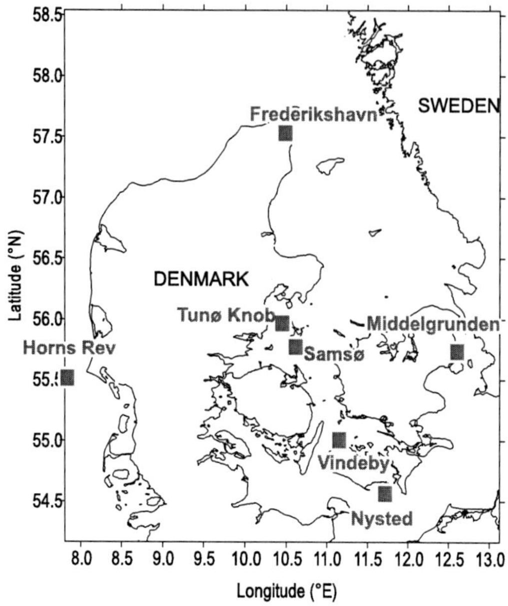

Figure 3. Location of Danish offshore wind farms (2006). See Table 2 for details of these installations.

2. RESOURCE ESTIMATION

Typical wind farms are assumed to have an expected lifetime of 20 to 30 years. To predict the wind resource over the lifetime of a prospective wind farm it is usually assumed that the past wind climate can be used an analogy for the future. Thus the wind resource is implicitly assumed to be stationary (i.e. time invariant at time-scales beyond a few years) and consequently the only requirement for assessing the resource is a historical record of wind speeds at the proposed site. The stationarity of the actual wind resource can be difficult to ascertain; indeed the potential impact of global climate change on renewable energy supplies must be considered, especially the long-term wind climate.

2.1 Prediction of the wind resource on long time scales: impact of climate change

2.1.1 Historic variability

It is difficult to characterise long-term variations in wind speeds effectively because, unlike other meteorological variables, such as temperature and precipitation, wind speeds have been measured using different types of instrument and often with poor accuracy and precision. Although innovative and robust measurement methods have been introduced (starting in the 1980's), trend analyses are confounded by the absence of long homogenised records. Additionally, even more than temperature and precipitation, wind speed measurements are strongly influenced by changes in site characteristics, such as the growth of trees or construction of buildings. For this reason there have been only a few studies of historic wind speed trends. These studies have tended to employ reanalysis data sets [23], derived using a state-of-the-art analysis/forecast system to model weather using relatively few past data as input. The aim is to generate historic global homogenised records with spatial resolutions of $\approx 2 \times 2\,^{\circ}$ and typical temporal resolution of 4 hours [24].

As an example of the potential impact of climate change on wind resource magnitude, we draw on previous research focussed on northern Europe and specifically the Scandinavian countries. This region has a relatively large penetration of carbon-neutral electricity from hydro and wind. For example, in Denmark over 18% of the annual electricity supply is derived from wind farms [18]. Many wind energy projects were developed in this region during the 1990's on the basis of wind climates during the mid-1980's onwards. However, average wind speeds over the Baltic significantly increased over the second half of the twentieth century, with the majority of the increase being in the southwest of the region [23] (Figure 4) and in the upper quartile of the wind speed distribution. This 'large wind-speed' quartile has the largest impact on the wind energy production

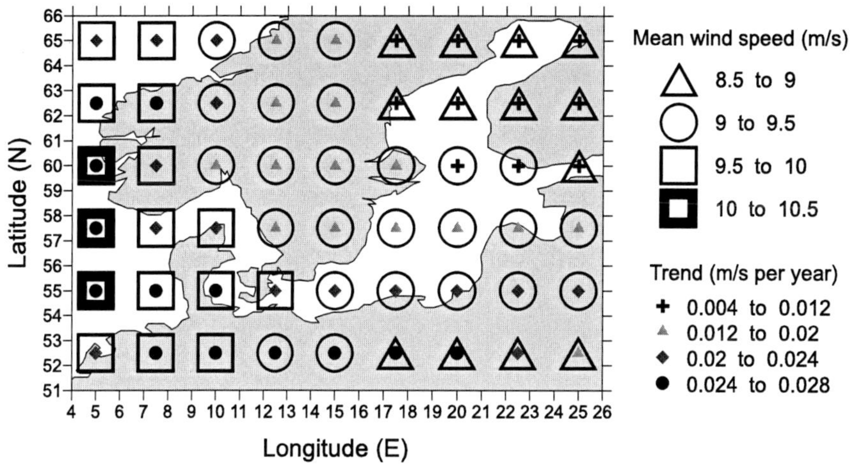

Figure 4. The mean wind speed and temporal trends in the 850 mb (approximately 1.5 km above mean sea-level) wind speed over Scandinavia from 1953-1999 [23] based on data from the NCEP reanalysis data set [24, 25].

These changes in wind speed are causally linked to variations in the synoptic scale circulation and the prevalence of a positive phase North Atlantic Oscillation (NAO) [26], which is itself linked to the evolution of the global climate. The NAO index is a teleconnection index (a teleconnection is a link between atmospheric circulation in different parts of the globe) that links to the global scale climate via its dependence on (i) aspects of the primary circulation patterns of the atmosphere, and (ii) the low pressure associated with the polar front. The latter divides the cold polar air from relatively warm mid-latitude air, and the sub-tropical 'Azores high' that is a component of the descending limb of the Hadley cell. The NAO is computed from the pressure difference between Iceland and the Azores. It describes, in broad terms, the steering and intensity of low-pressure transient circulation systems into Europe [27]. Positive NAO is associated with low-pressure systems (mid-latitude cyclones) moving northwards into the Baltic and can be characterised by westerly circulation with large wind speeds.

The Baltic region experienced a trend towards increased storminess and wind speeds starting in the 1960s, associated with increased prevalence of positive phase NAO that peaked in 1985-1990 and appears to have been broken in the mid-1990s [23, 28]. This example (shown graphically in Figure 5) illustrates that, although wind energy developers frequently use a ten year period to describe the long-term variations in site wind speeds, trends in wind speeds or wind speed cycles can exist on much longer time scales [23, 29].

Note that local/regional changes in flow regimes need *not* be linked to hemispheric dynamics, changes, but may also occur on more local scales. If areas which currently experience sea ice become ice free, it should be anticipated that changes in the thermal climate will also impact the local wind climate.

2.1.2 Future wind climates

Global Climate Models (GCMs) are the principal tools for understanding possible future climate states, and output from a large number of these models are now available with a daily time-step [30].

However, GCMs exhibit greatest accuracy at large scales and long averaging periods [31], while wind speeds exhibit rather low spatial autocorrelation [32, 33]. Accordingly GCMs are unable to replicate the historically observed magnitude and spatial variability of wind speeds [32] [34] [29], so alternative techniques are required to quantify the potential climate change impact of the wind energy resource. These include:

(i) Use of wind indices, where a normalization is applied to the GCM wind-speed time series to account for bias in the modelled wind speeds [29]. This approach has the advantage that it is already used in the wind energy community to quantify inter-annual variability and hence is well known to the end-user community.

(ii) Application of dynamical downscaling so a Regional Climate Model (RCM) is nested within a GCM to derive smaller scale climate realizations [35]. This approach has the advantage that it generates wind speeds, and hence wind energy estimates, for the entire area (model domain), regardless of data availability.

(iii) Application of empirical downscaling approaches where statistical relationships are developed between large-scale predictors of the climate system (derived from the GCM output) and local wind observations [36]. This approach has the advantage that it generates site-specific wind speed and energy density estimates.

We now briefly present the basis for these approaches and some results from research in northern Europe. Note that in these analyses we use daily output from different GCM using the SRES A2 emission scenario [37]. This scenario predicts a moderate to large greenhouse gas global cumulative emission for 1990 to 2100 as a result of projected population growth and fairly slow introduction of carbon-free technologies. The A2 emissions scenario thus provides a reasonable upper bound on likely climate change and hence a strong driving function when comparing current to future wind-flow climates.

2.1.2.1 Use of wind indices

Wind indices are a normalization tool to quantify variability in wind energy density. Recall that wind power is proportional to the cube of wind speed. A wind index is the average of the cube of the wind speeds in the time window of interest, divided by the average of the cubes of the wind speed during the normalization period. They are computed as follows:

$$Index = \frac{\overline{U^3_{j...n}}}{\overline{U^3_{i...k}}} * 100 \tag{16}$$

where the bar indicates an average
U are the wind speed observations
j..n denotes the time series of wind speeds in the time window of interest
i…k denotes the time series wind speeds during the normalization period

An overview of historical and projected annual wind indices for western Denmark is given in Figures 5a, b [29]. The indices were calculated using 10-m wind speed data from the ECMWF reanalysis data set for 1958-2001 and output from a simulation conducted using the HadCM3 GCM for 1990-2100 [38-40]. Here the grid cell average wind speeds directly from the GCM simulation are used. However, it should be noted that individual simulations conducted with GCM represent one realization of possible future atmospheric composition and climate [31] and that comprehensive analyses should be based on multiple simulations and ensembling techniques to produce more robust results [41].

Fig 5a and 5b (i) demonstrate significant variability on time scales longer than a decade, (ii) show that the inter-annual variability of wind indices over western Denmark (like much of Scandinavia) often exceeds 30%, and hence individual years may deviate by 30% from the long term average wind energy density.

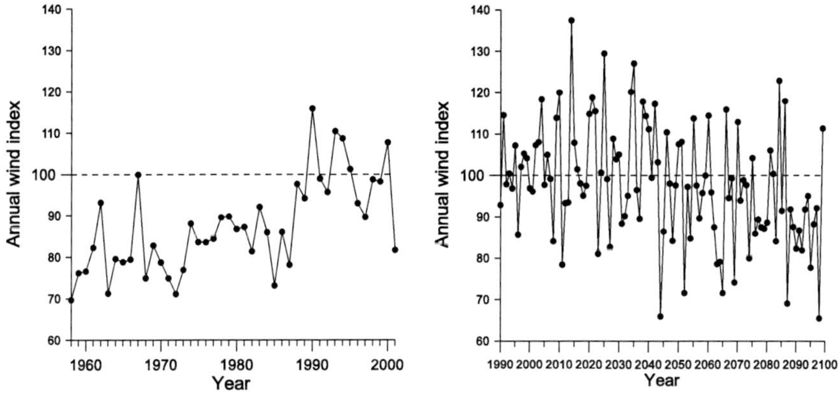

Figure 5a, b.. Annual wind indices. (a) for a grid cell located over western Denmark derived for 1958-2001 using data from the ECMWF reanalysis data set. (b) for 1990-2100 using output from HadCM3. For both sets of calculation, a normalization period of 1990-2001 was used to compute the wind indices.

Country specific wind indices, computed using the same HadCM3 simulations as in Figure 5a & b, are shown in Table 3. These simulations predict that the next 30 years will exhibit similar wind energy densities to those experienced during 1990-2001. However in all countries, except Finland, wind indices from this simulation of HadCM3 are projected to be substantially smaller at the end of the 21st century than during 1990-2001 or 2005-2034 and the decline in wind indices is most marked in Iceland [42].

Table 3. Country-wide annual wind indices from HadCM3 output calculated for 2005-2034, 2035-2064, and 2065-2094 with a normalization period 1990-2001. 'Mean' is the mean annual wind index. 'Std. Dev'. is the standard deviation of the annual wind indices.

	2005-2034		2035-2064		2065-2094	
Country	**Mean**	**Std. Dev.**	**Mean**	**Std. Dev.**	**Mean**	**Std. Dev.**
Denmark	94	14	97	11	95	11
Norway	100	10	100	10	99	8
Sweden	100	8	99	9	98	7
Finland	100	10	101	11	101	9
Baltic States	101	10	98	10	98	8
Iceland	96	10	91	8	87	8

2.1.2.2 Dynamical downscaling

In dynamical downscaling, Regional Climate Models (RCMs) use results of GCMs (or observations) as lateral boundary conditions and simulate (by use of essentially the same procedures as global models) scenario climates that are dynamically consistent with these lateral boundary conditions. Typical resolutions of these RCM simulations are ≈ 50 × 50 km.

Example results from analyses of wind speeds and energy density conducted using RCM simulations over the Baltic region [35] are shown in Figures 6a,b,c,d.. These indicate:

- The Rossby Centre coupled Regional Climate Model (RCAO) simulations with boundary conditions derived from the ECHAM4/OPYC3 and HadAM3H GCMs exhibit realistic wind climates during the control period (1961-1990).
- Simulations conducted for climate projection periods in the twenty-first century (C21st) show important variations in the simulated changes when the GCM is used to supply the lateral boundary conditions:
    - Simulations conducted using boundary conditions from the ECHAM4/OPYC3 GCM indicate increases in virtually all descriptive parameters of the wind speed distribution and energy density in the projected climate change simulations for C21st.
    - RCAO simulations of wind speeds and energy density over the Baltic region, (conducted using boundary conditions supplied from HadAM3H) indicate more spatially heterogeneous changes between the control run simulations and projected climates for 2071-2100.

The changes in wind speeds in the simulations conducted using boundary conditions supplied from the ECHAM4/OPYC3 GCM are linked to the modelled increase in north-south pressure gradient over the North Atlantic during the winter (the NAO). This feature is not present in the HadAM3H simulations, and accordingly the wind climate exhibits a much greater degree of consistency between the control run and the climate projections.

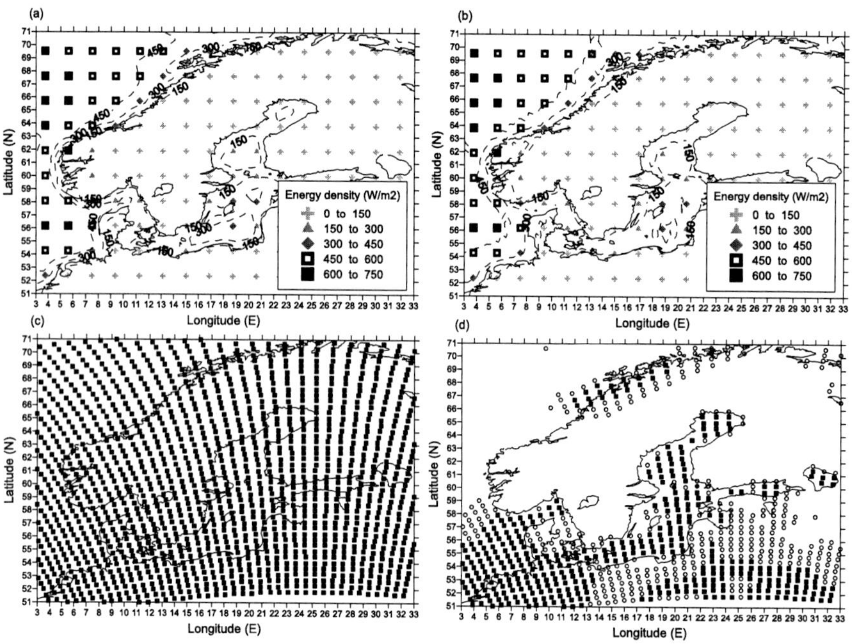

Figure 6. (a) and (b) show (in the contours) the energy density (at 10-m) from the RCAO simulations for 1961-1990 derived using lateral boundary conditions from (a) ECHAM4/OPYC3 and (b) HadAM3H. Symbols depict energy density derived from the NCEP reanalysis data.

Frames (c) and (d) show the change in energy density by RCAO grid cell in 2071-2100 relative to 1961-1990 for simulations conducted using the A2 emission scenario for (c) ECHAM4/OPYC3 and (d) HadAM3H. A solid black square indicates that the mean from 2071-2100 is larger than the 97.5th percentile in the control period. If the grid cell is blank, the mean from 2071-2100 is less than the 2.5th percentile in the control period. If the symbol is an open circle, the mean of the future period is within the 95% confidence intervals for the control period (1961-1990) [35].

2.1.2.3 Empirical downscaling

Empirical downscaling methods generally use GCM simulations of large-scale variables and atmospheric conditions. They relate those variables statistically to historical observations of the surface parameter of interest (temperature, precipitation, wind speed, etc.). The required transfer function can be developed using a range of statistical methods, ranging in sophistication from linear regression to highly non-linear techniques, such as neural networks [43]. Once the transfer function is determined, it is assumed to remain invariant under climate change, so it may be applied to the large scale parameters from a GCM climate change scenario.

Recent research has developed an innovative approach to downscaling, focussed on wind speeds and energy density [36, 44]. This approach starts with GCM projections of mean sea-level pressure and of the mean and standard deviation of vorticity. These projected values are compared to the Weibull parameters of the wind speed probability distribution (see section 1.4). This empirical downscaling approach has been demonstrated to capture the observed variability of 10-m wind speeds and energy density. At all but one of 46 observing stations across northern Europe, (i) the downscaled mean wind speed is within ± 5% of the independent observations, and (ii) energy density is within ± 20% of that calculated from observations (Figure 7, a,b,c). When applied to output from ten GCMs, the downscaling results indicate the range of percent changes in the mean and 90th percentile. Wind speed change is ≤ 20% for 2046-2065 and ≤ 35% during 2081-2100 at all stations. As with the changes in downscaled mean and 90th percentile wind speed, the results for energy density at each of the stations tend to span zero, with the downscaled results from some GCMs showing both increases and decreases (Figure 6).

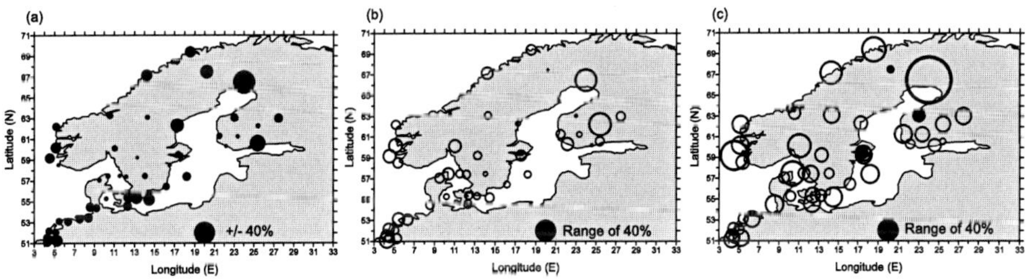

Figure 7. (a) The range of downscaled energy density from ten GCMs relative to independent observations during 1982-2000. A value of ± 40% means downscaled energy density from all ten GCMs lie within ± 40% of the observed values.

Frames (b) and (c) show the change in energy density presented in terms of the range of changes from downscaling of the ten GCMs; (b) for 2046-2065 (i.e. ((2046-2065) – (1961-1990))/(2046-2065)) and (c) for 2081-2100 relative to 1961-1990. If all the downscaled values indicated decline in the specified parameter, the symbol is solid. If the results from the downscaling of different GCMs span zero, the symbol is an open circle. No stations exhibited consistent increases in downscaled values from each of the ten GCMs. The diameter of the symbol used in each frame is linearly related to the data range.

2.1.2.4 Synthesis of research over the Baltic region

The various methods described above generate wind energy projections across the Baltic region under climate change scenarios. Comparing results, there are some commonalities and also some discrepancies. The major conclusions are are:

(i) The late 1980's and early 1990's seem to be somewhat atypical of the second half of the 20th century and of the likely climate states within the 21st century. According to reanalysis data sets, the mean wind energy density in the 1960's and 1970's was approximately 20% less than that of the 1990's. Wind speeds and energy density have generally declined since the peak in the early 1990's. Hence future resource assessments for this region should not rely solely on data from the 1990's in quantifying the available wind resource.

(ii) The modelling is inconsistent regarding an increase or decrease of the mean wind speed and energy density over the Baltic during C21st relative to the end of C20th. This uncertainty in decadal average wind energy density is mostly due to variations between simulations from different GCMs and is of comparable magnitude to historical estimates of inter-annual variability. Also downscaling results from different GCMs exhibit increased diversity at the end of C21st, which reduces confidence in the projections for the end of C21st relative to the middle of this century.

2.2 Prediction of the wind resource over the lifetime of a wind farm assuming a stationary climate

2.2.1 Sources of wind speed observations

In order to predict the wind resource over the lifetime of a wind farm, it is usually assumed that the past wind climate will continue into the future. Nevertheless, even with stationary climate inter-annual variability (Figure 5) and with seasonal fluctuations, there will be uncertainties in the projected wind climate. These uncertainties relate to the length of reference period. A long record of at least 30 years can be assumed to have an uncertainty of at least ± 5 %, whereas a reference period of one year has an uncertainty of at least ± 15 % [45]. These uncertainties relate solely to the climatological variability and not to any uncertainty introduced by the measurement techniques or to long-terms trends in the climate system.

Because offshore atmospheric stability varies seasonally offshore, rather than diurnally as over land surfaces, the importance of making at least one year of measurements to avoid seasonal bias has to be emphasised. If no site measurements are available, predicting the offshore wind resource becomes difficult and the resulting estimates are highly uncertain. Although large spatial scale datasets, such as the NCEP-NCAR data [25] and the POWER data set [46], are available, the major uncertainty lies in predicting the change of the resource at distances less than 50 km from the coast, where the majority of existing and planned wind farms are located. Since the large-scale datasets have spatial resolutions of 0.5° to 2° these can only be used as a general guide. In addition, these datasets have much reduced accuracy compared with onsite measurements.

An alternative technique to in situ measurements for resource assessment is application of remotely sensed data [47]. Scatterometer produced [48] wind speed datasets, such as QuikSCAT, are available with good time resolution worldwide, however the spatial resolution is relatively poor and wind speeds close to the coast are not well resolved. Synthetic Aperture Radar (SAR) data have better accuracy and resolution relative to Scatterometers, especially close to the coast, but poor time resolution, due to infrequent satellite over-passes [49]. Wind-speed accuracy in a single image for SAR is given as ± 2 m s^{-1} in the range 2 to 24 m s^{-1} [50] at a spatial resolution of ~ 300 by 300 m [51]. In general, comparison of satellite-derived wind-speeds from different sources, models and measurements give good results in case study analyses [50]. However, a relatively large number of satellite images is required to develop robust estimation of the entire wind-speed probability distribution which may prove prohibitively expensive in terms of analysis time and/or may be limited by availability of suitable images [15, 16]. A further drawback of large scale and satellite derived wind-speed data is that they are wind speeds at heights at, or below, 10-m above the sea surface. The same is true of course for data from buoys [51]. As discussed in section 2.3, extrapolation of wind speeds to hub-heights from near-surface measurements is a major source of uncertainty, particularly in the coastal zone. Hence, although remote sensing data can be employed for initial site exploration, they are not yet adequate for resource estimation.

By far the most accurate method of measuring site wind-speeds is to use standard equipment (i.e. good quality, individually calibrated, anemometers) at hub-height on a slim meteorological mast fixed to the seabed. Low-power instrumentation and satellite transmission of the resulting data means that accurate and reliable measurements for wind resource prediction or wind profile characterisation can be made even at remote offshore sites [52]. The height above water of a bottom-fixed mast is a compromise between: (i) uncertainty in the power prediction from extrapolating to hub-height, and (ii) the non-linear expense of increasing mast height to hub height above about 50 m. Care should be taken to avoid mast shadow effects, either by measuring on two sides of the mast or by using a slim open mast structure which is not cluttered by obstacles close to the anemometers.

2.3 Climatologically robust estimates of wind speeds and energy densities

Assuming one year of in situ meteorological data are available, there are essentially two methods of relating on-site measurements to a longer term reference data set to derive a climatologically adjusted resource estimate. One is to use a *physical model* and the other is to use *statistical analysis*. These alternatives are described below.

Physical models range in complexity from: (i) *mesoscale models* which attempt to incorporate physical processes across a range of scales in the atmosphere and are based on primitive equations [53, 54], to (ii) *linearised models*, such as the Wind Analysis and Application Program (WAsP), which employ certain semi-analytical approximations [55, 56] and have been designed specifically for wind energy.

While the sophistication of *mesoscale models* has many advantages in terms of the complexity of flow modelled, they are much more demanding in terms of input and computing resources. Mesoscale models are rarely run in 'time-series' mode for wind energy; instead,. a number of climatologically representative simulations are made, for example wind speed and wind direction classes, and then weighted according to the specific climate of the area. A further major advantage is that no site-specific data are required, although clearly the whole area must be described in adequate detail (orography, roughness etc). Mesoscale models have also been used to demonstrate the importance of coastal roughness, temperature and topography in dictating offshore wind speeds up to 200 km from the coastline [53], [57].

Linearised models require only wind speed and direction time series as input. Local effects such as obstacles, roughness changes and topography can be modelled. However neither type of model gives adequate results if flow separation occurs (such as wind flow over a cliff). As discussed above, wind flow in coastal areas tends to be non-equilibrium and to be affected by horizontal and vertical temperature gradients. These are not modelled specifically by linearised models. In WAsP, a mean stability correction is added to the vertical profile of wind speeds, but temporal or specific site variations in atmospheric stability are not taken into account. Also linearised models cannot capture mesoscale thermal circulations like low-level jets or wind channelling. WAsP predictions for offshore are mainly related to the roughness change at the coastline. This occurs in conditions which are close to near-neutral within about 20 km of the coast. The clear implication is that if WAsP is initialised using wind speeds from a land site, it will not determine further change in wind speed beyond about 20 km from the coast, depending on the height under consideration. The major advantages of use of the WAsP model is the computational efficiency (a site assessment can be generated in a few minutes) and the ability of the model to determine the uncertainty in the resource estimate based on the input data.

The most common statistical method for a climatological wind speed and energy density prediction for a potential windfarm site is 'measure-correlate-predict' (MCP). MCP uses a short-term wind dataset from the site itself and relates this to a longer time series from a relatively nearby location. There are a number of variations of MCP [58, 59] but all derive a statistical relationship between wind speeds at both the prediction site and the longer-term site, using data from the overlap period. This relationship is then used to derive the wind speed/direction statistics at the site for the longer period, assuming the relationship between the two sites remains consistent. The two sites should therefore be under the same synoptic influence and, as a rule-of-thumb, not be separated by more than about 50 km. MCP may under-predict wind resources at offshore sites if predictions are made from land-based observations and if the change in the wind speed distribution (not just the mean wind speed) are not accounted for.

Comparisons of these techniques for developing a site-specific wind resource assessment have been made for a number of offshore sites [45]. The results indicate merit in both statistical and modelling approaches. For instance, a study of wind-speed prediction over the Baltic compared model results from WAsP and from the Uppsala University mesoscale model (MIUU). As shown in Figure 8, wind speeds derived from the two methods are in good agreement in the central Baltic, but in coastal areas the differences between the wind-speeds derived from the two methods can be substantial. This is probably due to air-sea temperature differences (atmospheric stability variations) close to the coastline.

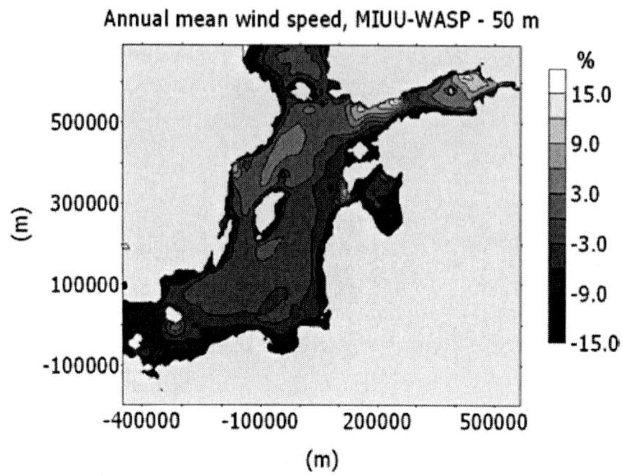

Figure 8. A comparison of 50 m wind speeds predicted by WAsP and the MIUU mesoscale meteorological model for the Baltic [54].

2.4 Extrapolation of vertical wind speed profiles

The extrapolation of measured wind speeds to turbine hub-heights involves significant uncertainty for wind energy prediction. Ideally, measurements would be at hub-height and clearly the shorter the extrapolation distance, the less the error. If extrapolation is required, a simple model is usually used, such as the power law [60] or logarithmic wind profile (with stability corrections) [4, 61]. Figure 9 shows the difference between the logarithmic and stability-corrected profile at a height of 100 m for different values of the Monin-Obukhov length, which relate to the stability (see equation 2).

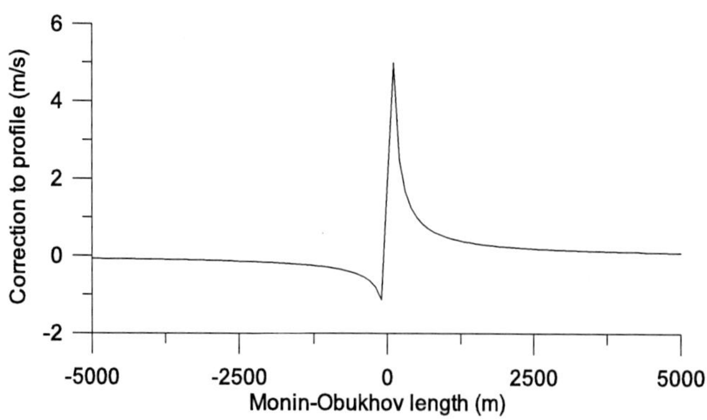

Figure 9. Corrections to the logarithmic wind speed at 100 m height for different values of the Monin-Obukhov length

Figure 10 shows predicted wind speed profiles for different stabilities defined according to the Monin-Obukhov length. The central thick solid line is the logarithmic profile prediction from an initial wind speed of 8 m s^{-1} at 10 m height. It is important to note that corrections under stable conditions (0 m < L<1000 m) are typically larger than for negative L in unstable conditions (0 m > L> -1000 m).

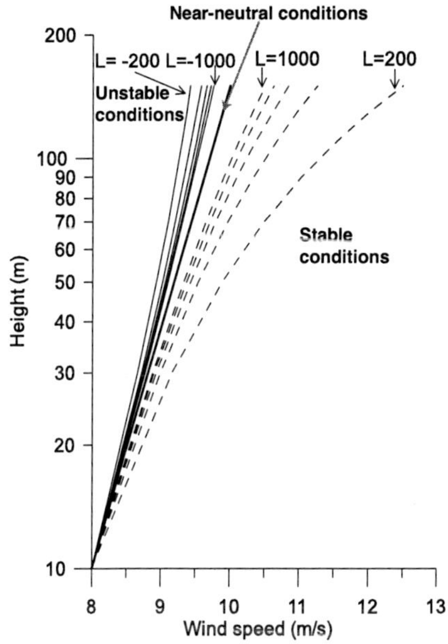

Figure 10. Example of wind speed profiles under different stability conditions with different values of the Monin-Obukhov length. The central thick solid line represents near-neutral conditions, the dashed lines represent stable conditions and the thin solid lines, unstable conditions.

Recently both *sodar* and *Doppler lidar* anemometry have been used on a fixed platform to investigate wind speed profiles up to about 200 m close to the Nysted offshore wind farm (see Table 1 for a description of this wind farm) [62]. Deploying these usually ground-based remote sensing technologies on a boat or moored-buoy is possible [63], although corrections for yaw and tilt are complex, especially in strong wind.

The varying roughness of the sea surface does not affect resource predictions significantly if the lowest height from which the extrapolation to hub-height wind speeds is above 10 m. However, for measurements from a buoy or using an anemometer at less than 10 m height, the extrapolation factor is large and subject to large uncertainties [64].

As explained in section 1.3, atmospheric stability has a major impact of wind speed profiles and corrections to the profile [65]. On average, a correction assuming 'slightly stable' conditions to the logarithmic profile (as is applied in the WAsP model) has been found to give a good fit to offshore observations in northern Europe, but the majority of these are at less than 70 m height [45]. However, [66] and [67] indicate (i) that similarity theory (used to compute L and relying on a constant flux layer assumption) is not valid above about 50 m from the sea surface, and (ii) that above this height the wind-speed profile becomes linear, which is why the stable correction gives a reasonable fit. The suggestion is that the atmospheric Ekman layer begins at 15 to 45 m, and that below this height, theory linking the atmosphere and the wave surface is required [68]. This tendency towards linear profiles at greater heights has also been advanced for land surfaces [67]. If the constant flux layer theory does not apply, an additional scaling parameter may be required, such as the boundary-layer height.

3. FROM RESOURCE TO WIND ENERGY: IMPACTS OF WAKES

3.1 Wind turbine wakes

As energy is extracted by a wind turbine rotor, previously unperturbed air passing through the rotor plane has its speed reduced and its turbulence increased. In a windfarm, the separation between otherwise identical turbines is seldom large enough for the airflow to return fully to its unperturbed state, i.e. for the wake to 'recover' and disappear. So assuming flat terrain, the wind speed at the first downwind turbine will be less than for the initial turbine, so its energy production will be less. The reduced wind speed and enhanced turbulence region of the airflow is known as the *wind turbine wake*. Smoke injected as a tracer, also shows that the wake has angular momentum, which helps define the region. Energy losses of otherwise potential energy relate to the properties of oncoming wakes and the design characteristics of specific wind turbines. In effect, these losses are inversely proportional to the power-extraction efficiency of the turbine, known as the *power coefficient* C_p, which itself is a function of the tip-speed ratio (speed of blade tip divided by speed of oncoming wind). In essence therefore, the more that is understood about wakes, the greater the opportunity for wind turbine and windfarm designers to improve energy capture. Wake recovery depends on both ambient turbulence intensity and wake-generated turbulence intensity. Under conditions of small ambient turbulence intensity, i.e. very stable conditions when vertical energy transfer is limited, wake recovery requires longer distances than in near-neutral conditions when there is vertical mixing with the higher, less perturbed, airflow. In unstable conditions, wake recovery occurs over shorter distances than for near-neutral conditions, due to increased momentum transfer with the higher airflow. Analysis of wake losses by simulation, is usually limited to near-neutral conditions, since most wake models cannot (yet) account for stability variations.

3.2 Introduction to wake models

Given the trend towards large offshore wind farms, accurate prediction and analysis of wind farm wakes is a major issue in determining power output. As indicated above, wake losses are more important offshore than onshore because turbulence in the airflow is less and wake recovery longer. The corollary is that the transfer of energy from higher unperturbed airflow into the reduced-energy wake is reduced compared to identical turbines on land sites, so allowing wakes offshore to propagate further downstream than onshore.. Wake models extend in complexity from (a) engineering solutions which approximate physical processes and are very fast to run (e.g. WAsP [56], to (b) a number of models based on the Ainslie solution [69] to the Navier Stokes equations [70], to (c) computational fluid dynamics models [71]. For a detailed description of wind turbine wakes and wake models see e.g. [72]. For the influence of wakes on turbulence and loads within wind farms see [73].

Engineering models. The fundamental basis for most engineering models is the wake expansion model [74], [75]:

$$D_w \propto X^{1/3} \tag{17}$$

$$D_w = D(\beta + \alpha \cdot s)^{1/2} \tag{18}$$

where X is the downwind distance in rotor diameters (D)
D_w is the wake diameter
α is the wake decay constant (engineering model)
β is a constant related to the initial wake diameter and to α
s = X/D is the non-dimensional distance from the first wind turbine

This general approach is followed in Risø's Wind Analysis and Application Program (WAsP) which is widely used by the wind energy community. The wake model in WAsP [55] is based on a mathematical model of the wake behind a wind turbine [76] [77] (Figure 11). The wake is assumed to have a 'top hat' form, and the spread of the wake (D_w) is assumed to be symmetric in the vertical and lateral directions, as in Figure 11:

$$D_w = D + 2kX \tag{19}$$

Where k is the wake decay coefficient (WAsP definition).
The velocity deficit at hub-height, ΔU is calculated:

$$\Delta U = U_{freestream} - U_{wake} \tag{20}$$

and the wind speed in the wake (U_{wake}) is:

$$U_{wake} = U_{freestream}\left[1 - \left(1 - \sqrt{1 - C_t}\right)\left(\frac{D}{D + 2kX}\right)^2\right] \tag{21}$$

where C_t is the thrust coefficient (see Fig 2).

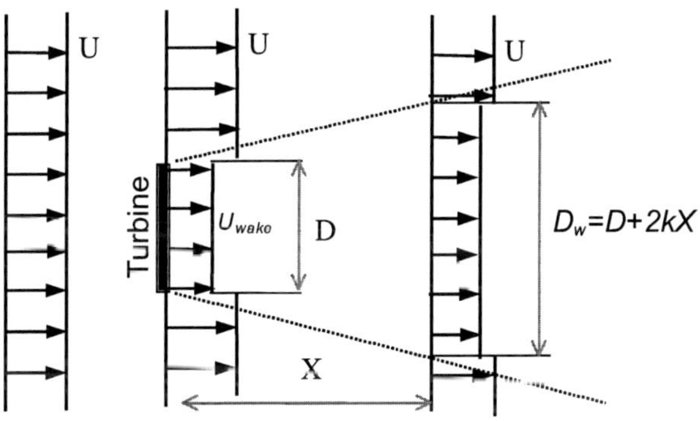

Figure 11. Schematic of the wake model used in WAsP

In the WAsP model the suggested wake decay coefficients (see equation 20) are k = 0.075 for land sites and k = 0.04 for small offshore wind farms. Current research suggests that for large offshore wind farms (more than 2 rows) the WAsP wake decay coefficient (k) should be less, corresponding to increased wake losses for the same turbine separation with identical turbines.

In WAsP, the model assumes the centre-line of the expanding wake follows the terrain (flat for offshore), and the different hub heights and rotor diameters are taken into account by the overlapping fraction of a wake with a down-wind rotor plane. Since the near-wake is not modelled specifically, the algorithms for WAsP wakes are valid only for distances in excess of 3D to 4 D, which, as shown below, have been evaluated over the entire range of downwind distances.

3.3 Evaluation of the wake magnitude and wake models in single wake cases

The performances of state of the art wake models for offshore, including WAsP, have been compared and evaluated [78, 79] using data from the Vindeby wind farm, principally for single wake conditions [63]. The first study used results from a sodar experiment at Vindeby [63] (Table 4). The second study used a set of systematic scenarios developed for the full-scale evaluation of wake models within the ENDOW project [78], [79]. The former represent the wake behaviour at a range of downwind distances, and the latter allow a more comprehensive representation of ambient conditions.

Figure 12 shows predictions of the wake magnitude from the WAsP wake-algorithm along with the sodar-measured free-stream and wake profiles. As shown in Figure 12 and Table 4, the direct calculation of U_{wake} gives good results for distances 2D to 4 D but under-predicts the wake wind-speed in the near and far wake. The direct U_{wake} calculations have a Root Mean Square Error (RMSE) from measurement of 0.7 m s^{-1}. However, these results are broadly comparable with those from a range of wake models [44].

Table 4. Measured and modelled velocity deficit from the sodar experiment conducted at the Vindeby offshore wind farm and computed using WAsP. The experiment numbers are as shown in Figure 12.

Experiment #	Freestream U (m s^{-1}) at 48 m	D	ΔU (m s^{-1}) Sodar (measured)	ΔU (m s^{-1}) WAsP (modelled)
4	6.90±0.59	1.7±0.3	4.26	2.71
7	5.74±0.20	2.8±0.3	2.28	2.28
9	7.54±0.45	2.9±0.5	2.57	2.01
10	6.37±0.25	3.4±0.5	1.48	2.11
11	8.19±0.46	3.4±0.3	2.28	1.98
12	6.12±0.74	7.4±0.5	0.61	1.21

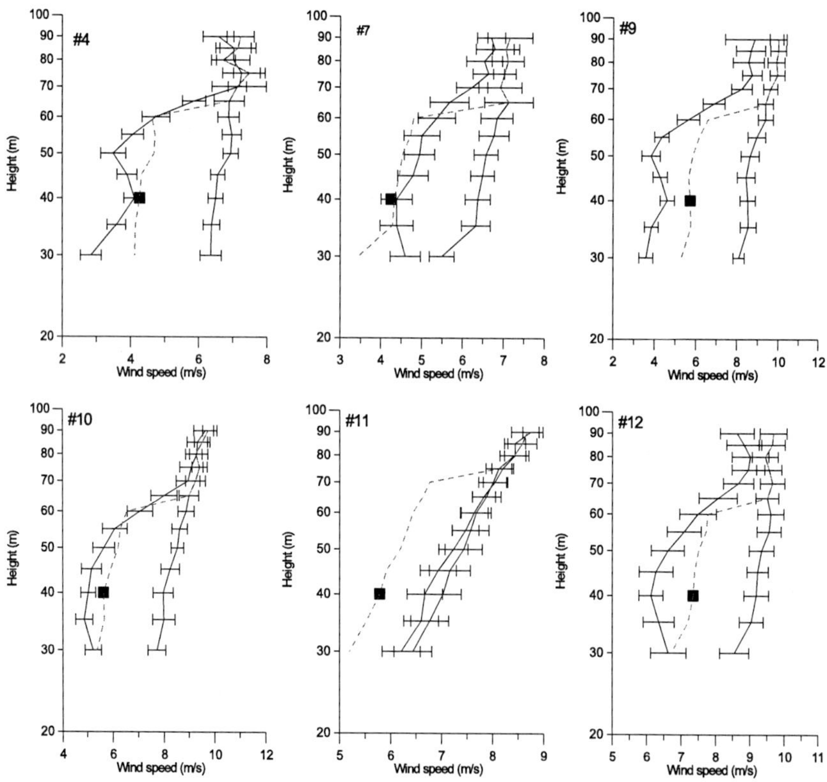

Figure 12. Comparison of sodar wake and free-stream wind speeds (full lines) and WAsP algorithm predicted wake wind speeds at hub-height (square symbol). Error bars calculated from the sodar data are ±0.5 standard deviation. The dashed line shows the wake profile predicted assuming the velocity deficit is applied to the free-stream wind profile within a vertical/horizontal spread calculated from equation 10 (a top-hat profile).

A limitation of this comparison is that only a few observational conditions are presented. In a second evaluation of the WAsP wake algorithms, data from a fixed mast were used, and hence the comparison focussed on a fixed distance downstream of a turbine with different flow conditions. Data from the meteorological masts at Vindeby were selected only for when the upstream-turbine to met mast direction was ± 5° of the direct wake direction. For single wakes, wind speed profiles in the wake were determined from the sea mast (south) while free-stream wind profiles were taken from simultaneous observations at the near-shore land mast. As shown in Table 5, predicted wind speeds at hub-height are approximately equal to those observed. The RMSE was 0.37 m s^{-1} in the three smallest wind-speed scenarios (up to 10 m s^{-1}) at a distance of 9.6D. However, the WAsP wake-algorithms underpredict the wake horizontal speed magnitude at large wind-speeds, possibly indicating that the calculated wake recovery of the algorithms is too rapid, although this discrepancy may also relate to the influence of stability on turbulence profiles and hence on wake decay.

Table 5. Wake calculations (wind-speed) from the WAsP algorithms at a range of downstream distances. The wake wind-speed increases as the airflow recovers with downstream distance. Also shown are the wake wind speeds derived from the observations at the meteorological mast at a distance of 9.6 rotor diameters downstream. The wake wind speeds are for the hub-height in m s^{-1}.

Distance as rotor diameters	3 WAsP	5 WAsP	7 WAsP	9.6 WAsP	9.6 observed	Difference at 9.6 U_{WAsP}-$U_{freestream}$	10 WAsP
Distance (m)	106.5	177.5	248.5	340.8			355
Freestream wind speed (m/s)							
5.02	2.89	3.42	3.77	4.08	4.33	-0.25	4.12
7.27	5.03	5.59	5.96	6.29	6.42	-0.13	6.32
9.75	7.40	8.02	8.40	8.73	8.80	-0.07	8.78
13.70	11.50	11.96	12.23	12.46	11.74	0.69	12.49

3.4 Quantifying multiple wakes

A major requirement in the development of large offshore windfarms is design for wake effects within the particular windfarm and the propagation of total windfarm wakes between adjacent windfarms. Hence there is a need to predict multiple wakes both within and downwind of large offshore windfarms. A first order approach for the integrated downwind impact of a large windfarm is to treat the windfarm as a 'topographical' element of large roughness [80]. As shown in Table 6 (rows 1,2,3) for calculations using WAsP, the choice of the roughness length (z_0) assigned to the windfarm affects the downwind distance required for the windspeed at hub-height to recover to 98% of its upwind freestream value. It is also possible to use WAsP to predict the same condition using multiple wake superposition; the results are on Table 6, rows 4 and 5. It is clear that the results of the two sets of simulations vary significantly.

Table 6. Recovery distances behind a large (100 turbine) wind farm, shown as distance (km) after the wind farm at which the hub-height wind speed recovers to 98% of its upwind free-stream value

Table row number	Model	Distance (km)
1	WAsP z_0(block) 0.1 m	6
2	WAsP z_0(block) 0.5 m	7
3	WAsP z_0(block) 1.0 m	8
----------	----------	----------
4	WAsP wake decay 0.075	2
5	WAsP wake decay 0.05	3

One possible reason for the divergent results in Table 6 is that turbulence generated by large windfarms is likely to alter the structure of the atmospheric boundary-layer and so change momentum transfer affecting wake recovery. This effect is not important over small windfarms so is not analysed in most wake models. Note that a relatively new model within WAsP (the analytical model), which integrates the momentum deficit from individual wakes in order to calculate the whole wind farm wake [81], is now being evaluated [82].

4. WINDFARM POWER OUTPUT AT SHORT TIME SCALES: IS OFFSHORE DIFFERENT?

The primary concern of this chapter so far has been the prediction of power output over the lifetime of a wind farm, so long-term effects have been studied. However, wind climates vary over a range of time scales, including scales relevant for turbine control (from milliseconds to seconds), and for the integration of wind power in the electrical grid (from minutes to days). Short-term forecasting involves many of the same issues as long-term forecasting, especially relating larger scale model data to site-specific energy output..

Knowing weather conditions and with previous experience, electricity power demand can be predicted with a high degree of accuracy; usually about ±1.5% for 24 hours ahead and about ±5% for one week ahead [83]. Windfarm energy production cannot be forecast with such accuracy, although the greater the number of dispersed turbines, the greater the accuracy of predication for the total output using national meteorological information.

Hence there is considerable emphasis on developing more accurate wind forecasts which can be used to estimate windfarm power output and to bid for a price in the electricity market up to, say, 48 hours ahead. Most methods use Numerical Weather Prediction (NWP) models to predict site specific wind speeds and hence power output, using similar downscaling methods to those described in section 2. The NWP models typically have resolutions of 5 to 55 km in the horizontal, have up to 40 vertical height zones and use a six hour data assimilation cycle. The grid-cell averaged NWP model output is related to small scale variability using physical/dynamical, empirical/statistical or hybrid approaches [83] to generate location specific power production. For short-term forecasting, statistical methods have some advantages, since in the first six hours persistence-type models give good results. Statistical methods are fast and require limited computing resources and power output can be predicted directly, avoiding the extra step of utilising the turbine power curve [83] to relate the predicted wind speed at hub-height to the power produced by the turbine. As described in section 1, use of a manufacturer's standard power curve gives uncertainty in power predictions. Nevertheless in short-term prediction, the largest error in prediction of short-term fluctuations in power output is due to bias and uncertainties in the NWP model simulations [83].

In general, short-term forecasting should be more straightforward offshore than onshore. This is because offshore wind-speeds in the power-producing range (i.e. more than cut-in speed) are more persistent, and the duration of calms is less, than for adjacent land sites [84]. Indeed improved accuracy in offshore wind speed predictions, as compared with nearby onshore, has been obtained [66]. A further advantage for offshore sites, in terms of the accuracy of power production, is that the average diurnal variability of wind speeds is of minor importance [10], especially in comparison with land sites in the summer. At typical land sites, wind speed maxima are obtained in the early afternoon whereas the maxima appear to occur later at offshore sites which are influenced by nearby land. Minimum wind speeds appear to occur in the early morning at coastal offshore sites. However, this seems to be location dependent and maybe due to advection of stronger wind speeds from land, thermal gradients or interaction with larger scale circulation, e.g.. sea breezes. The major issues remaining for offshore short-term forecasting are large horizontal gradients in predicted wind speeds in coastal areas [85]. For some sites there is relatively little information from upwind locations giving a large range of accuracy of the forecasts. Errors in the predicted wind-speed profiles may be due to use of simple prediction models, e.g. logarithmic or stability corrected profiles. The effects of this will be greatest for large turbines with taller hub-heights and larger rotor diameters. The trend towards larger windfarms means that wake losses have a greater impact on the overall power prediction.

5. SUMMARY

Accurate prediction of power output from wind farms requires an assessment of the hub-height wind speed which will be valid for the next 20-30 years. Variability of wind speed is usually assessed based on past climates, where the inter-annual variability has usually been at least ± 5% at northern European sites. However, the variability has been as large as ± 30% for individual years, especially if sites are positioned in heterogeneous locations (e.g. the coastal zone). For site measurements, in general terms, the shorter the record used to assess the wind climate, the more uncertainty will be associated with it. However, even using 10 years data will not fully capture the total variability, especially under non-stationary climate conditions.

In terms of future wind speeds under climate change, recent modelling with Global Climate Models (GCMs) suggests that, at least in northern Europe, major shifts in the wind climate are not anticipated over the coming decades. However, the range of GCM wind-speed projections remains large (i.e. there is large uncertainty), although this can be reduced with appropriate downscaling techniques. Best available estimates suggest the climate signal is moderate and of the same order as the uncertainty in GCM modelling of the 1961-1990 wind climate.

For site wind-speed estimation, use of remote sensing data derived from satellite-borne instrumentation provide a good overview of the likely range and spatial distribution of wind speeds. However, these techniques are not yet fully developed to provide good temporal and spatial resolution data at the required accuracy for site assessment. Mesoscale model data are also becoming available and appear to give a good representation of the wind climate. However, at present, by far the most accurate method of determining the potential power output is to measure site wind speeds at hub-height on a fixed meteorological mast. Assuming these are short-term measurements (of the order of a few years) they can be related to longer-term wind speeds using (i) mesoscale, (ii) linearised models, or (iii) well-known statistical techniques, such as measure-correlate-predict. If extrapolating from onshore to offshore, care needs to be taken that the whole wind speed distribution, and not just the mean wind speeds, are transformed.

If hub-height measurements are not available, wind speeds will need to be extrapolated vertically. Standard methods, e.g. the power law, logarithmic profile and stability corrected profile, all seem to be inadequate offshore at heights above about 50 m. A major issue is the lack of observations at heights above 50 to 70 m, to use for model validation.

Wake effects from large offshore windfarms are expected to cause power reductions in the range 10 to 20%. Present models give acceptable results for single wakes and small windfarms, but none of the state-of-the-art models include feedback to and from the boundary-layer which is probably needed for large windfarms. A new analytical model has been specifically designed to calculate the effect of wakes within and down-wind of large windfarms which requires further evaluation.

For short-term forecasting, errors in wind speeds from Numerical Weather Prediction models are typically the largest source of uncertainty in prediction of power production up to 48 hours ahead. For short-term forecasting offshore, the additional major uncertainties appear to be the strong horizontal gradients of wind speed in coastal areas and the accurate prediction of vertical profiles offshore..

It should be noted that we have focussed exclusively on meteorological and climatological aspects of wind energy developments in offshore environments. Other issues such as losses within the electric cables are also significant and are covered in other chapters within this volume.

6. ACKNOWLEDGEMENTS

This work has been funded in part by the National Science Foundation grant #0618364, Danish PSO (FU 4103) and Research Ministry of Science Research and Innovation (2104-04-0005) projects, the 'Impacts of Climate Change on Renewable Energy Sources and their Role in the Energy System: 2003–2006' project funded by Nordic Energy Research and National and the Edinburgh Research Partnership. We would also like to acknowledge one of the editors John Twidell for his clarifications.

7. REFERENCES

1. Wieringa, J., *Updating the Davenport roughness classification*. Journal of Wind Engineering and Industrial Aerodynamics, 1992. **41**: p. 357-368.
2. Charnock, H., *Wind stress on a water surface*. Quarterly Journal of the Royal Meteorological Society, 1955. **81**: p. 639-640.
3. Barthelmie, R.J., *The effects of atmospheric stability on coastal wind climates*. Meteorological Applications, 1999. **6**(1): p. 39-48.
4. Motta, M., R.J. Barthelmie, and P. Vølund, *The influence of non-logarithmic wind speed profiles on potential power output at Danish offshore sites*. Wind Energy, 2005. **8**: p. 219-236.
5. Stull, R.B., *An introduction to boundary layer meteorology*. ISBN 90-277-2768-6 ed. 1988, Dordrecht: Kluwer Publications Ltd. 666 pp.
6. Stull, R.B., *Meteorology for scientists and engineers*. 2000, Pacific Grove, CA: Brooks/Cole. 502 pp.
7. Van Wijk, A.J.M., A.C.M. Beljaars, A.A.M. Holtslag, and W.C. Turkenburg, *Evaluation of stability corrections in wind speed profiles over the North Sea*. Journal of Wind Engineering and Industrial Aerodynamics, 1990. **33**: p. 551-566.
8. Barthelmie, R.J. *Monitoring offshore wind and turbulence characteristics in Denmark*. in *Proceedings of the 21st British Wind Energy Association Conference*. 1999. Cambridge: Professional Engineering Publishing.
9. Garratt, J.R., *The stably stratified internal boundary layer for steady and diurnally varying offshore flow*. Boundary-Layer Meteorology, 1987. **38**(4): p. 369-394.
10. Barthelmie, R.J., B. Grisogono, and S.C. Pryor, *Observations and simulations of diurnal cycles of near-surface wind speeds over land and sea*. Journal of Geophysical Research (Atmospheres), 1996. **101**(D16): p. 21,327-21,337.
11. Smedman, A.S., U. Hogstrom, and H. Bergstrom, *Low level jets – a decisive factor for off-shore wind energy siting in the Baltic Sea*. Wind Engineering, 1996. **20**(3): p. 137-147.
12. Simpson, J.E., *Sea breeze and local winds*. 1994, Cambridge: Cambridge University Press. 234 pp.
13. Badger, J., R. Barthelmie, S. Frandsen, and M. Christiansen. *Mesoscale modelling for an offshore wind farm*. in *European Wind Energy Association Conference*. 2006. Athens, Greece, February 2006. .
14. Pavia, E.G. and J.J. O'Brien, *Weibull statistics of wind speed over the ocean*. Journal of Climate and Applied Meteorology, 1986. **25**: p. 1324-1332.
15. Barthelmie, R.J. and S.C. Pryor, *Can satellite sampling of offshore wind speeds realistically represent wind speed distributions?* Journal of Applied Meteorology, 2003. **42**: p. 83-94.
16. Pryor, S.C., M. Nielsen, R.J. Barthelmie, and J. Mann, *Can satellite sampling of offshore wind speeds realistically represent wind speed distributions? Part II: Quantifying uncertainties associated with sampling strategy and distribution fitting methods*. Journal of Applied Meteorology, 2004. **43**: p. 739-750.
17. Vikkelsø, A., J.H.M. Larsen, and H.C. Sørensen, *The Middelgrunden offshore wind farm. A popular initiative*. 2003, Copenhagen Environment and Energy Office CEEO: Copenhagen. p. 28.
18. IEA, *IEA Wind Energy Annual Report 2004*. 2005, Kendall Printing Company, USA. p. 264.
19. Dyre, K. *Vindeby off-shore wind farm – the first experiences*. in *'The Potential of Wind Farms', EWEA Special Topic Conference '92, September 1992*. 1992. Herning, Denmark.
20. Frandsen, S., L. Chacon, A. Crespo, P. Enevoldsen, R. Gomez-Elvira, J. Hernandez, J. Højstrup, F. Manuel, K. Thomsen, and P. Sørensen, *Measurements on and modelling of offshore wind farms*. 1996, Risø National Laboratory: Denmark.
21. Barthelmie, R.J., M.S. Courtney, J. Højstrup, and S.E. Larsen, *Meteorological aspects of offshore wind energy – observations from the Vindeby wind farm*. Journal of Wind Engineering and Industrial Aerodynamics, 1996. **62**(2-3): p. 191-211.
22. Jensen, L. *Wake measurements from the Horns Rev wind farm*. in *European Wind Energy Conference*. 2004: EWEA (on CD).
23. Pryor, S.C. and R.J. Barthelmie, *Long term trends in near surface flow over the Baltic*. International Journal of Climatology, 2003. **23**: p. 271-289.

24. Kistler, R., E. Kalnay, W. Collins, S. Saha, G. White, J. Woollen, M. Chelliah, W. Ebisuzaki, M. Kanamitsu, V. Kousky, H. van den Dool, R. Jenne, and M. Fiorino, *The NCEP-NCAR 50 year reanalysis: Monthly mean CD-ROM and documentation*. Bulletin of the American Meteorological Society, 2001. **82**: p. 247-267.
25. Kalnay, E., M. Kanamitsu, R. Kistler, W. Collins, D. Deaven, L. Gandin, M. Iredell, S. Saha, G. White, J. Woollen, M. Chelliah, Y. Zhu, M. Chelliah, W. Ebisuzaki, W. Higgins, J. Janowiak, K.C. Mo, C. Ropelweski, J. Wang, A. Leetmaa, R. Reynolds, R. Jenne, and D. Joseph, *The NCEP/NCAR 40 reanalysis project*. Bulletin of the American Meteorological Society, 1996. **77**: p. 437-471.
26. Marshall, J., Y. Kushnir, D. Battisti, P. Chang, A. Czaja, R. Dickson, J. Hurrell, M. McCartney, R. Saravanan, and M. Visbeck, *North Atlantic climate variability: Phenomena, impacts and mechanisms*. International Journal of Climatology, 2001. **21**: p. 1863-1898.
27. Hurrell, J., Y. Kushnir, G. Ottersen, and M. Visbeck, *An overview of the North Atlantic Oscillation*. The North Atlantic Oscillation: Climatic Significance and Environmental Impact. 2003: Geophysical Monograph 134, American Geophysical Union. 35 pp.
28. Alexandersson, H., H. Tuomenvirta, T. Schmith, and K. Iden, *Trends of storms in NW Europe derived from an updated pressure data set*. Climate Research, 2000. **14**: p. 71-73.
29. Pryor, S.C., R.J. Barthelmie, and J.T. Schoof, *The impact of non-stationarities in the climate system on the definition of 'a normal wind year': A case study from the Baltic*. International Journal of Climatology, 2005. **25**: p. 735-752.
30. Meehl, G., C. Covey, M. Latif, B. McAvaney, J. Mitchell, and R. Stouffer, *Soliciting participation in climate model analyses leading to IPCC Fourth Assessment Report*. EOS, 2004. **85**: p. 274.
31. IPCC, *IPCC Third Assessment Report. Climate Change 2001: The scientific basis*, ed. J. Houghton, et al. 2001, Cambridge, U.K.: Cambridge University Press. 881 pp.
32. Pryor, S.C., J.T. Schoof, and R.J. Barthelmie, *Winds of Change? Projections of near-surface winds under climate change scenarios*. Geophysical Research Letters, 2006. **33**(L11702): p. doi:10.1029/2006GL026000.
33. Robeson, S.M. and K.A. Shein, *Spatial coherence and decay of wind speed and power in the north-central United States*. Physical Geography, 1997. **18**: p. 479-495.
34. Breslow, P.B. and D.J. Sailor, *Vulnerability of wind power resources to climate change in the continental United States*. Renewable Energy, 2002. **27**: p. 585-598.
35. Pryor, S.C., R.J. Barthelmie, and E. Kjellstrom, *Analyses of the potential climate change impact on wind energy resources in northern Europe using output from a Regional Climate Model*. Climate Dynamics, 2005. **25**: p. 815-835.
36. Pryor, S.C., J.T. Schoof, and R.J. Barthelmie, *Empirical downscaling of wind speed probability distributions*. Journal of Geophysical Research, 2005. **110**(D19109): p. doi: 10.1029/2005JD005899.
37. IPCC, *Emissions Scenarios, 2000*, ed. N. Nakicenovic and R. Swart. 2000: Cambridge University Press, UK. 570 pp.
38. Stratton, R.A., *A high resolution AMIP integration using the Hadley Centre model HadAM2b*. Climate Dynamics, 1999. **15**: p. 9-28.
39. Pope, V., M. Gallani, P. Rowntree, and R. Stratton, *The impact of new physical parameterizations in the Hadley Centre climate model: HadAM3*. Climate Dynamics, 2000. **16**: p. 123-146.
40. Johns, T.C., R.E. Carnell, J.F. Crossley, J.M. Gregory, J.F.B. Mitchell, C.a. Senior, S.F.B. Tett, and R.A. Wood, *The second Hadley Centre coupled ocean-atmosphere GCM: model description, spinup and validation*. Climate Dynamics, 1997. **13**: p. 103-134.
41. Giorgi, F. and L.O. Mearns, *Probability of regional climate change based on the Reliability Ensemble Averaging (REA) method*. Geophysical Research Letters, 2003. **30**(12): p. 1629, doi:10.1029/2003GL017130.
42. Pryor, S.C., R.J. Barthelmie, and J.T. Schoof, *Inter-annual variability of wind indices over Europe*. Wind Energy, 2006. **9**: p. 27-38.
43. Giorgi, F., B. Hewitson, J. Christensen, M. Hulme, H. von Storch, P. Whetton, R. Jones, L. Mearns, and C. Fu, *Chapter 10: Regional climate information – Evaluation and projections*, in *Climate Change 2001: The Scientific Basis*. 2001, Cambridge University Press: Cambridge. p. 583-638.

44. Pryor, S.C., J.T. Schoof, and R.J. Barthelmie, *Potential climate change impacts on wind speeds and wind energy density in northern Europe: Results from empirical downscaling of multiple AOGCMs*. Climate Research, 2005. **29**: p. 183-198.
45. Barthelmie, R.J., M. Courtney, B. Lange, M. Nielsen, A. Sempreviva, M., J. Svenson, F. Olsen, and T. Christensen. *Offshore wind resources at Danish measurement sites*. in *Proceedings of the 1999 European Wind Energy Conference and Exhibition*. 1999. Nice: James and James Scientific Publishers Ltd.
46. Watson, G., J. Halliday, J. Palutikof, T. Holt, R. Barthelmie, J. Coelingh, L. Folkerts, G. Wiegerinck, E. van Zuylen, J. Cleijne, and G. Hommel. *POWER -A methodology for the prediction of offshore wind energy resources*. in *Proceedings of OWEMES (Offshore Wind Energy in Mediterranean and Other European Seas) conference*. 2000. Sicily: ATENA/ENEA.
47. Wu, J., *Sea surface winds – a critical input to oceanic models but are they accurately measured?* Bulletin of the American Meteorological Society, 1995. **76**: p. 13-19.
48. Liu, W.T., *Progress in scatterometer application*. Journal of Oceanography, 2002. **58**: p. 121-136.
49. Hasager, C.B., M. Nielsen, P. Astrup, R. Barthelmie, E. Dellwik, N.O. Jensen, B. Jørgensen, S. Pryor, O. Rathmann, and B. Furevik, *Offshore wind resource assessed from satellite SAR wind field maps*. Wind Energy, 2005. **8**: p. 403-419.
50. Hasager, C.B., R. Barthelmie, M. Christiansen, M. Nielsen, and S. Pryor, *Quantifying offshore wind resources from satellite maps: study area the North Sea*. Wind Energy 2006. **9**(1-2): p. 63-74.
51. Kerbaol, V., B. Chapron, and P.W. Vachon, *Analysis of ERS-1/2 synthetic aperture radar wave mode imagettes*. Journal of Geophysical Research: Oceans, 1998. **103**: p. 7833-7846.
52. Barthelmie, R., O. Hansen, K. Enevoldsen, J. Højstrup, S. Larsen, S. Frandsen, S. Pryor, M. Motta, and P. Sanderhoff, *Ten years of meteorological measurements for offshore wind farms*. Journal of Solar Energy Engineering, 2005. **127(2)**: p. 170-176.
53. Bergström, H., *Boundary-layer modelling for wind climate estimates*. Wind Engineering, 2002. **25(5)**: p. 289-299.
54. Bergström, H. and R.J. Barthelmie. *Offshore boundary-layer modelling*. in *2002 Global Windpower*. 2002. Paris: EWEA.
55. Mortensen, N.G., L. Landberg, I. Troen, and E.L. Petersen, *Wind Analysis and Application Program (WASP)*. 1993, Risø National Laboratory: Roskilde, Denmark.
56. Mortensen, N.G., D. Heathfield, L. Landberg, O. Rathmann, I. Troen, and E. Petersen, *Getting started with WASP 7.0*. 2000, Risø National Laboratory: Roskilde.
57. Kallstrand, B., H. Bergstrom, J. Højstrup, and A.S. Smedman, *Mesoscale wind field modifications over the Baltic Sea*. Boundary-Layer Meteorology, 2000. **95**: p. 161-188.
58. Rogers, A., J. Rogers, and J.F. Manwell, *Comparison of the performance of four measure-correlate-predict algorithms*. Journal of Wind Engineering and Industrial Aerodynamics, 2005. **93**: p. 243-264.
59. Bunn, J.C. and S.J. Watson. *A new matrix method of predicting long-term wind roses with MCP*. in *European Union Wind Energy Conference, May 1996*. 1996. Goteborg, Sweden: H.S. Stephens and Associates, Bedford, UK.
60. Hsu, S.A., *Coastal Meteorology*. ISBN 0-12-357955-4 ed. 1988, London: Academic Press Inc. 260 pp.
61. Hsu, S.A., *Determining the Power-law wind-profile exponent under near-neutral stability conditions at sea*. Journal of Applied Meteorology, 1994. **33**(6): p. 757-765.
62. Antoniou, I., H. Jørgensen, T. Mikkelsen, S. Frandsen, R. Barthelmie, C. Perstrup, and M. Hurtig. *Offshore wind profile measurements from remote sensing instruments*. in *European Wind Energy Association Conference*. 2006. Athens, Greece, February 2006.
63. Barthelmie, R.J., L. Folkerts, F. Ormel, P. Sanderhoff, P. Eecen, O. Stobbe, and N.M. Nielsen, *Offshore wind turbine wakes measured by SODAR*. Journal of Atmospheric and Oceanic Technology, 2003. **30**: p. 466-477.
64. Barthelmie, R.J., *Evaluating the impact of wind induced roughness change and tidal range on extrapolation of offshore vertical wind speed profiles*. Wind Energy, 2001. **4**: p. 99-105.

65. Van Wijk, A.J.M., A.C.M. Beljaars, A.A.M. Holtslag, and B. Turkenburg, *Evaluation of stability corrections in wind speed profiles over the North Sea*. Journal of Wind Engineering and Industrial Aerodynamics, 1990. **33**: p. 551-566.
66. Tambke, J., M. Lange, U. Focken, J. Wolff, and J. Bye, *Forecasting offshore wind speeds above the North Sea*. Wind Energy, 2005. **8**: p. 3-16.
67. Larsen, S.E., S.E. Gryning, N.O. Jensen, H.E. Jørgensen, and J. Mann. *Profiles of mean wind and turbulence in the atmospheric boundary layer above the surface layer*. in *In: EUROMECH colloquium 464b: Wind energy. International colloquium on fluid mechanics and mechanics of wind energy conversion*. 2005. Oldenburg (DE), 4-7 Oct 2005: ForWind, Oldenburg.
68. Bye, J., *Inertially coupled Ekman layers*. Dynamics of Atmospheres and Oceans, 2002. **35**: p. 27-39.
69. Ainslie, J.F., *Calculating the flow field in the wake of wind turbines*. Journal of Wind Engineering and Industrial Aerodynamics, 1988. **27**: p. 213-224.
70. Lange, B., H.P. Waldl, A. Guerrero, and R. Barthelmie, *Improvement of the wind farm model FLaP (Farm Layout Program) for offshore applications*. Wind Energy, 2003. **6**: p. 87-104.
71. Schepers, J.G., *ENDOW: Validation and improvement of ECN's wake model*. 2003, ECN: Petten, The Netherlands. p. 113.
72. Crespo, A., J. Hernandez, and S. Frandsen, *Survey and modelling methods for wind turbine wakes and wind farms*. Wind Energy, 1999. **2**: p. 1-24.
73. Frandsen, S., *Turbulence and turbulence-generated fatigue loading in wind turbine clusters*. 2005, Risoe National Laboratory: Roskilde. p. 128
74. Schlichting, H., *Boundary layer theory (6th edition)*. 1968, New York: McGraw-Hill. 748 pp.
75. Pope, S.A., *Turbulent flows*. 2000: Cambridge University Press. 806 pp.
76. Jensen, N.O., *A note on wind turbine interaction*. 1983, Risø National Laboratory: Roskilde, Denmark. p. 16.
77. Katic, I., J. Højstrup, and N.O. Jensen. *A simple model for cluster efficiency*. in *European Wind Energy Association*. 1986. Rome.
78. Rados, K., G. Larsen, R. Barthelmie, W. Schelz, B. Lange, G. Schepers, T. Hegberg, and M. Magnusson, *Comparison of wake models with data for offshore windfarms*. Wind Engineering, 2002. **25**: p. 271-280.
79. Schlez, W., A.E. Umaña, R. Barthelmie, S. Larsen, K. Rados, B. Lange, G. Schepers, and T. Hegberg, *ENDOW: Improvement of wake models within offshore windfarms*. Wind Engineering, 2002. **25**: p. 281-287.
80. Crespo, A., L. Chacon, J. Henandez, F. Manuel, and R. Gomez Elvira. *Modelization of offshore wind farms. Effect of the surface roughness of the sea*. in *European Union Wind Energy Conference*. 1996. Goteborg: H.S. Stephens and Associates, Bedford, UK.
81. Frandsen, S., R. Barthelmie, S. Pryor, O. Rathmann, S. Larsen, J. Højstrup, and M. Thøgersen, *Analytical modelling of wind speed deficit in large offshore wind farms*. Wind Energy 2006. **9**(1 2): p. 39-53
82. Rathmann, O., R.J. Barthelmie, and S.T. Frandsen. *Wind turbine wake model for wind farm power production*. . in *European Wind Energy Conference*. 2006. Athens.
83. Giebel, G., R. Brownsword, and G. Kariniotakis, *The state-of-the-art in short-term prediction of wind power. A literature overview*. 2003, http://anemos.cma.fr/download/ANEMOS_D1.1_StateOfTheArt_v1.1.pdf, 2003.08.12.: Roskilde. p. 36.
84. Pryor, S.C. and R.J. Barthelmie, *Comparison of potential power production at on- and off- shore sites*. Wind Energy, 2002. **2001:4**: p. 173-181.
85. Barthelmie, R.J. and G. Giebel. *Prediction of wind speed profiles for short-term forecasting in the offshore environment*. in *European Wind Energy Conference*. 2006. Athens.

Chapter 5:

Electrical Aspects of Wind Turbines

Z. Chen and F. Blaabjerg

I. INTRODUCTION

Global wind energy capacity is increasing rapidly at about 30%/year, outpacing all other electrical power technologies. Initially, wind power production exported to the local grid was too small to have any serious impacts on the operation and control of the total power system. However, now large penetration of wind power in certain areas plays an active part in the grid, such as offshore windfarms with capacities of hundreds of MW, as being developed in Europe and other regions around the world.

Early electrical technology used in wind turbines was based on squirrel-cage induction generators directly connected to the grid. Consequently, power variations in the wind were directly transferred to the grid without control of the active and reactive power, which, in principle, are the key control parameters for power system frequency and voltage. However, as the power capacity of new wind turbines increases, so their controllability has significantly improved.

Although turbines with standard induction generators are still used, the trend is to use more advanced generators, e.g. doubly-fed induction or fully decoupled synchronous generators. Such systems use solid-state power electronics to (i) allow the turbines to have variable speed and so be more efficient and less mechanically stressed in changing winds, and (ii) to strengthen, rather than weaken, the reactive power impact on the local grid [1].

This chapter describes the basic electrical aspects of wind power conversion systems, including (i) the methods of controlling the input power to the system, (ii) the generators and power electronic converters, (iii) the various wind turbine configurations, and (iv) some control methods. The discussion applies to grid-connected wind turbines in general; both on land and offshore wind farms.

The main electrical power components that may be included in a wind turbine system are illustrated in Fig. 1, including turbine rotor, gearbox, generator, power electronics and a transformer for grid connection. In essence, the turbine rotor converts the fluctuating wind energy into mechanical energy which is converted into electrical power by the generator, and then transferred into the grid through a transformer and transmission lines.

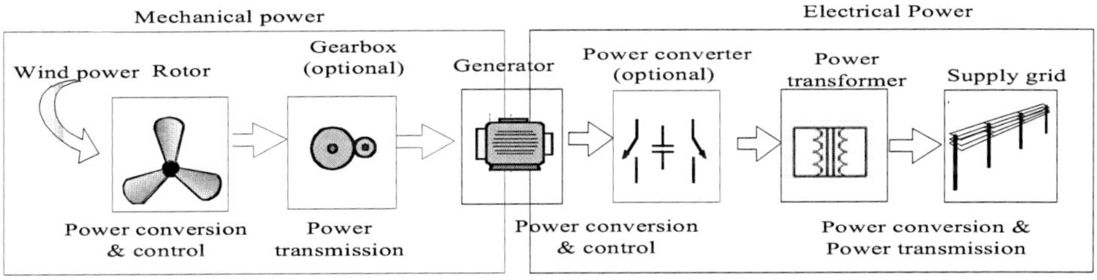

Figure 1. Main components of a wind turbine system

Wind turbines capture the power from the wind by means of aerodynamically designed blades and convert it to rotating mechanical power. The number of blades is normally three and the rotational speed will *decrease* as the radius of the blade *increases*. For large multimegawatt-scale wind turbines, the rotational speed will usually be 10 to 15 rpm (i.e. rotation period 4 to 6 seconds) . The traditional way to convert such low-speed, high-torque mechanical power to electrical power is with a gear-box to increase the shaft speed to a 4 or 6 pole-pair standard generator rotating at about 1,500 rpm or 1,000 rpm. However, the gear box may not be necessary for multi-pole generators.

The generator converts the mechanical power to electrical power, which is usually, fed through a transformer into a grid directly or through power electronic converters. In addition, there will be circuit breakers and electricity meters. The connection of wind turbines to the local circuitry or grid is possible at the voltage of distribution system level or at the high and the extra high voltage of national transmission since in practice the transmittable power of an electrical system increases with increase of voltage. Most onshore wind farms are connected to distribution systems, but, large offshore wind farms are expected to be connected to the high and extra high voltage grids of national transmission.

The electrical losses within a wind farm include the losses within generators, transformers, power cables and power electronic converters. In addition, the onsite power for computation, lights and heating etc are a form of loss. Each turbine has its own adjacent transformer to raise voltage from the generator terminal voltage (usually under 1,000 V) to distribution system site voltage. Most wind farms have a dedicated substation for monitoring and further voltage increase for grid compatible connection. At this 'point of common coupling' (PCC) with the grid, there is a circuit breaker for the disconnection if faults arise in either the grid (which are relatively common) or within the wind plant.

2. TURBINE POWER CONTROL

The conversion of wind power to mechanical power is done aerodynamically by the turbine rotor. It is important to be able to limit the input mechanical power during high-speed wind that would otherwise cause the system damage. The *aerodynamic power control* is either (a) with fixed blades designed to stall aerodynamically in strong winds at rated power, or (b) with blades or blade tips that can be turned actively at all power conditions. The purpose of blade pitching is to produce either stall at large angle of attack or less power at small angles of attack. In practice, method (a) is used with constant speed rotors (as with simple induction generators), and methods (b) are mostly used with variable speed rotors linked to doubly-fed generator systems.

Passive stall on fixed-speed turbines is the simplest control method, since the blade-fixing at the rotor hub does not have to change. At higher-speed wind, the angle of attack of the relative wind onto the blade increases, so reducing lift and increasing drag, thereby reducing the output power. The disadvantage of passive stall is that erratic aerodynamic behaviour occurs after stall, which produces variations of power at and above rated wind-speed.

Pitch control limits power above rated wind-speed by rotating all, or part, of each blade about its axis in the direction of reducing angle of attack (blade feathering). In comparison with passive stall, *active pitch control* allows the power to be controlled smoothly and limited to the rated value at rated wind-speed and above. In addition, the aerodynamic braking facility of pitch control can reduce extreme loads on the turbine. The blade pitch angle is typically kept at a small angle of a few degrees near rated operation, and is adjusted to limit the input of wind power at wind-speeds higher than the rated wind-speed. The blades can be pitched towards 90° ('full feather') for aerodynamic braking. Pitch-regulation requires the fast response of *closed-loop control*, which is not required for a stall-regulated machine. Therefore, additional costs are incurred, including the cost of maintenance.

Active stall control limits the power input by pitching the blades initially into stall at larger angles of attack, i.e., in the opposite direction to pitch control. An advantage of active stall control is that the blade remains stalled above the rated wind-speed, so gusts produce less fluctuations in blade loads and power output. Also only small changes of pitch angle are required to maintain the power output at rated, so pitch rates do not need to be as large as that for pitch control. Moreover, full aerodynamic braking requires only a small pitch angle so the travel of the pitch mechanism is significantly reduced compared with pitch control. However, it may be difficult to predict aerodynamic behaviour accurately in stall conditions, with consequent erratic power generation.

Fig. 2 shows the power curves of different types of turbine rotor power limitation methods [2]. It can be seen that the power may be limited smoothly by rotating the blades either by pitch or active stall control, while the power from the stall-controlled turbine shows a small overshoot and then less power output for higher wind-speed.

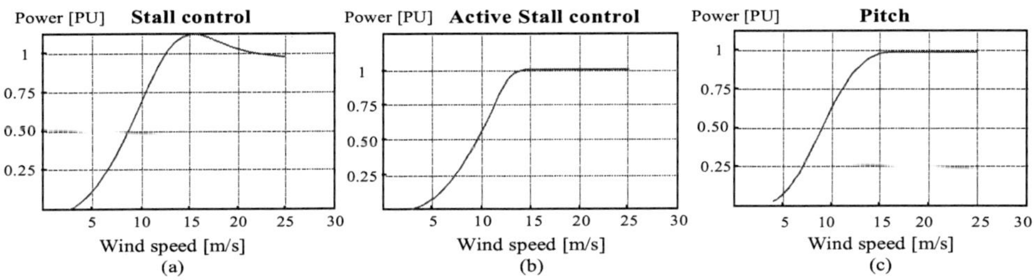

Figure 2. Power characteristics of fixed speed wind turbines. [2]
a) Stall control b) Active stall control c) Pitch control

3. GENERATORS FOR WIND TURBINES

Generators convert mechanical power into electrical power, and are either induction or synchronous generators. Both synchronous and induction generators have similar winding arrangements on the stator, but the rotors of the two machines are quite different. A synchronous machine has permanent magnets or independently excited electromagnets mounted on the rotor (and electrically connected to the exciter by slip rings), so the frequency of the generated electricity is locked into the field produced by the stator. If grid connected, the mechanical rotor must always rotate in-phase with the grid frequency, which is a difficult condition for a wind turbine in varying wind-speed. For 50 Hz electricity, the synchronous speed, which is determined by the number of magnetic pole pairs, is exactly 1500 r.p.m. (4 pole), 1000 r.p.m. (6 pole) or 750 r.p.m. (8 pole).

However, the rotor of an induction machine has a 'squirrel cage' or a wound-rotor winding, so currents are induced as the rotor cuts the magnetic field produced by the stator at a non-zero relative speed. Hence, an induction generator can only develop torque and generate power at a rotational speed *different* from that of the stator field. We should note that a simple induction generator may also be a motor powered from the grid, so its generalised name is 'induction machine'. The change in the ratio of grid frequency, f_g, to the rotor current frequency, f_r, (i.e. $1 - f_g/f_r$) is the 'slip', which is usually in the range of 0% to 5%. Since the slip is small and reasonably constant, the turbine rotational speed is effectively constant, and so the turbine is said to be of 'fixed speed'. A disadvantage of simple induction generators is the reactive component of generation which may be detrimental to the grid supply.

The internal voltage of all generators is limited, usually at the relatively low voltage below 1000 V. Therefore connection to local grids or wind farm interconnections is through transformers. Generators need cooling, usually by air, however some wind turbines use liquid cooling with a heat exchanger to the environment.

3.1 Induction Generators

Induction generators are much less used than synchronous generators for large scale conventional power generation. This is because the inductive magnetisation results in relatively larger energy losses in the rotor. The reactive power necessary to energize the magnetic circuits of induction machines must be supplied from the network, or, especially for small machines, by parallel capacitor banks at the machine terminal. The latter allows self-excitation for autonomous operation or when connection to the network is lost. For simple induction machines, the terminal voltage and reactive power cannot be directly controlled. Simple induction generators in wind turbines may suffer from voltage instability which is a significant concern if used with large-scale wind farms.

A wound-rotor induction machine has a rotor with copper wire windings, which can be connected to an external resistor or to ac systems via power electronic systems. Such a system provides a partial variable speed operation with a small power electronic converter, therefore increased energy capture and reduced mechanical load to the system. This type of system is an economical way to obtain variable speed and supply reactive power and to increase energy yield at low wind-speeds. Also reduced torque loads in the drive train make the whole mechanical construction simple and reliability increased.

3.2 Synchronous Generators

Synchronous generators are used in large central power plants, so controlling the frequency on the grid. For power generation applications, electromagnets excited by an externally applied direct current are used on the rotor. The rotor operates at *exactly* the same 'fixed' speed as the stator magnetic field but it leads the stator field by an angle depending on the applied torque. Synchronous generators are often not practical to provide the degree of damping required for wind turbine applications. What are called 'Fixed-speed'[1] wind turbines normally use induction generators rather than synchronous generator because of the requirement for significant damping in the drive train due to the cyclic variations in the torque developed by the aerodynamic rotor. Basically the damping provided by induction generators is necessary for fixed-speed wind turbines to reduce dynamic stresses in the transmission system, especially the gear box. Therefore, in practice, synchronous generators powered by wind turbines cannot be directly connected to an ac grid. However, if an indirect coupling is used (e.g. rectifier/inverter solid state electronics 'converter'), synchronous generators can be used beneficially, since the turbine rotor can then have variable-speed and the blade rotation can be matched more efficiently to the variable wind-speed. Hence, synchronous generators can be used, if coupled to the grid through full-rate power-electronic converters.

3.3 Direct-drive generators

There is considerable interest in the application of generators driven directly by the wind-turbine rotor without a gearbox and incorporating multi-pole generators with decoupled electronic interface with the grid. The power output of a rotating electrical machine is propositional to the machine length, rotor area and rotor speed. Thus if a smaller rotational speed is adopted, it is necessary either to lengthen the generator or to increase the diameter to provide the same level of power. It is cheaper to increase the diameter, as this raises the power by the square rather than linearly. Thus, direct-drive generators for wind turbines tend to have rather large diameters. This also allows many pole-pairs to be fitted on the large diameter.

Induction generators require a rather small air gap to ensure an adequate air-gap magnetic flux density as the excitation is provided from the stator. In contrast, synchronous generators have excitation systems on the rotor and so can operate with larger air-gaps. It is difficult to manufacture large diameter electrical machines with small air gaps for mechanical and thermal reasons. Hence direct-drive wind turbines use synchronous generators (either with permanent magnet excitation or with an electromagnet). The use of a synchronous generator, in turn, leads to the requirement for power electronic conversion system to de-couple the generator from the network.

In wind turbine applications, the machine can be made as multiple pole and are directly connected to the turbine rotor without a gearbox or with a low turn ratio gear box. A field excitation system used to control the generator terminal voltage. Permanent magnet synchronous machines are also used for wind turbine systems, especially for multipole machines. In this case, slip rings for field excitation are eliminated, together with the power loss of an excitation system; however, the machine terminal

[1] Small variations in slip are ignored, hence the rotor speed is effectively 'fixed' and constant.

voltage cannot be easily controlled with permanent magnet excitation.

In a low speed direct drive system, the turbine and the generator are within a compact and structurally integrated unit. For a medium speed system, the permanent magnet generator may be constructed with integral turbine main bearing and have a single-stage gear box. Low or medium-speed permanent magnet generators may have simple and robust low-speed rotor-design; this results in less wear, reduced maintenance requirements, reduced life-cycle costs, and a long lifetime.

4. MODERN POWER ELECTRONICS AND CONVERTER SYSTEMS

Power electronics plays an important role in the field of modern electrical engineering; it is an essential part for the integration of wind power generation unit to achieve high efficiency and high performance in power systems. The solid-state power electronic converters are used to match the characteristics of the wind turbines and the requirements of the grid connections, including frequency, voltage, control of active and reactive power, harmonic minimization etc.

Many types of wind turbines use power electronic systems as interfaces to operate at variable rotor speed. The aim is to keep the speed of the blade-tip a constant ratio to the unperturbed wind speed, i.e. to maintain constant tip-speed ratio. In this way, the angle of the relative wind to the blade surface is kept constant when the unperturbed wind speed varies. Since the generator shaft is fixed to the rotor shaft, if the wind turbine operates at variable rotational speed, then the electrical frequency of the generator will similarly vary accordingly and must therefore be decoupled from the frequency of the grid. This can be achieved by a power electronic converter system. Even in a fixed speed system where wind power generators are directly connected to the grid, thyristors are used as soft-starters. The modern power electronics play an important role in wind power systems.

Power electronics has changed rapidly during the last thirty years and the number of applications has been increasing, mainly due to the developments of semiconductor devices and microprocessor technology. Improved performance is steadily obtained per unit area silicon chip, whilst the price of devices is continuously falling. The important issues are reliability; efficiency and cost. Today, the cost of power semiconductor devices is decreasing 2-5 % every year for the same output performance. In order to improve reliability and reduce cost still further, the number of components needed is being reduced by a higher level of integration. For example, Fig. 3 shows some key self commutated devices and the area where the development is still going on.

Power electronic converters are constructed by power electronic devices, driving, protection and control circuits. A converter, depending on the topology and application, may allow both directions of power flow and can interface between the load/generator and the grid. There are two different types of converter systems: grid commutated and self commutated converter systems. The grid commutated converters are mainly thyristor converters, 6 or 12 or even more pulse. This type of converter produces

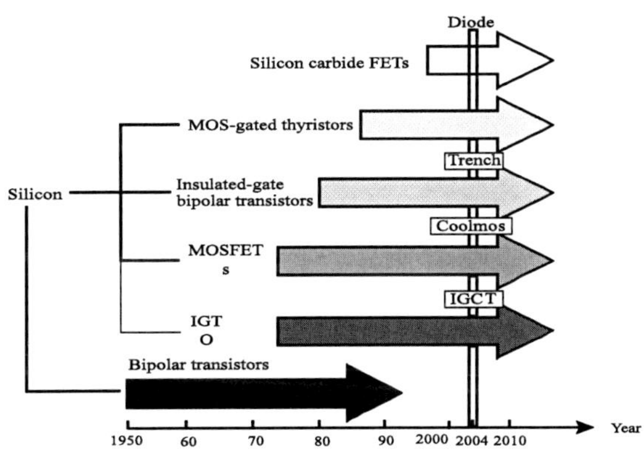

Figure 3. Development of power semiconductor devices in the past and in the future [3].

integer harmonics which in general requires harmonic filters [4, 5]. Also thyristor converters are not able to control the reactive power and consume inductive reactive power, and hence thyristor converters are mainly used for very high voltage and power applications such as conventional HVDC systems.

Self-commutated converter systems are mainly pulse-width modulated (PWM) converters, where IGBTs (Insulated Gate Bipolar Transistor) are mainly used. This type of converter can control both active power and reactive power [6, 7]. That means the reactive power demand can be delivered by a PWM-converter. The high frequency switching of a PWM-converter may produce harmonics and interharmonics. In general these harmonics are in the range of some kHz. Due to the high frequencies, the harmonics are relatively easier to be removed by small size filters. Fig. 4 shows a typical power electronic converter consisting of self commutated semiconductors such as IGBTs and Fig. 5 shows the waveforms of different operation modes.

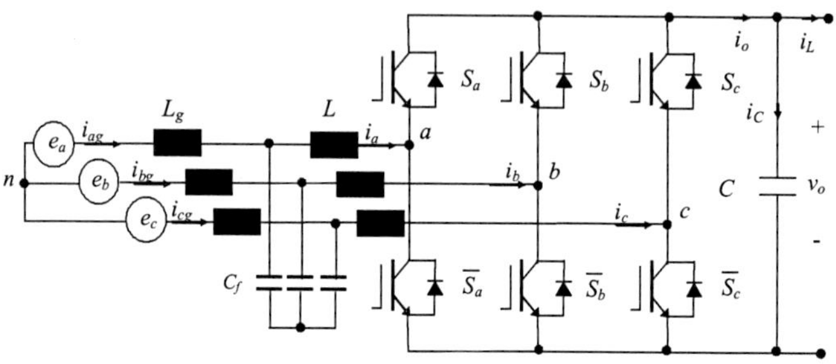

Figure 4. Circuit diagram of a voltage source converter (VSC) with IGBTs

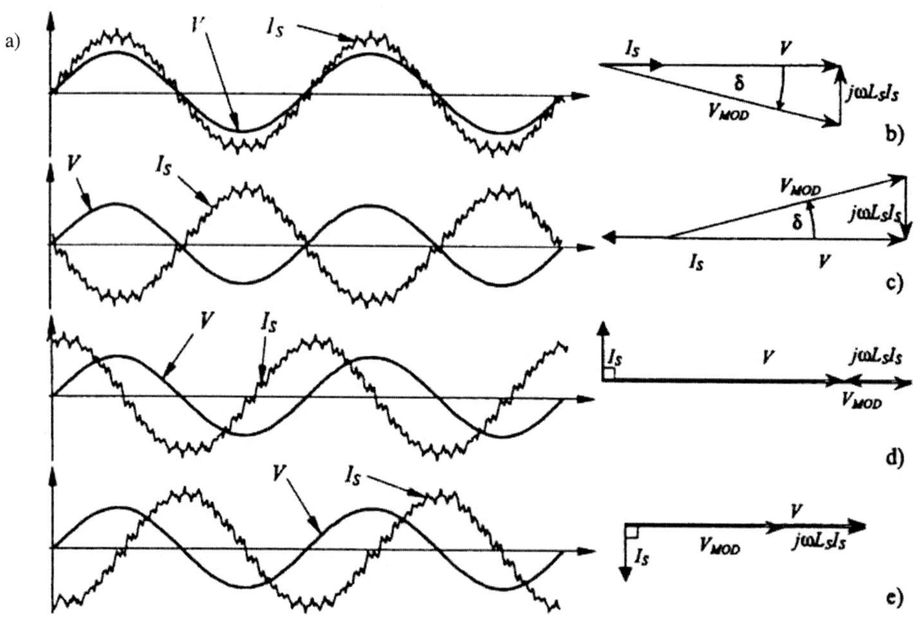

Figure 5. Bi- directional active and reactive power of a VSC
(a) Waveforms of different operation conditions
(b) active power flow from the ac system to the converter dc side
(c) active power flow from the converter dc side to the ac system
(d) the converter generating reactive power
(e) the converter consuming reactive power

Other types of circuit configurations also exist, including the multi-level converter, as shown in Fig. 6, and the matrix converter, as shown in Fig. 7.

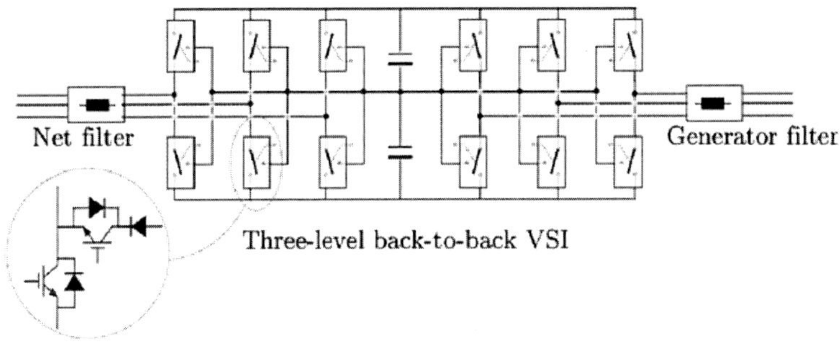

Figure 6. Neutral point clamped back-to-back three-level VSC

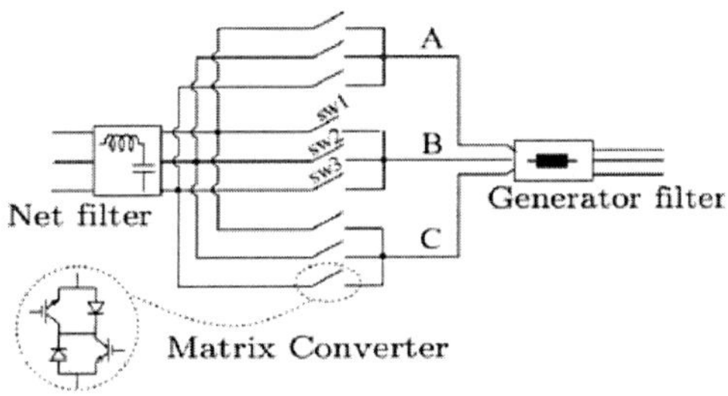

Figure 7. Circuit configuration of a matrix converter

Based on the previously discussed generators and power electronic converters, many possible technical solutions of the electrical system for wind turbines are available. Fig. 8 shows a technological roadmap starting with wind energy/power and converting the mechanical power into electrical power. It involves solutions with and without a gearbox, as well as solutions with or without power electronic conversion. In the following sections, main wind turbine configurations will be presented and explained.

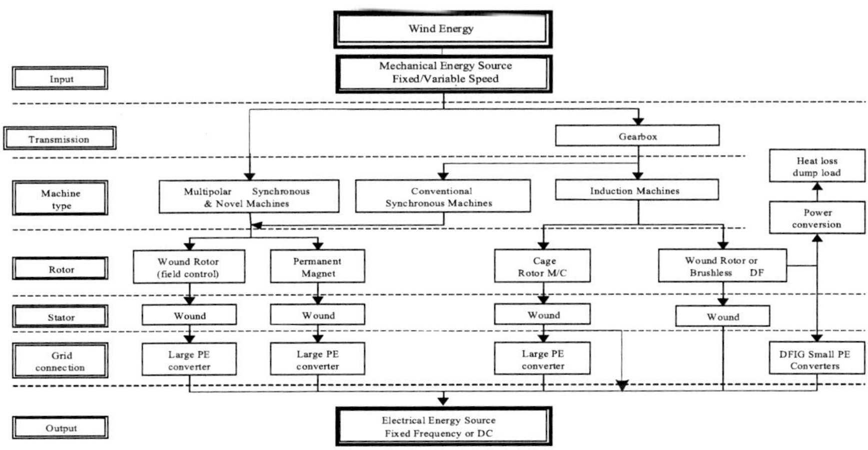

Figure 8. Road-map for wind energy conversion. PE: Power Electronics. DF: Doubly-fed [8], [9].

5. ELECTRICAL POWER CONVERSION SYSTEMS OF WIND TURBINES

Both induction generators and synchronous generators are used in wind turbines, including the grid-connected turbines of this book. Induction generators are of three general types, (i) squirrel cage rotor for fixed speed[2] operation and direct grid connection (Fig. 9), (ii) wound-rotor with variable resistance for enhanced slip and increased variability of speed (Fig. 12), and (iii) doubly-fed for variable speed operation with the rotor grid-connected through an electronic interface (Fig. 13). Variable speed turbines with synchronous generators (Fig. 14) require a power electronic interface with the grid for 100% of the power.

The cage rotor induction machine can be directly connected into an ac system and operates at a fixed speed or uses a full rated power electronic system to operate at variable speed. The wound-rotor generator with rotor resistance-slip control is normally directly connected to ac system, but the slip control provides the ability of changing the operation speed in a certain range. The doubly-fed induction generator is able to provide wide range of speed variation depending on the characteristics of the power electronic converter.. Below we discuss (1) fixed speed systems without power electronics except the thyristor soft-starter, (2) variable speed wind turbine systems including those with partially rated power electronics, and (3) full-scale power electronic interfaced wind turbine systems.

5.1 Fixed speed wind turbines

The "Danish Concept" of directly connecting a wind turbine to the grid is widely used for power ratings up to 2.3 MW. The scheme consists of a squirrel-cage induction generator[3] (SCIG), connected via a transformer to the grid. The wind turbine systems using cage rotor induction generators almost operate at a fixed speed (variation of 1-2%). The power can be limited aerodynamically either by stall, by active stall or by pitch control. The basic configurations of three different fixed speed concepts are shown in Fig. 9. The advantages of wind turbines with induction generators are the simple and cheap construction. In addition no synchronization device is required. These characteristics are attractive due to low-cost and high-reliability. However, directly connected induction machines cannot provide fast (within a few ms) control of the active power. Some drawbacks are: the wind turbine has to operate at constant speed, it requires a stiff power grid to enable stable operation and it may require more expensive mechanical construction to absorb large mechanical stress when wind gusts cause torque pulsations in the drive-drain and the gear-box.

The real power, reactive power and torque characteristics of a typical induction machine are shown in Fig. 10 (a) (b) and (c) [10]. It can be seen that, in the range of stable operation at so-called fixed speed, when the slip increases, the electrical power and torque of the machine also increase. Therefore, the variation of the slip and speed is directly related to the change in generated power from the induction machine.

The large starting currents of induction generators are usually limited by a thyristor soft-starter, and the demand for reactive power is normally met by capacitor banks that may be switched in or out according to the real-power production. In Fig. 10 (b), the reactive power characteristic with a fixed capacitor is also shown. As illustrated, the reactive power requirement can be improved by the capacitor bank. It is usually done by continuously switching capacitor banks following the production variation (5-25 steps).

Directly connecting induction generators to the grid produces transients that are of short duration, but with very large inrush currents. These cause disturbances to the grid and high-torque spikes in the drive train. Unless special precautions are taken, the inrush currents are normally 5-8 times the rated current of the generator after a very short period (less than 100 ms). A transient like this disturbs the grid and limits the acceptable number of wind turbines at the connection point to the grid. All three systems of directly connected induction generators in Fig. 9 use a thyristor controller, the soft-starter,

---

[2] Any small variation in slip is not considered significant.

[3] Generally called a 'machine', since the same device can also be a motor if power is taken from the grid.

to reduce the inrush current [11]. The current limiter or soft starter based on thyristor technology typically limits the RMS value of the inrush current to less than twice the generator rated current. The soft starter has limited thermal and continuous-current capacity, so it is short circuited by a mechanical contactor which carries the full load current after the connection to the grid has been completed. In addition to reducing the impact on the grid, the soft starter also effectively dampens the torque peaks associated with the peak currents and hence reduces the loads on the gearbox.

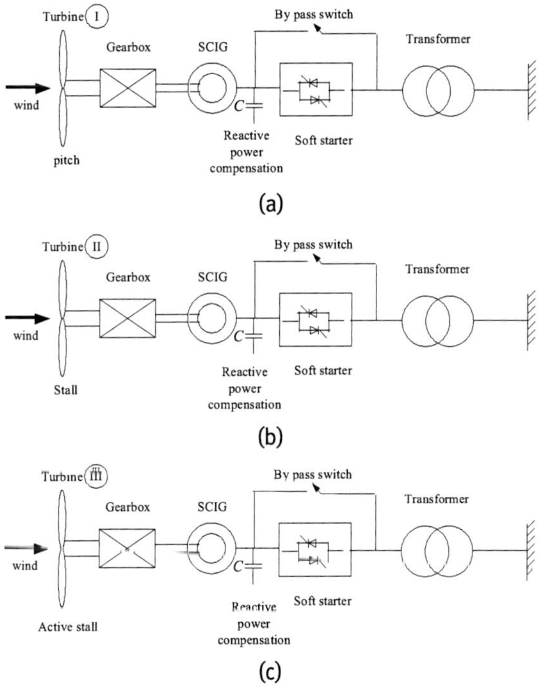

(a) Pitch controlled (System I), (b) Stall controlled (System II) (c) Active stall controlled (System III)
Figure 9. Wind turbine systems without power converter but with aerodynamic power control

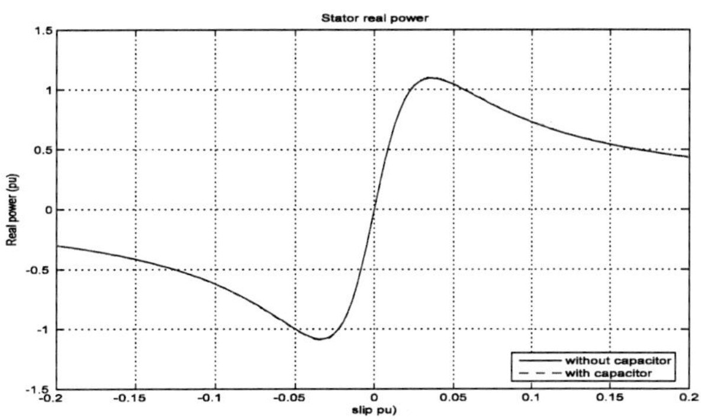

(a) real power characteristics
Fig. 10. *continued over*

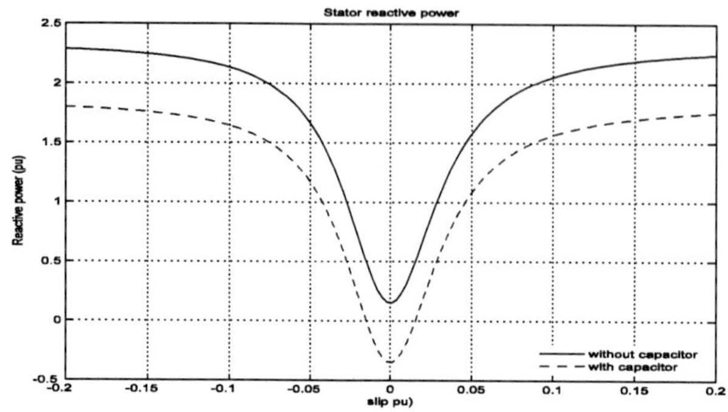

(b) reactive power characteristics

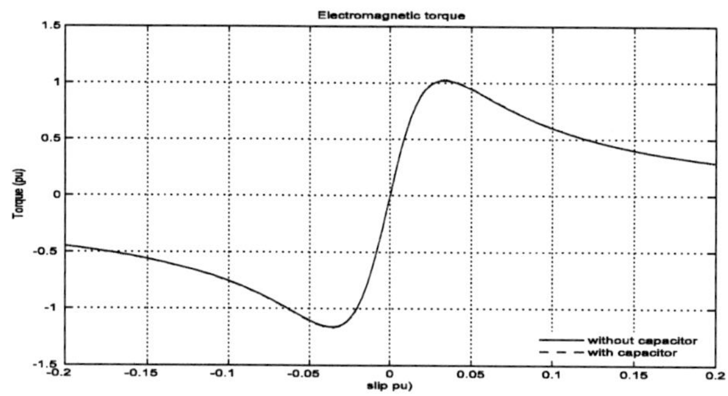

(c) torque characteristic

Figure 10. A cage induction generator directly connected to the grid [10]

An example is shown here to illustrate the start-up of a soft-starter-fed induction generator [12]. The induction machine has 2 MW rated power, 690 V / 1700 A rated phase-voltage and rated line current, respectively (delta connection). The induction machine is connected via a soft-starter to the supply voltage below synchronous speed (1450 rpm). The starting firing angle for the soft-starter is 120°. The equivalent diagram of this system is shown in Fig. 11(a).

The electromagnetic torque and the rotational speed of the high-speed shaft during the start-up are presented in two cases: direct start-up and using a soft-starter. Fig. 11(b) shows the simulation results for the direct start-up, while Fig. 11(c) shows the results when the machine is connected to the grid via a soft-starter. When the induction machine is connected directly to the grid, high starting torque is observed. Large oscillations in the shaft speed can be seen in Fig. 11(b). Using a soft-starter the inrush currents and therefore the high-starting torque are limited and the shaft speed is smoothed as shown in Fig. 11(c).

Fig.11 the start-up of a fixed-speed wind turbine [12]

5.2 Variable speed wind turbines

In variable speed turbines, the generator is normally connected to the grid by a power electronic system. For synchronous generators and for induction generators without rotor windings, a full rated power electronic system is connected between the stator of the generator and the grid, where the total power production must be fed through the power electronic system [13, 14].

For induction generators with rotor windings, the stator of the generator is connected to the grid directly. In addition, the rotor current is controlled by various techniques, for instance (a) by a series

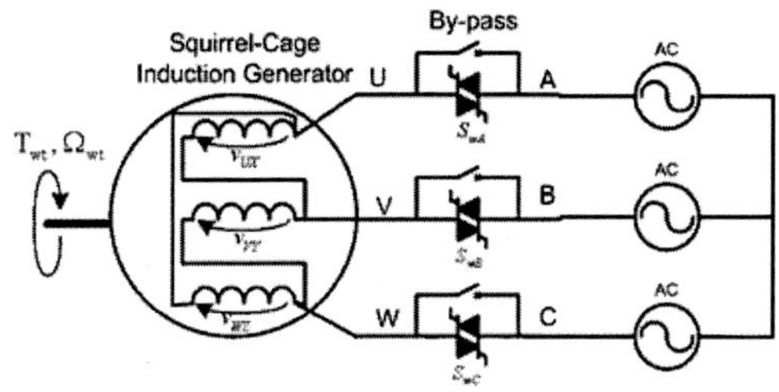

(a) Equivalent diagram of a fixed-speed wind turbine to show the start-up.

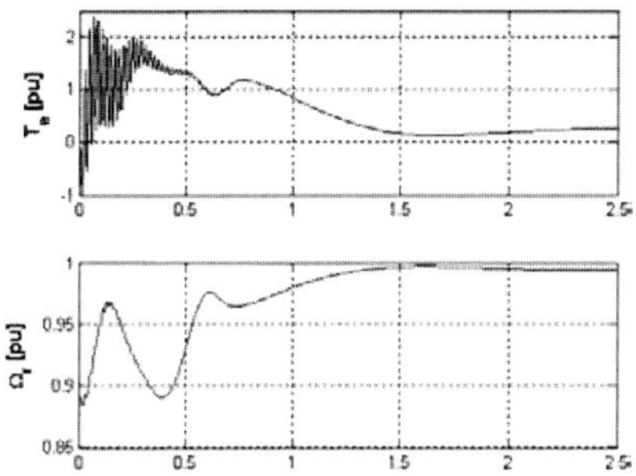

(b) Electromagnetic torque and shaft speed during the direct start-up of a 2 MW induction machine

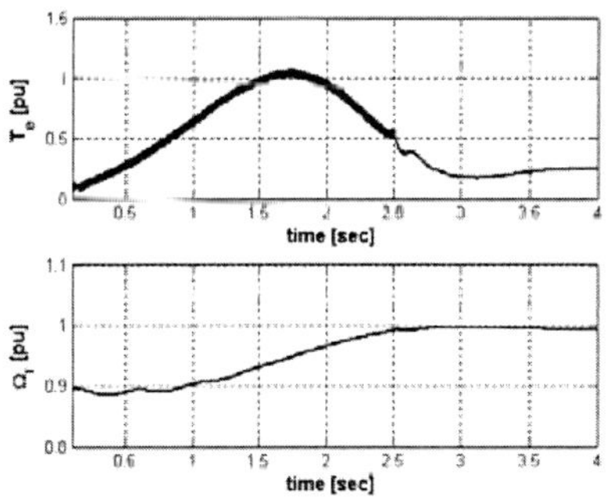

(c) Electromagnetic torque and shaft speed during start-up of a 2 MW soft-starter-fed induction machine

Figure 11. the start-up of a fixed-speed wind turbine [12]

resistance connected with the winding, or (b) by a grid-connected external circuit through power electronic converters.

Case (a) above is called a variable slip induction generator, since the active control of the resistance in series with the rotor windings allows the torque/speed relationship to be modified. By means of closed-loop control based on the measured currents, variable speed operation is effectively performed. Below rated power it behaves just like a normal induction generator. In case (b) the power electronics controls the power flow in the rotor circuit. Thus the power electronic converter is for the rotor circuit only and so needs to be rated to handle only about 30% of the total power, although the larger this proportion, the larger will be the achievable speed range.

Variable speed operation of a wind turbine system has several advantages compared with fixed-speed, but costs more. The fundamental benefit is that the blade tip-speed ratio can be kept constant at the optimum value (about 7 for most blades), so increasing the turbine efficiency in changing wind-speed conditions. Also, the electrical to mechanical link is softer, which means less wear on the gearbox and other components in the drive train, and structure, and reduced power transients on the grid.

5.2.1 Variable speed wind turbines with partially rated power converters

Consider turbines with an induction generator with a wound-rotor and so with partially rated power converters to obtain the improved control performance. The wind turbine systems with wound-rotor induction generators are in Fig. 12 and Fig. 13 [15, 16].

Wound-rotor induction generator with rotor resistance control (Dynamic slip control)

For a wound-rotor induction machine, the larger resistance of the rotor windings, the larger is the slip. Therefore, one way of varying the slip is to vary the resistance of the rotor winding. In this scheme, the rotor winding is connected to variable resistors through slip-rings, so the equivalent resistance in the circuit can be adjusted by an electronic control system as shown in Fig. 12 (a). In this way the generator speed can be varied over a limited range, typically a speed range of between 2% and 5 %. This is named *dynamic slip control*. The characteristics are shown in Fig. 12 (b) (c) and (d) [10]. The connection with brushes and slip rings is a drawback in comparison with the simple technical design of a cage rotor induction machine. It also requires more parts, which raise the maintenance requirements. The power converter for the rotor resistance control is for low voltage and high currents. In addition, the extra control is useful at stronger wind-speeds to maintain constant output power. This solution still needs a softstarter (not shown in Fig. 12(a)).

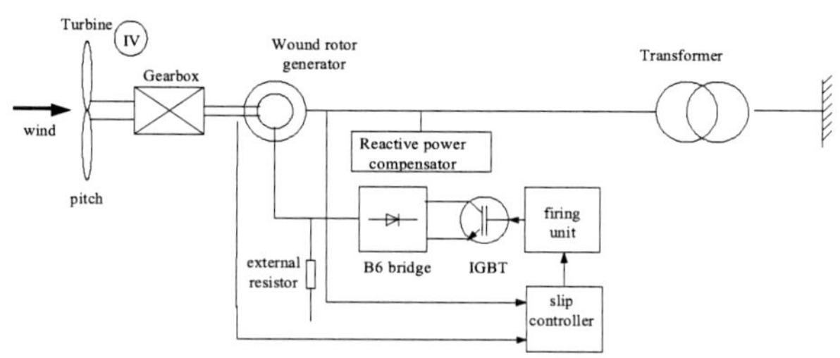

(a) System configuration diagram

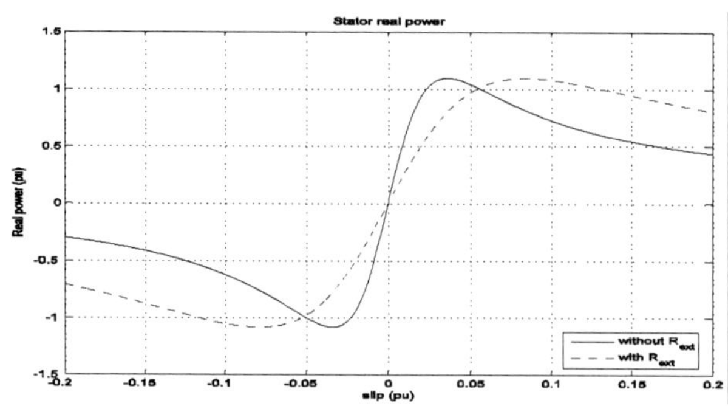

(b) real power characteristics

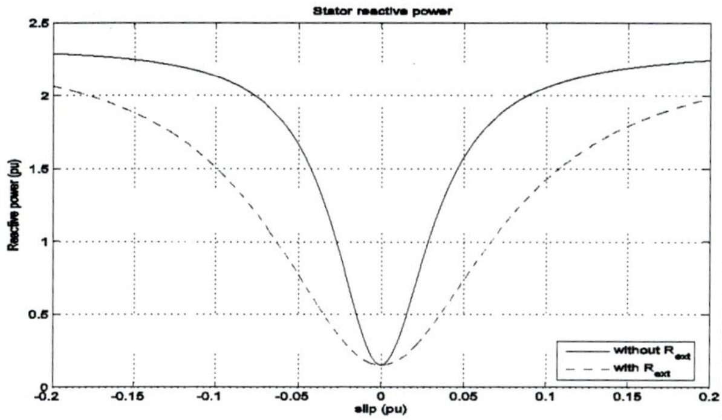

(c) reactive power characteristics

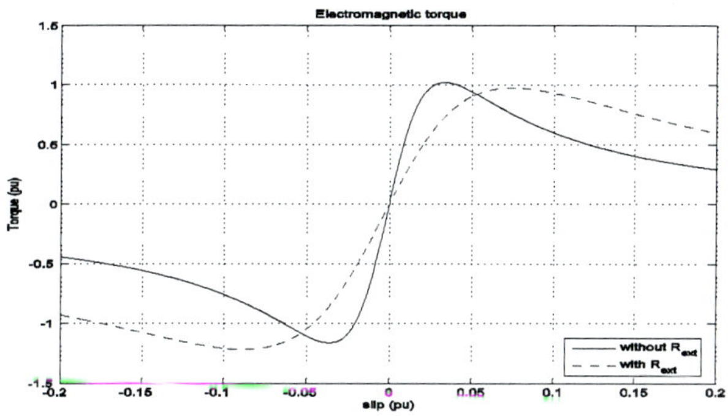

(d) torque characteristic

Figure 12. A wound-rotor induction generator with rotor-resistance converter (System IV) [10].

As an alternative, the slip rings, brushes, and associated maintenance may be eliminated by mounting the resistors and associated electronic control system on the rotor itself (Vestas OptiSlip). The requirement of changing the slip is communicated to the rotor without physical contact and therefore without slip-rings by using optical communication.

Both (i) cage-induction generators and (ii) rotor-resistance controlled wounded induction generators, need to operate at a super-synchronous speed to generate electricity. Both types draw reactive power, which is either supplied from the grid or from installed compensation equipment, such as capacitor banks or additional power electronic equipment. In order to reduce costs, capacitor banks are normally used. Modern MW-class turbines have thyristor-switched capacitors allowing for a more dynamic compensation. A Static Var Compensator (SVC), or similar technology, may be needed to improve the dynamic responses of the windfarm.

Doubly-fed induction generator

The stator of a doubly-fed induction generator is connected to the grid directly, while the rotor of the generator is connected to the grid via electronic converters and slip rings, as shown in Fig. 13. The generator can deliver energy to the grid at both super-synchronous ($s < 0$, $n > n_s$) and sub-synchronous ($s > 0$, $n < n_s$) speeds. The slip may vary considerably, as the rotor power passes through the power electronic circuit to the grid. The advantage is only a part of the power production is fed through the power electronic converter. That means the nominal power of the power electronic converter system can be less than the nominal power of the wind turbine. In general the nominal power of the converter may be about 30 % of the power of the wind turbine, enabling a rotor speed variation in the range of about ±30 % of the nominal speed. By the control of active power of the converter, it is possible to vary the rotational speed of the generator and thus of the rotor of the wind turbine.

Self commutated converter systems, such as Insulated Gate Bipolar Transistors (IGBT) Pulse Width Modulated (PWM) converters, are normally used for this type of system. The power electronic converters for variable speed generators have the ability to control both the active and reactive power delivered to the grid. This optimizes the grid integration with respect to steady state operation conditions, power quality, voltage and angular stability. The reactive power to the grid from the generation unit can be controlled either as zero or to a value required by the system operator within the converter rating limit. In general, the harmonics generated by the converter are in the range of some kHz, so filters are necessary to reduce harmonic amplitudes. At such high frequencies, the construction of the filters is simple.

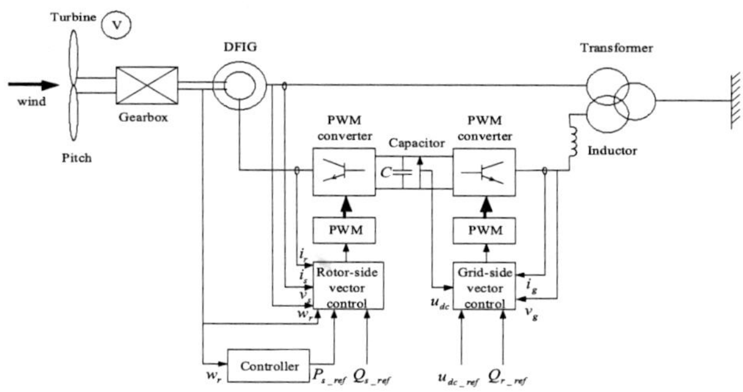

Figure 13. Wind turbine topologies with Doubly-fed induction generator (System V)

Doubly-fed induction generators also allow special operation strategies and provide high quality power to the grid. Acoustic noise from the turbines is reduced, since the rotor slows as the winds peed decreases. The dynamic response and controllability is better than with traditional induction generators, due to the power electronics.

The doubly-fed induction generator solution may need neither a soft-starter nor a reactive power compensator. The solution is naturally a little more expensive compared to the classical solutions shown before in Fig. 9 and Fig. 12. However, it is possible to save money on the safety margin of gearboxes and reactive power compensation units, and, moreover, more energy is captured from the wind.

3.2.2. Full-scale power electronic converter integrated systems

Wind turbines with a full-scale power converter between the generator and grid allow excellent technical performance. Fig. 14 shows four possible solutions with full-scale power converters.

Since the generator is decoupled from the grid, it can operate over a wide variable frequency range at optimal efficiency. The generated power enters the grid through the grid-side converter, which can be used for controlling both the active and reactive power [6, 7], and hence, the dynamic response is improved. However, the initial cost of a full converter system is usually more expensive than otherwise.

The solutions shown in Fig. 14 (a) and Fig. 14 (b) have a gear box. The wind turbine with a cage rotor induction generator and full rated power electronic converters is shown in Fig.14 (a). Usually, a back-to-back voltage source converter is used to achieve full control of the active and reactive power.

The synchronous generator shown in Fig. 14(b) needs an additional small power converter for field excitation. Multi-pole systems with the synchronous generator without a gearbox are shown in Fig. 14(c) and Fig. 14(d). The latter uses permanent magnets, which are becoming cheaper and thereby attractive. All four solutions have almost the same controllable characteristics, since the generator is decoupled from the grid by a dc-link. The power conversion to the grid enables the system to control active and reactive power very fast, however, this requires a more complex system with more sensitive electronic parts.

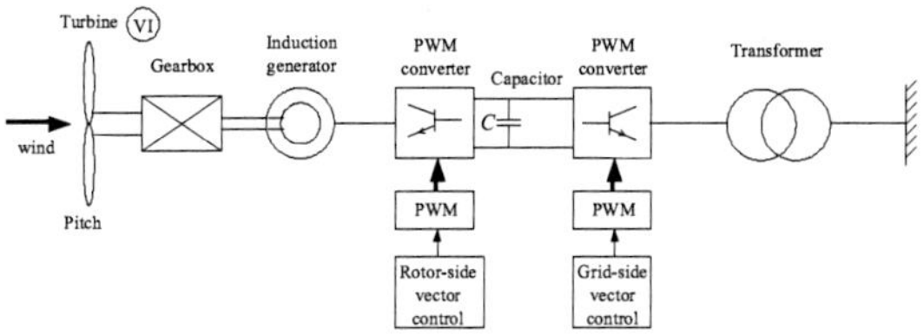

a) Induction generator with gear (System VI)

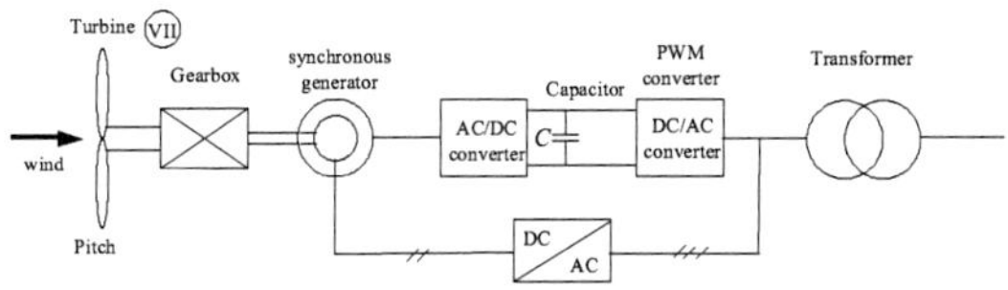

b) Synchronous generator with gear (System VII)

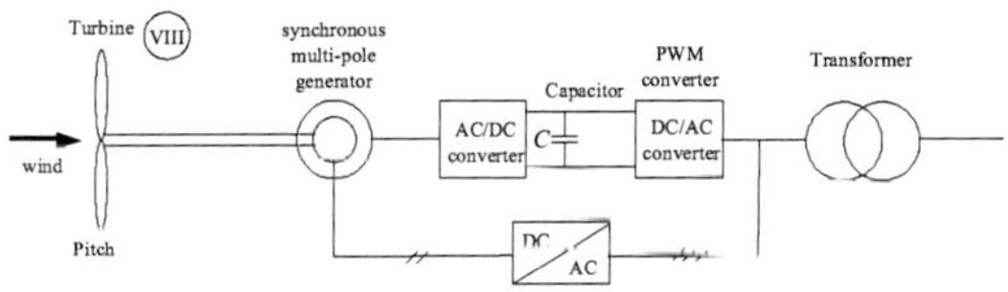

c) Multi-pole synchronous generator (System VIII)

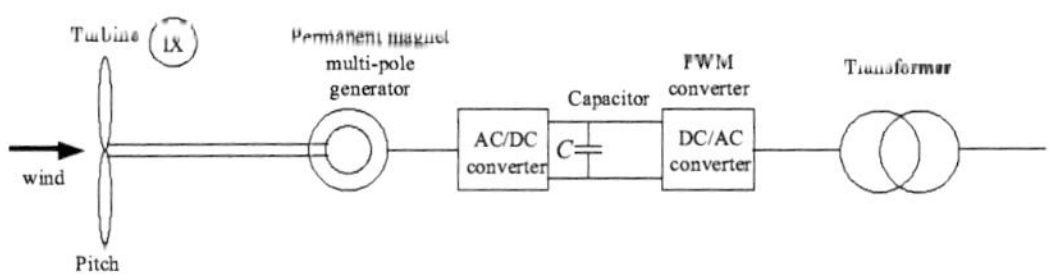

d) Multi-pole permanent magnet synchronous generator (System IX)

Figure 14. Wind turbine systems with full-scale power converters.

5.3 Summary of wind turbine- generator systems

By introducing power electronics, wind turbines may perform similarly to conventional power plants. Of course, the amount of produced real power depends on the available wind. A further service in some arrangements, e.g. weak grids, is that reactive power can be delivered to benefit the grid without having any wind.

Table 1 summarizes the systems discussed so far. Each of these topologies has benefits and drawbacks. A fixed speed wind turbine is relatively simple, so the price tends to be less than the price for variable speed turbines. Nevertheless, if the rotor speed cannot be varied, these turbines must be more robust than the other designs due to the larger structural loads.

Table 1 Comparison of different wind turbine systems. (see Figs 10, 12, 13, 14 for definition of types)

System comparison of wind turbines

System type	I	II	III	IV	V	VI	VII	VIII	IX
Variable speed	No	No	No	No	Yes	Yes	Yes	Yes	Yes
Control active power	Limited	No	Limited	Limited	Yes	Yes	Yes	Yes	Yes
Control reactive power	No	No	No	No	Yes	Yes	Yes	Yes	Yes
Short circuit (fault-active)	No	No	No	No	No/Yes	Yes	Yes	Yes	Yes
Short circuit power	contribute	contribute	contribute	contribute	contribute	limit	limit	limit	limit
Control bandwidth	1-10 s	1-10 s	1-10 s	100 ms	1 ms	0.5-1 ms	0.5-1 ms	0.5-1 ms	0.5-1 ms
Standby function	No	No	No	No	Yes +	Yes ++	Yes ++	Yes ++	Yes ++
Flicker (sensitive)	Yes	Yes	Yes	Yes	No	No	No	No	No
Softstarter needed	Yes	Yes	Yes	Yes	No	No	No	No	No
Rolling capacity on grid	Yes, partly	No	Yes, partly	Yes, partly	Yes	Yes	Yes	Yes	Yes
Reactive compensator (C)	Yes	Yes	Yes	Yes	No	No	No	No	No
Island operation	No	No	No	No	Yes/No	Yes/No	Yes/No	Yes/No	Yes
Investment	++	++	++	++	+	o	o	o	o
Maintenance	++	++	++	++	o	+	+	+	+

A variable-speed wind turbine generates more energy for a given wind-speed especially at low wind-speed. Moreover, the active and reactive power can be easily controlled and there is less mechanical stress. The direct driven (variable speed) machines are normally more expensive. However, costs may be saved with a multipole synchronous generator because there need be no gearbox. The major drawback of the direct-driven topologies is the large and relatively heavy generator. Moreover, the power converter has to be designed to handle the full rated power.

In Table 1, some other important issues for wind turbines acting as a real power source in the grid are also listed. *Rolling capacity* on the grid, which means that a wind turbine keeps active and grid-connected when a fault appears in the grid, allows power production to be reduced even though more power is available in the wind and thereby the turbine is able to have a 'rolling capacity' for the benefit of the network. *Island operation* means that the turbine remains generating even if the grid supply voltage reduces or fails (collapses); the ability of operating in such situations reduces the risk of total area blackout.

6. CONTROL OF WIND TURBINES

The variation of generator speed is determined by the difference of the input aerodynamic torque and the output electromagnetic torque. Therefore the operation point and the efficiency of the energy conversion system can be adjusted by controlling the torques. The aerodynamic torque can be controlled by adjusting blade pitch with the aim of maximizing power production in the available wind or limiting the power at the rated value. In a variable speed system with power electronics, the electromagnetic torque may also be controlled to achieve the desired operation point. The power control system should also be able to limit the power. Controlling a wind turbine involves both fast and slow control. Some typical control methods associated with different wind power generators are described in this section.

6.1 Active stall wind turbine with cage rotor induction generators

The main difference between a stall turbine and an active stall turbine, is that the latter has a pitch system for variable pitch angles, which allows the stall effect to be controlled. An active stall wind turbine has to pitch in a negative direction (i.e. to increase the angle of attack) to limit the power when the electrical power of the wind turbine exceeds nominal power. Large windfarms such as Nysted (170 MW installed capacity) have been built with active stall wind turbines.

The generator of an active stall turbine may be a simple squirrel cage induction generator directly connected to the grid. In order to compensate for the output power factor a capacitor bank can be used. A soft-starter is used only during the start-up sequence of the generator in order to limit the in-rush currents.

The electromagnetic torque developed by a fixed-speed squirrel cage induction generator is determined by the slip speed. As the aerodynamic torque varies, the rotor speed varies by a very small amount such that the electromagnetic torque changes to match the aerodynamic torque. The electromagnetic torque cannot be actively controlled in a system with a squirrel cage induction generator directly connected to the grid.

The maximum power output of the active stall turbines can be maintained at a constant value. In addition, the aerodynamic efficiency, C_p, can be optimized to a certain extent. The active stall control can improve the efficiency of the overall system. The adjustable coupling angle of the blades to the hub also facilitates emergency stopping and start up. One drawback of the active-stall controlled wind turbine compared to the passive stall one is the higher price, due to the pitching mechanism and its controller.

Depending on the pitch system, less than optimum efficiency due to slow control may be justified by reduced stress and wear in the pitch system and reduced fatigue loads on the wind turbine. This applies both to the power optimization region, where the controller strives for maximum power yield by using the moving average of the wind-speed signal to find the appropriate pitch angle in a lookup table, and to the power limitation region, where the power output is controlled in a closed control loop. With a slow control system, substantial overpower may occur, which is avoided by an overpower protection feature.

An example

In Fig. 15, the following conditions are illustrated: the mean wind-speed is 11 m/s until the simulation time reaches 60 s, between 60 s and 160 s it is ramped up from 11 m/s to 16 m/s. This corresponds to a slope in mean wind-speed of 3 m/s per minute.

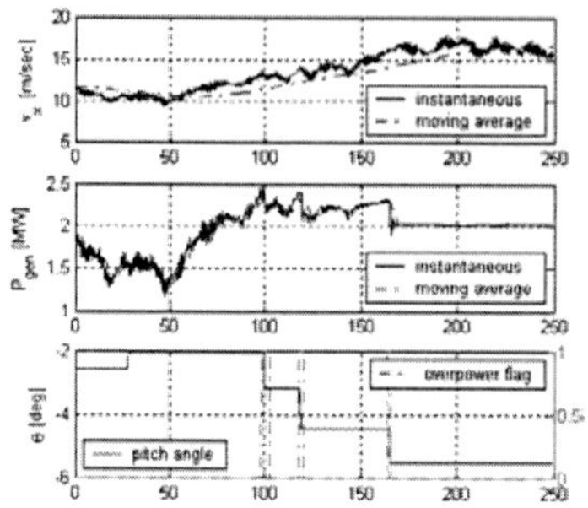

Figure 15. Simulation results for an active stall controlled wind turbine [12]

The 2 MW turbine starts off in power optimization mode where an increase in pitch angle takes place. As the wind-speed increases the turbine speed enters power limitation mode; a further increase in wind-speed causes the average power to exceed 2300 kW (300 kW beyond nominal power, which is the maximum allowed level of tolerable overpower). As soon as the overpower is detected, the pitch angle is adjusted. For steadily increasing wind-speed, as shown in the figure, it takes three adjusting operations to finally limit the power at the nominal level as shown in Fig. 15.

6.2 Variable pitch angle control with doubly-fed generators

The variable speed doubly-fed induction generator wind turbine is the widely used concept today. The control system of a variable speed wind turbine with DFIG mainly functions to:

Control the power drawn from the wind turbine in order to track the wind turbine optimum operation point;

Limit the power in the case of high wind-speeds;

Control the reactive power exchanged between the wind turbine and the grid.

Two related hierarchical control levels have different bandwidths, namely DFIG control level and wind turbine control level. An example of an overall control scheme of a wind turbine with a doubly-fed generator system is shown in Fig. 16.

The DFIG control, with a fast dynamic response, contains the electrical control of the power converters and of the doubly-fed induction generator. The wind turbine control, with slow dynamic response, supervises both the pitch system of the wind turbine and the active power set point of the DFIG control level.

A vector control approach is adopted for the DFIG control, while two cross-coupled controllers are used to control the wind turbine. These controllers are speed and power limitation controllers, intended to track the wind turbine optimum operation point, to limit the power in the case of high wind-speeds and to control the reactive power exchanged between the wind turbine generator and the grid.

Below maximum power production, the wind turbine will typically vary the speed proportional with the wind-speed and keep the pitch angle nearly fixed. At very low wind speed, the speed of the turbine will be fixed at the maximum allowable slip in order not to have overvoltage. A pitch angle controller limits the power when the turbine reaches the nominal power. The generated electrical power is controlled by the doubly-fed generator through the rotor-side converter. The control of the grid-side converter simply just keeps the dc-link voltage fixed. Internal current loops in both converters are used with typical PI-controllers.

The significant feature of the control method is that it allows the turbine to operate with the optimum power efficiency over a wide range of wind-speeds. Moreover, due to the design of this control method, small changes in generator speed do not lead to large power fluctuations and unnecessary transitions between power optimization and power limitation mode. A gain scheduling control of the pitch angle is also implemented in order to compensate for the non-linear aerodynamic characteristics.

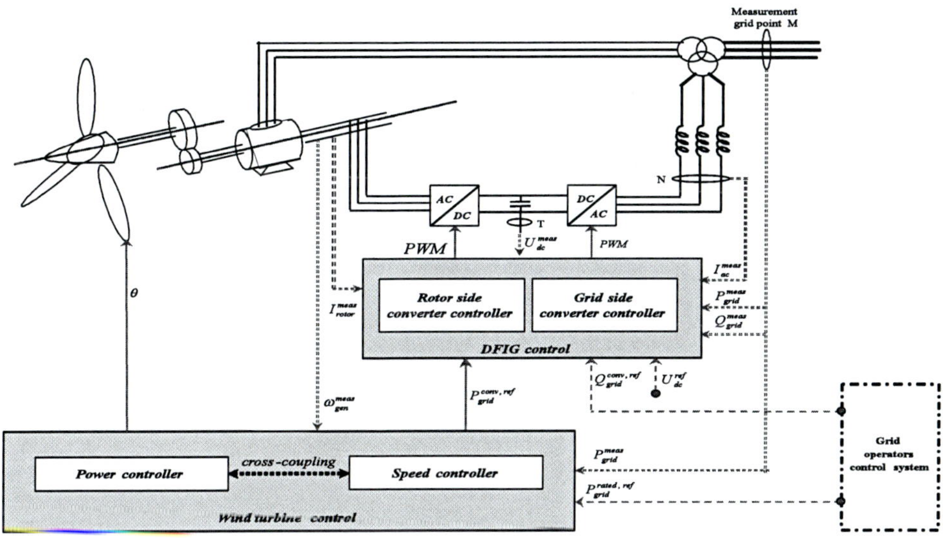

Figure 16. Control of wind turbine with doubly-fed induction generator system [18].

An example

The example below illustrates the performance both of the DFIG controller and of the overall control of the variable speed /variable pitch wind turbine. The variable speed wind turbine has a rated power of 2 MW. The rated wind-speed is at 11.5 m/s and the rated generator speed is 1686 rpm. Fig. 17 shows the simulation results when a wind-speed at a mean value of 22 m/s with a turbulence intensity of 10%. Typical quantities, including wind-speed, the pitch angle, the generator power to the grid and the generator speed, the reference of the generator speed, are shown in Fig. 17.

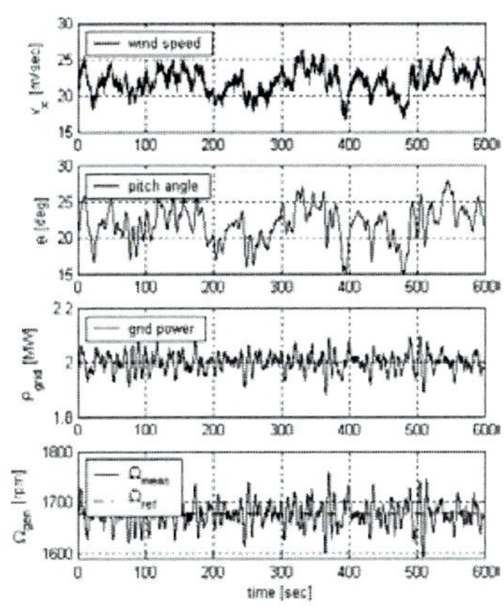

Figure 17. Simulation results for a variable speed/pitch wind turbine with DFIG under turbulent wind speed [12]

The wind-speed corresponds to the power limitation strategy, where both the speed control loop and power control loop are active. The power control loop is fast, while the speed control loop is much slower, allowing dynamic variations of the generator speed in a predefined range. It can be seen that the power on the grid is limited to 2 MW; its variations being less than 2% of the rated power. The reference of the generator is maintained at the nominal speed, while the generator speed varies as the electrical power. The pitch angle reacts to the slow variations in the wind-speed.

6.3 Full rated power electronic interface wind turbine systems

Cage induction generators and synchronous generators can be integrated into the system by full rated power electronic converters. As shown in Fig. 18, a passive rectifier, for example, a multi-module diode rectifier system shown in Fig. 19 [19, 20], may be used for a multipole permanent magnet generator. A boost converter is then used to regulate the voltage at the dc terminal of the grid side converter [6, 7], and to control the active power from the generator. A grid inverter interfaces the dc-link to the grid. Here it is also possible to control the reactive power to the grid. The system is able to control reactive and active power very fast and thereby the turbine may take part in the power system control.

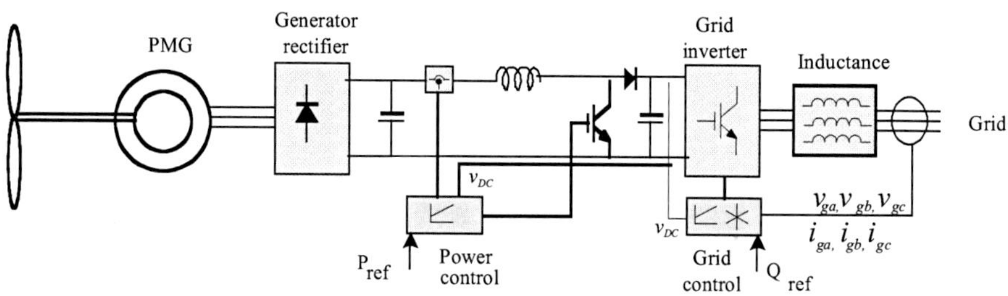

Figure 18. Basic control of active and reactive power in a wind turbine [6].

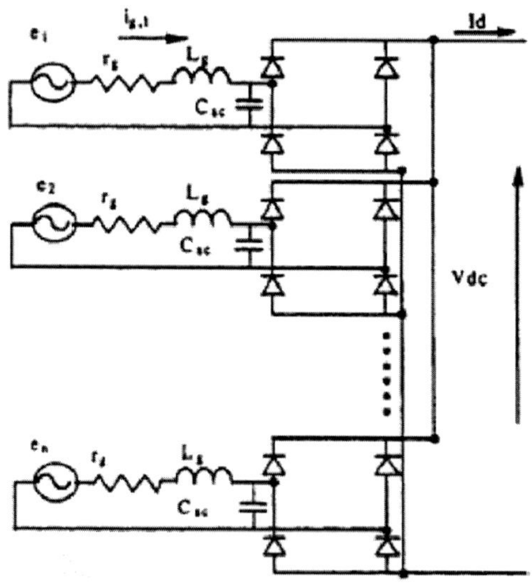

Figure 19. Module stator coil and rectifier connection [20].

7. NETWORK TOPOLOGIES OF WINDFARMS

Many countries have, or are planning to have, a significant penetration (say more than 20%) of wind energy into the national grid from large onshore and, where possible, offshore windfarms with inbuilt power electronics. It is important to note that such windfarms need not present overall difficulties to the network operators, but can both be a significant power contribution, and also improve power quality and control.

Consequently, high technical demands are required for these generation units, such as frequency and voltage control, regulation of active and reactive power, quick responses under power system transient and dynamic situations. For example, it may be required to reduce the power from full rated power to 20 % power within 2 seconds. The power electronic technology is important, both for the network support and the control of the windfarms themselves. Some possible electrical configurations of windfarms are shown in Fig. 20, the requirements of integration wind turbines and windfarms into power system are to be discussed in the next section.

A windfarm equipped with power electronic converters, as shown in Fig. 20(a), can perform both real and reactive power control and also operate the wind turbines in variable speed to maximize the energy captured as well as reduce the mechanical stress and noise. Such a system is in operation in Denmark as a 160 MW off-shore wind power station.

Fig. 20(b) shows a windfarm with induction generators; a STATCOM controls the reactive power, supports voltage control and provides the reactive power demand of the induction generators.

For long distance transmission of power from an offshore windfarm, HVDC may be an option. In HVDC transmission, the low or medium AC voltage at the windfarm is converted into high-voltage dc on the transmission side. The dc power is then transferred onshore, where the dc is converted back into ac, Fig. 20(d). For certain power levels, a HVDC transmission system, based on voltage source converter technology, may be used instead of the conventional thyristor technology. The topology may even be able to vary the rotational speed of the wind turbine rotors in the complete windfarm. Another possible dc transmission system configuration is shown in Fig. 20(c), where each wind turbine has its own power electronic converter, so it is possible to operate each wind turbine at an individual optimal speed. A comparison of the topologies is given in Table 2. As it can be seen the windfarms have interesting features in order to act as a power source to the grid, with some having better abilities than others. The overall considerations will include production, investment, maintenance and reliability.

There are also other possibilities, such as field-excited synchronous machines or permanent magnet synchronous generators, as shown in Fig. 20 (c) and (d), (with a multipole generator, the gearbox may not be necessary).

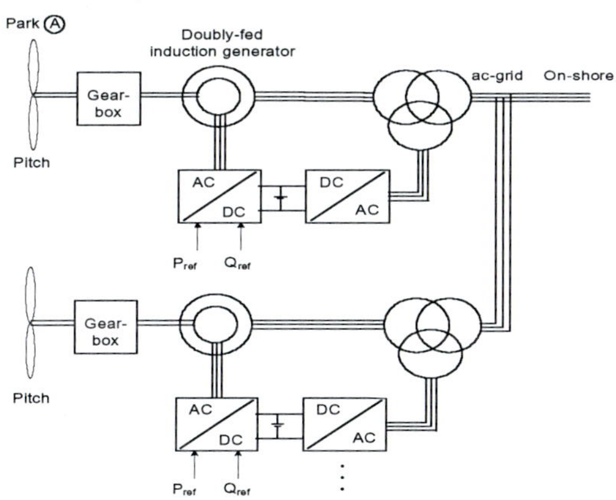

(a) Doubly-fed induction generator system with ac-grid (System A)

Figure 20. *continued over*

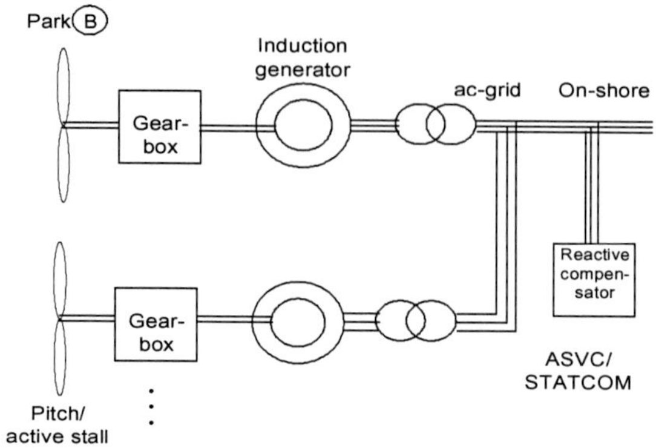

(b) Induction generator with ac-grid (System B)

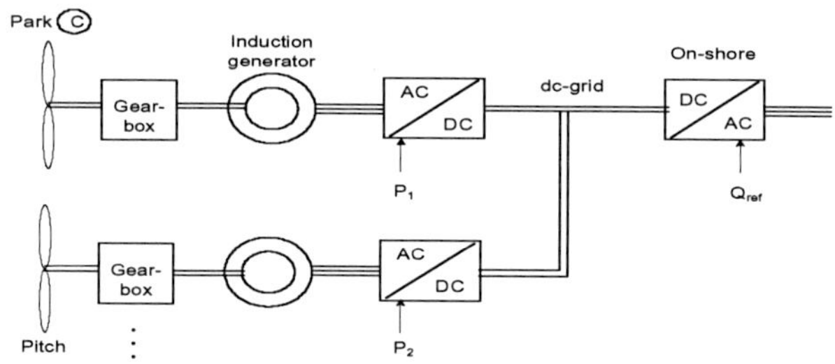

(c) Speed controlled induction generator with common dc-bus and control of active and reactive power (System C)

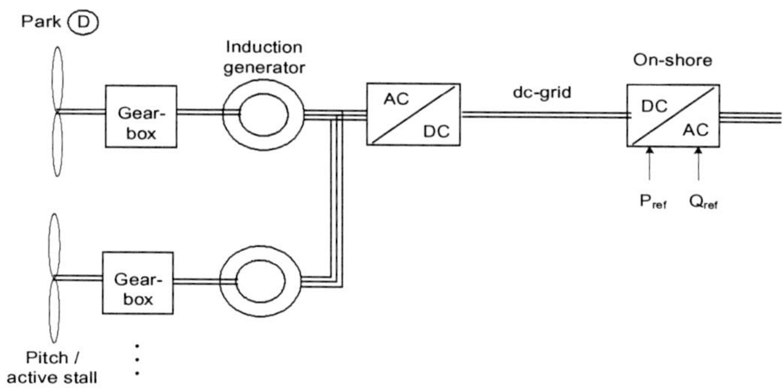

d) Speed controlled induction generator with common ac-grid and dc transmission (System D)

Figure 20. Wind farm solutions.

8. INTEGRATION OF WIND TURBINES INTO POWER SYSTEMS

Large scale integration of wind turbines may have significant impacts on power system operation. Traditionally, wind turbines are not required to participate in frequency and voltage control. However, in recent years, [20-24]. some grid codes specify the requirements that wind turbines must meet to be connected to the grid. Examples include the capabilities of contributing to frequency and voltage control by continuously adjusting active power and reactive power supplied to the power systems, and the power regulation rate that a windfarm must provide, etc. Some of the requirements can be dealt with by implementing control schemes in certain types of wind turbines, such as reactive power control with wind turbines having power electronic converters. Many research activities have been conducted in this area [26-41].

Table 2. Wind turbine topologies market in 2001. (Source: BTM consult, [17])

Turbine concept (refer to Table 1)	World-Market Share	Market Share Germany
Fixed speed (Stall or active stall, gearbox), System I, II, III	23%	22%
Dynamic slip control (Limited variable speed, pitch, gearbox), System IV	11%	0%
Doubly-fed generator (Variable speed operation, pitch control, gearbox), System V	50%	49%
Direct-driven (variable speed operation, pitch control), System VIII	16%	29%
Total	**100%**	**100%**

8.1 Requirements of wind turbine grid integration

Frequency and active power control

Early windfarms have much smaller capacity than the centralized plants, so their input cannot effect the system frequency significantly. However, the peak output from large windfarms, including offshore windfarms, may perturb the power system and so these windfarms need to be capable of performing control. A windfarm may have to operate at a lower power level than the available power to be able to increase the power output for frequency control, which means reduced utilization of the wind energy resources. One way to improve the situation may be the use of "energy storage" technologies, such as batteries, pump storage and fuel cells, though the speed of response will vary depending on the energy storage technology. So far large scale, cost effective energy storage technologies are yet to be developed.

Short circuit power level and voltage variations

Considering a central point, A, in the network. Let Z_k. be the equivalent impedance between A and a remote location, B, where U_k is the voltage. The term $U_k{}^2/Z_k$ is called the short circuit power S_k (MVA) between A and B. Let B be the Point of Common Connection (PCC) of the windfarm with the grid, if the impedance Z_k is small (i.e. the grid is strong), the voltage variations at B will be small, if Z_k is large (i.e. the grid is weak) the voltage variations at B will be large. The short circuit power at a given point in an electrical network is a measure of its strength. Although it is not a direct parameter of voltage quality, it has a significant influence on the ability of the grid to absorb disturbances.

Fig 21 shows the equivalent circuit of the wind power plant connected to a network with short circuit impedance, Z_k. The network voltage at the assumed remote busbar (A) and the voltage at the Point of Common Coupling (PCC) are U_s and U_g, respectively. The output power and reactive power of the generation unit are P_g and Q_g, which corresponds to a current I_g.

$$I_g = \left(S_g/U_g\right)^* = \frac{P_g - jQ_g}{U_s} \tag{1}$$

The voltage difference, DU, between the system and the connection point is given by

$$
\begin{aligned}
U_g - U_s = \Delta U = Z_k I_g &= (R_k + jX_k)\left(\frac{P_g - jQ_g}{U_g}\right) \\
&= \frac{R_k P_g + X_k Q_g}{U_g} + j\frac{P_g X_k - Q_g R_k}{U_g} = \Delta U_p + j\Delta U_q
\end{aligned} \tag{2}
$$

The voltage difference, ΔU, is related to the short circuit impedance, the real and the reactive power output of the wind power plant. It is clear that the variations of the generated power will result in variations of the voltage at PCC.

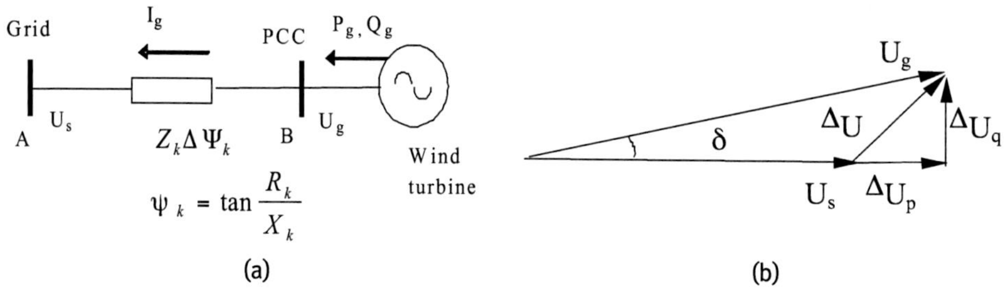

Figure 21. A simple system with an equivalent wind power generator connected to a network. a) System circuit b) Phasor diagram [27]

Equation (2) indicates the relationship between the voltage and power transferred into the system. The voltage difference, ΔU, can be calculated with load flow methods and other simulation techniques. The voltage at the PCC should be maintained within utility regulatory limits. Operation of wind turbines may affect the voltage in the connected network. If necessary, the appropriate methods should be taken to ensure that the wind turbine installation does not bring the magnitude of the voltage outside the required limits.

Reactive power control

Reactive power may be understood as the oscillation of energy stored in capacitive and inductive components in a power system. Reactive power is produced in capacitive components and consumed in inductive components. A synchronous generator can either produce or consume reactive power by controlling the magnetizing level of the generator, i.e. a high magnetizing level results in high voltage and production of reactive power.

The current associated with the reactive power flow causes a system voltage drop and power loss. Furthermore, large reactive currents flowing in a power system may cause voltage instability in the network, due to the associated voltage drops in the transmission lines. Therefore reactive power control is important to prevent these unwanted effects. This is especially important for wind turbines with basic induction generators, which consume reactive power. To minimize the power losses and to increase voltage stability, such wind turbines are compensated according to the requirements of the local utility or distribution company. For wind turbines with PWM converter systems, the reactive power can be controlled by the converter. Thus these wind turbines can have a power factor of 1.00 (power factor is defined as cos ψ, where the angle ψ can be found with tan $\psi = Q/P$) and so the voltage may be controlled by controlling the reactive power by the generation or consumption of reactive power.

Flicker

Voltage variations caused by fluctuating wind power generation may affect voltage quality.

Fluctuations in the system voltage (in terms of rms value) may cause perceptible light flicker depending on the magnitude and frequency of the fluctuation. This type of disturbance is called 'voltage flicker', often shortened as 'flicker'.

Rapid variations in the power output from a wind turbine, such as from generator and capacitor switching, can also result in unwanted variations in the RMS value of the voltage. For instance, certain rates and magnitudes of these variations may cause perceived flickering of incandescent electric lights. Wind power should not cause such conditions.

Flicker evaluation based on IEC 1000-3-7 gives guidelines for emission limits of fluctuating loads in medium voltage and high voltage networks. The basis for the evaluation is a measured curve giving the threshold of visibility for rectangular voltage changes applied to an incandescent lamp. The level of flicker is quantified by the short-term flicker severity , which is normally measured over a ten-minute period. Disturbances just visible are said to have a flicker severity factor of $P_{st} = 1$. Furthermore, a long term flicker severity factor P_{lt} is defined where P_{lt} is measured over two hour periods.

Determination of flicker emission can be done on the basis of measurement. IEC 61000-4-15 specifies a flickermeter which can be used to measure flicker directly. The flicker emissions may also be estimated with the coefficient and factors, $c_f(\Psi_k, v_a)$ and $k_f(\Psi_k)$ obtained from the measurements, which are usually provided by wind turbine manufacturers [27].

Harmonics

Harmonics are a phenomenon associated with the distortion of the voltage and current waveforms. Any periodical function may be expressed as a sum of sinusoidal waveforms with different frequencies, including the fundamental frequency and a serious of integer multiples of the fundamental component. Depending on their harmonic order, they may causes various damage or interference to different type of electrical equipment. All harmonics cause increased currents and possible destructive overheating in capacitors, because the impedance of a capacitor decreases in proportion to the increase in frequency. The higher harmonics may further give rise to increased audible noise in nearby analogue telephone circuits. The harmonic distortion is expressed as Total Harmonic Distortion (THD). THD and individual harmonics should meet the system requirements.

Note that Pulse Width Modulation (PWM) switching converters are used in most variable speed wind turbine technologies today. The switching frequency is typically around a few kHz. The high frequency harmonics are small in magnitude and are easier to be removed by filters.

Stability

The problem of network stability is often associated with different types of faults in the network, such as tripping of transmission lines (e.g. overload), loss of production capacity and short circuits. Tripping of transmissions lines due to overload or component failure disrupts the balance of power (active and reactive) flow. Consequently, a period of low voltage, possibly followed by complete loss of power, often occurs. Many of the faults are cleared by the relay protection of the transmission system, either by disconnection and fast reclosure, or by disconnection of the equipment in question after a few hundred milliseconds. In all the situations, the result is a short period with low or no voltage followed by a period when the voltage restores. A large windfarm in the vicinity will 'see' this event and may be disconnect from the grid if no appropriate control has been implemented. This leads to the situation "loss of production capacity". Loss of production capacity may cause a large power unbalance, at least momentarily, with perhaps a large frequency and voltage drop followed by complete loss of power. The disconnection of the windfarm will further aggravate the situation, therefore, in some grid codes, wind turbines and windfarms are required to have the ability of 'ride through' such disturbances. Wind turbines may have various control methods to cope with such transients [15, 28-39].

8.2. Voltage quality assessment

The assessment of the impacts from integrating wind turbines may be performed according to the methods given in the IEC 61400-21 /2/ to determine the acceptability of such integration regarding steady-state voltage, flicker and harmonics

(a) Steady-state voltage

The grid and wind turbine voltage should be maintained within the utility limits. Operation of a wind turbine may affect the steady-state voltage in the network. It is recommended that load-flow analyses be conducted to assess this effect to ensure that the wind turbine installation does not bring the magnitude of the voltage beyond the required limits of the network, In general, some extreme case of the loads and the wind turbine production may be checked for compatibility, such as

- low-loads and low-windpower
- low-loads and high-windpower
- high-loads and low-windpower
- high-loads and high-windpower

Depending on the scope of the load-flow analysis, a wind turbine installation may be assumed as a PQ node, which may use ten minute average data (P_{mc} and Q_{mc}), 60 s average data (P_{60} and Q_{60}) or 0.2 s average data ($P_{0.2}$ and $Q_{0.2}$).

A windfarm with multiple wind turbines may be represented with its output power at the PCC. Ten minute average data (P_{mc} and Q_{mc}) and 60 s average data (P_{60} and Q_{60}) can be calculated by simple summation of the output from each wind turbine, whereas 0.2 s average data ($P_{0.2}$ and $Q_{0.2}$) may be calculated according to equations (3) and (4) below.

$$P_{0.2\Sigma} = \sum_{i=1}^{N_{wt}} P_{n,i} + \sqrt{\sum_{i=1}^{N_{wt}} \left(P_{0.2,i} - P_{n,i}\right)^2} \tag{3}$$

$$Q_{0.2\Sigma} = \sum_{i=1}^{N_{wt}} Q_{n,i} + \sqrt{\sum_{i=1}^{N_{wt}} \left(Q_{0.2,i} - Q_{n,i}\right)^2} \tag{4}$$

where: $P_{n,i}$ and $Q_{n,i}$ are the rated real and reactive power of the individual wind turbines, N_{wt} is the number of wind turbines in the group.

(b) Voltage fluctuations

There are two types of flicker emissions, the flicker emission during continuous operation and the flicker emission due to generator and capacitor switchings. Often, one will be dominant. The flicker emissions from a wind turbine installation should be limited to comply with the flicker emission limits. However, different utilities may have different flicker emission limits. The assessments of the flicker emissions are described below.

1) Continuous operation

The flicker emission from a single wind turbine during continuous operation may be estimated by:

$$P_{st} = c_f(\psi_k, v_a)\frac{S_n}{S_k} \tag{5}$$

where $c_f(\Psi_{k,} v_a)$ is the flicker coefficient of the wind turbine for the given network impedance phase angle, Ψ_k, at the PCC, and for the given annual average wind-speed, v_a, at hub-height of the wind turbine.

A table of data produced from the measurements at a number of specified impedance angles and wind-speeds can be provided by wind turbine manufacturers. From the table, the flicker coefficient of the wind turbine for the actual Y_k and v_a at the site may be found by applying linear interpolation.

The flicker emission from a group of wind turbines connected to the PCC is estimated using equation (6)

$$P_{st\Sigma} = \frac{1}{S_k}\sqrt{\sum_{i=1}^{N_{wt}}\left(c_{f,i}(\psi_k, v_a)S_{n,i}\right)^2} \tag{6}$$

where $c_{f,i}(Y_{k,} v_a)$ is the flicker coefficient of the individual wind turbines; $S_{n,i}$ is the rated apparent power of the individual wind turbine; N_{wt} is the number of wind turbines connected to the PCC.

If the limits of the flicker emission are known, the maximum allowable number of wind turbines for connection can be determined.

2) Switching operations

The flicker emission due to switching operations of a single wind turbine can be calculated as

$$P_{st} = 18 \times N_{10}^{0.31} \times k_f(\psi_k)\frac{S_n}{S_k} \tag{7}$$

where $k_f(\Psi_k)$ is the flicker step factor of the wind turbine for the given Ψ_k at the PCC.

The flicker step factor of the wind turbine for the actual Ψ_k at the site may be found by applying linear interpolation to the table of data produced from the measurements by wind turbine manufacturers. The flicker emission from a group of wind turbines connected to the PCC can be estimated from:

$$P_{st\Sigma} = \frac{18}{S_k}\left(\sum_{i=1}^{N_{wt}} N_{10,i}\left(k_{f,i}(\psi_k)S_{n,i}\right)^{3.2}\right)^{0.31} \tag{8}$$

where $N_{10,i}$ and $N_{120,i}$ are the number of switching operations of the individual wind turbine within 10 minute and 2 hour period respectively; $k_{f,i}(\Psi_k)$ is the flicker step factor of the individual wind turbine; $S_{n,i}$ is the rated apparent power of the individual wind turbine;

Again, if the limits of the flicker emission are given, the maximum allowable number of switching operations in a specified period, or the maximum permissible flicker emission factor, or the required short circuit capacity at the PCC may be determined.

(c) Harmonics

A wind turbine with an induction generator directly connected to the grid is not expected to cause any significant harmonic distortions during normal operation. Only wind turbines with power electronics need to be checked concerning harmonics. The harmonic current emission of such wind turbine system is normally given in the power quality data sheet, while the limits for harmonics are often specified for harmonic voltages. Then harmonic voltages should be calculated from the harmonic currents of the wind turbine, which requires the information of the grid impedances at different frequencies.

9. IMPROVEMENT OF WIND TURBINE PERFORMANCE IN POWER SYSTEMS

9.1. Voltage flicker minimization

Voltage variation and flicker emission of grid connected wind turbines are related to many factors, including

- Mean wind-speed, v
- Turbulence intensity, In
- Short circuit capacity ratio, $SCR = S_k/S_n$

where S_k is the short circuit capacity of the grid where the wind turbines are connected, S_n is the rated power of the wind turbine.

The voltage difference in (2) may be approximated as:

$$\Delta U \approx \Delta U_p = \frac{P_g R_k + Q_g X_k}{U_g} \tag{9}$$

With the grid impedance angle ψ_k and the wind turbine power factor angle ψ being defined as

$$\begin{aligned} tan\psi_k &= X_k / R_k \\ tan\psi &= Q_g / P_g \end{aligned} \tag{10}$$

Equation (9) can be written as

$$\Delta U_p = \frac{P_g R_k (1 + tan\psi_k \cdot tan\psi)}{U_g} = \frac{P_g R_k \, cos(\psi - \psi_k)}{U_g \, cos\psi_k \cdot cos\psi} \tag{11}$$

It can be seen from equation (11) that when the difference between the grid impedance angle ψ_k and the wind turbine power factor angle ψ approaches 90 degrees, the voltage fluctuation is minimized [27]. Equation (11) also indicates that the reactive power may be regulated with the real power generation to minimize voltage variation and flicker.

Variable speed wind turbines with doubly-fed induction generator are capable of controlling the output active and reactive power respectively. Normally the output reactive power of the wind turbine is controlled as zero for unity power factor. It is possible to control the output reactive power appropriately with the variation of the output real power, so the voltage changes caused by the real power flow may be cancelled by the reactive power flow. For example, the reactive power can be controlled in proportion to the wind turbine output active power, so that the power factor angle is close to the value of ψ_k + 90 degrees.

Based on IEC standard IEC 61000-4-15, a flickermeter model is built to calculate the short-term flicker severity P_{st} as shown in Fig. 22. Fig. 23 shows the wind-speed and output power produced by

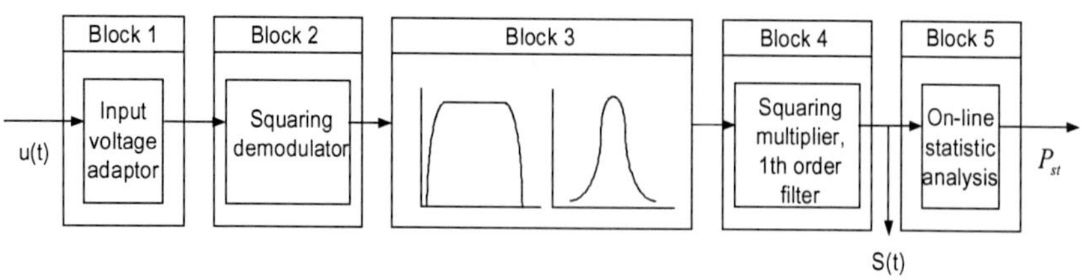

Figure 22. Flickermeter model according to IEC 61000-4-15

the simulation model and the corresponding power spectra are shown in Fig. 24 where the 3p effect can be clearly seen.

A simulation study has been carried out to investigate the flicker minimization by using the reactive power compensation. The results, as shown in Fig. 25, indicates that the flicker is significantly reduced if the angle difference () is regulated to be 90 degrees by controlling the reactive power flow.

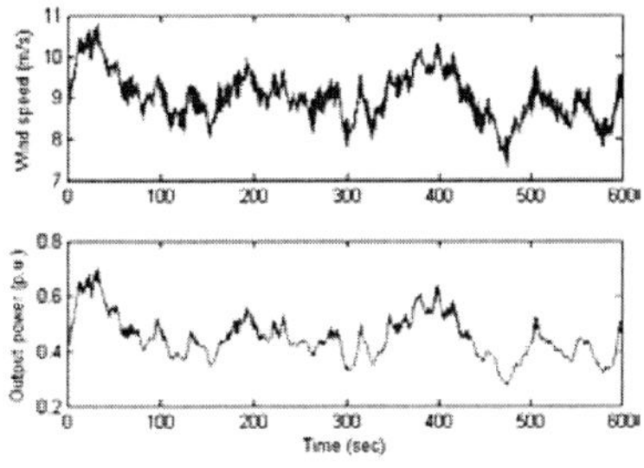

Figure 23. Wind speed and output power of a wind turbine [29]

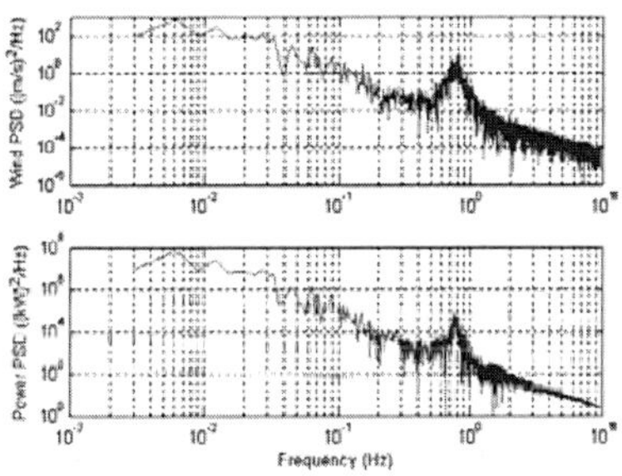

Figure 24. Spectrum of wind speed and output power of the wind turbine [29]

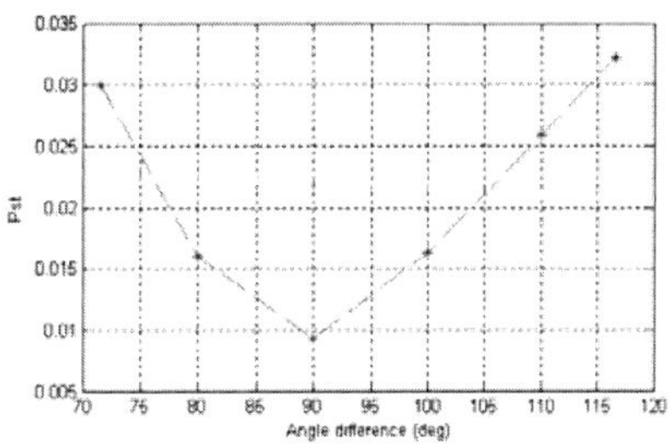

Figure 25. Short-term flicker severity, P_{st}, variation with angle difference, ψ-ψ_k, (*v=9m/s, In=0.1, SCR=20ψ_k =63.4°*) [29]

Fig. 26 – Fig. 29 illustrate the flicker level differences between (i) output reactive power control, i.e. the angle difference is regulated to be 90 degrees, and (ii) normal operation. The comparison is done with different parameters, such as mean wind-speed, turbulence intensity, and short circuit capacity ratio. The relationships between the short-term flicker severity and the other parameters in the two cases have a similar pattern. However, compared with normal operation, when the angle difference is controlled to be 90 degrees by controlling the reactive power, the flicker is significantly reduced.

From Fig. 26 to Fig. 29, it can be concluded that regulating the angle difference, by controlling the wind turbine output reactive power, is an effective way for flicker mitigation regardless of mean wind-speed, turbulence intensity and short circuit capacity ratio. To implement this method for flicker mitigation, some reactive power may be absorbed from the grid. It may not be necessary to realize the full compensation, (i.e. controlled to be 90 degrees), which may cause a significant amount of reactive power to be absorbed from the grid. If only a small amount of reactive power is absorbed from the grid, (may become nearer to 90 degrees), the flicker level could be effectively reduced

The flicker emission effects of variable speed wind turbines with doubly-fed induction generators during continuous operation are shown in Fig 29. The wind turbine output reactive power can be controlled to vary with the output active power, so the difference between the grid impedance angle and the power factor angle may approach 90 degrees, which leads to reduced flicker. The wind turbine output reactive power control provides an effective means for flicker mitigation regardless of mean

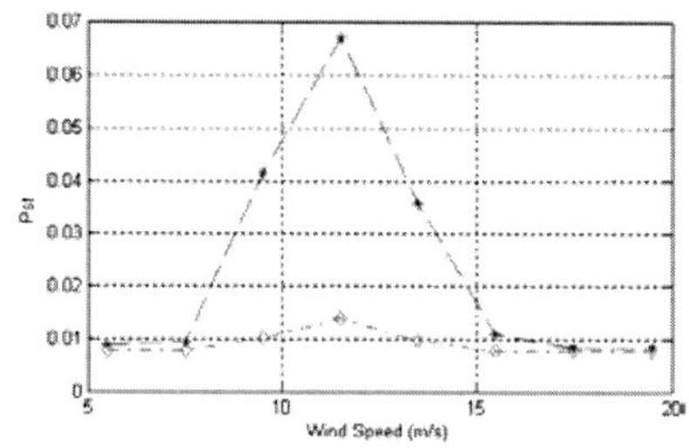

Figure 26. Short-term flicker severity P_{st} variation with mean wind speed (*In = 0.1, SCR = 20,* ψ_k = *63.4349°*) normal operation (asterisk), with output reactive power control (diamond)) [29].

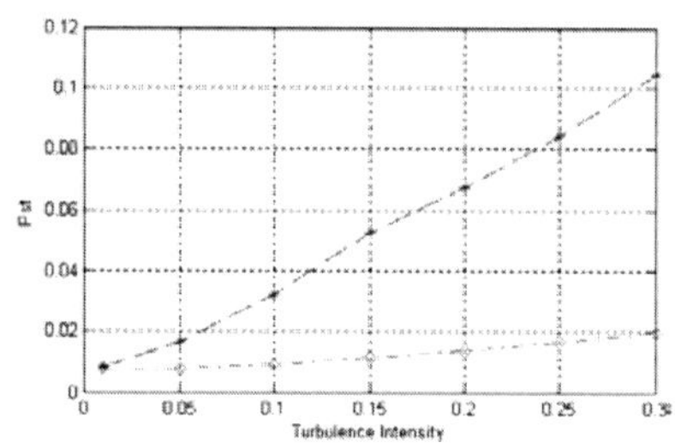

Figure 27. Short-term flicker severity P_{st} variation with turbulence intensity in low wind speeds (*v = 9ms, SCR = 2-,* ψ_k = *63.4349°*) normal operation (asterisk), with output reactive power control (diamond)) [29].

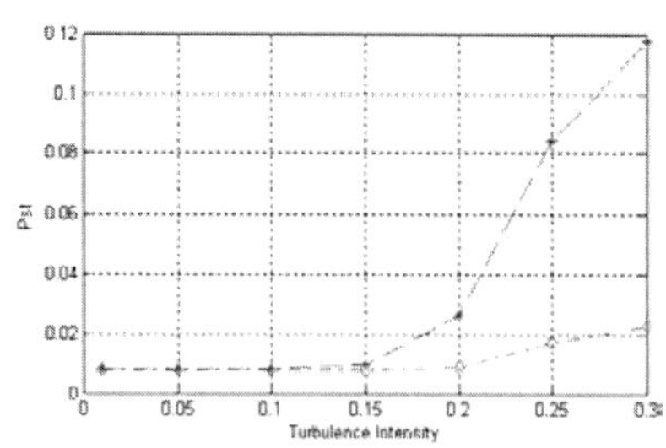

Figure 28. Short-term flicker severity P_{st} variation with turbulence intensity in high wind speeds (*v = 18ms, SCR = 20,* ψ_k = *63.4349°*) normal operation (asterisk), with output reactive power control (diamond)) [29].

wind-speed, turbulence intensity and short circuit capacity ratio. It is clear that variable speed wind turbines with power electronic interfaces are capable of controlling the output active and reactive power independently, therefore, power quality can be improved.

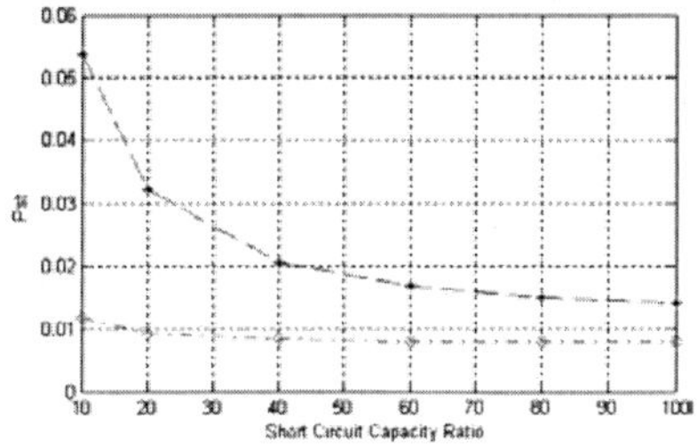

Figure 29. Short-term flicker severity P_{st} variation with short circuit capacity ratio (*v = 9 ms, In = 0.1,* ψ_k *= 63.4349°*) normal operation (asterisk), with output reactive power control (diamond)) [29].

9.2 Stability improvement in power systems

(a) Steady state voltage stability

In general, voltage instability problems and collapse are typically associated with power systems that are heavily loaded and are not able to meet the demand for reactive power. Therefore, voltage stability is sometimes called load stability. In the centralised power plants, most synchronous machines are equipped with excitation control systems for the supply and control of reactive power. They, therefore, may increase the voltage stability limit of the system. For wind power integration, however, the wind power generation may be viewed as a complex load, but with a negative real part in contrast to normal load. The characteristic of windfarms depend on the types of wind turbines and their operation and control strategies.

The connection of induction machines may reduce the voltage stability limit. This is because the induction machines cannot control reactive power (without power electronic converter systems or other reactive power compensation devices), but take reactive power from the system. Furthermore, the number of large central generators with voltage control ability may be reduced, since part of the power demand is supplied by wind turbine units. This may lead to less reactive power support in the system.

Fig. 30 shows the relationship between the real power and reactive power of a cage induction generator, where the negative real power means generation, and the positive reactive power represents absorption. A characteristic with a fixed capacitor bank is also shown in the figure. It can be seen the capacitor bank compensates for the reactive power to give a unit power factor at the rated real power generation. However, the reactive power absorption will increase quickly if the real power generation is greater than the rated value. Fig. 31 shows the system voltage drops with the increase of wind power penetration, until load flow convergence fails. The system used for this case study is a simplified equivalent system from reference [20]. The capacitance compensation for the cage rotor induction generator is fixed at a value so (i) there is unit power factor at the rated power, and (ii) each turbine operates at about 1.1 pu of the rated power. It is clearly seen that the high penetration of wind power may cause system voltage instability. It is also shown that capacitor reactive power compensation can increase the allowed wind power penetration capacity. It is expected that other reactive power compensation devices, such as SVC and STATCOM, are better for improving the voltage stability due to their controllability.

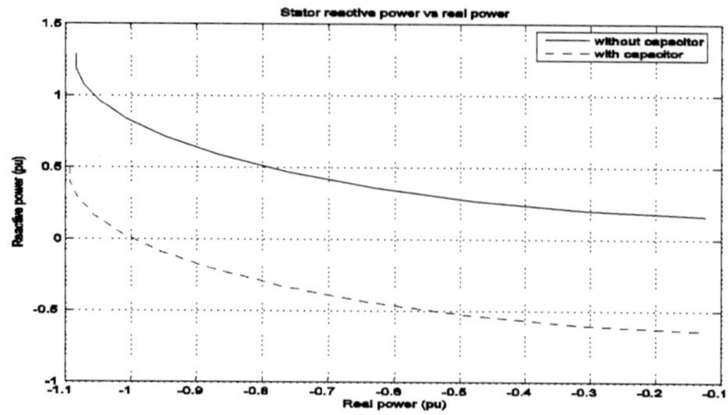

Figure 30. Relation between real power and reactive power of a cage induction generator [10]

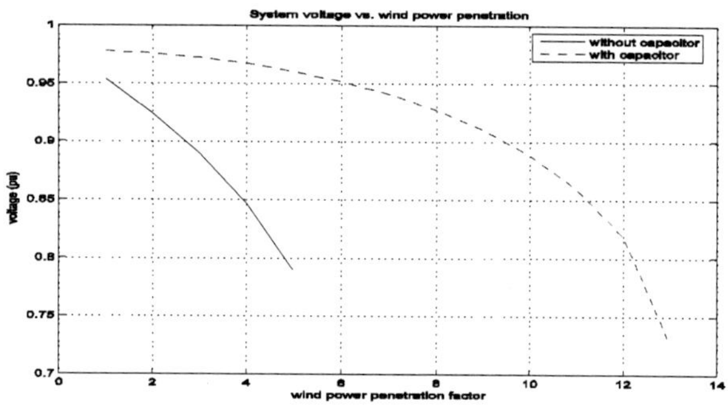

Figure 31. Relationship between system voltage and wind turbine power generation [10]

(b) Dynamic stability

An important issue of integrating large-scale windfarms is the impact of the wind power technologies on the stability and transient behaviour of the power systems [33-41].

When a fault occurs, the system voltage will drop, which will reduce the output electrical power from the wind turbines and decrease the electromagnetic torque applied on the turbine shaft, while mechanical torque may still be applied on the turbine as for normal operation. Consequently, the turbine and generator system will accelerate due to the unbalanced torque. After the clearance of the fault, the voltage of the power system will recover, and the output power and electromagnetic torque will increases again. If there is not sufficient electrical power to balance the mechanical power, the power surplus then leads to acceleration, and an increase in machine speed. The generator may have to be tripped out by protection systems. If the combined torque (electromagnetic torque and mechanical torque) can bring the generator back to normal operation speed, the turbine-generator system may restore normal operation, which is very important for overall power system stability. Many power system operators have grid codes to require the wind turbines remain connected in the system during a power system fault, in order to support the system recovery from the fault and prevent power system from collapse caused by the further loss of generation capacity. This becomes more significant as wind power penetration is increased. The main issues are voltage recovery and torque control. For mechanical torque control, turbines having blades with pitch adjustment have the advantage of being able to actively reduce the input mechanical torque so limiting the acceleration of the generator. For satisfactory generator restoration, a quick recovery of voltage and electromagnetic torque is crucial.

In the example of Fig. 32 [10], a load is at bus 2 and a windfarm, with wound-rotor induction generators, is represented by a single machine at bus 3. A three-phase short-circuit fault occurs at the middle of one of the two parallel circuits. It begins at 2 s and the line is tripped after 150 ms. The voltage at the wind turbine drops during the fault period, which leads to reduction in the electromagnetic torque and acceleration of the generator. The results with different controls are respectively presented in Fig. 33, Fig. 34 and Fig. 35. From Fig. 33 and Fig. 34, it can be clearly seen that neither rotor slip control nor pitching control alone can restore the system normal operation in this case; the generator terminal voltage cannot be recovered after the fault and the generator will be tripped by the over-speed protection. In Fig. 35, the rotor resistance control in combination with the pitching control restores normal operation of the system. Note that a single control, pitching or rotor resistance control alone, improves the situation but may not be enough in some cases, where the combination of pitching and rotor resistance control can be more effective to rebuild the voltage and may bring the generator back to normal operation after the short-circuit fault.

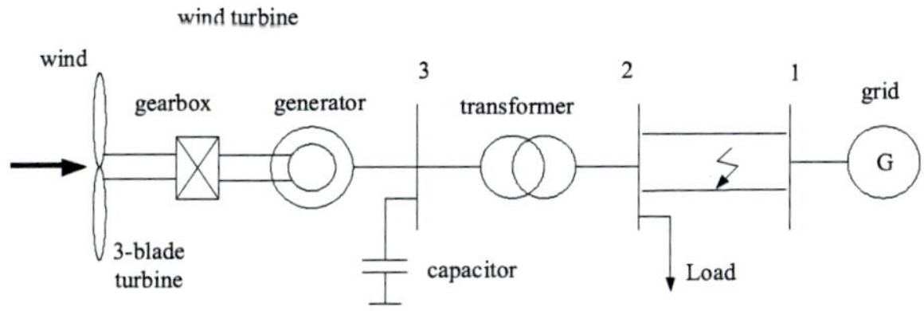

Figure 32. Block diagram of a wind turbine connected to a grid [10].

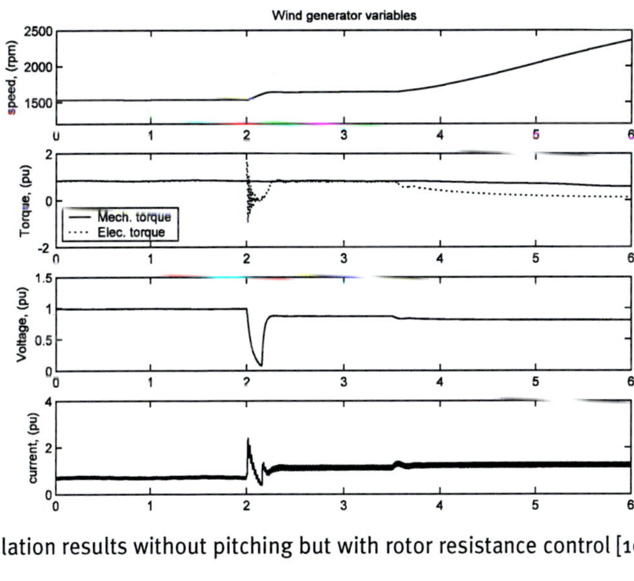

Figure 33. Simulation results without pitching but with rotor resistance control [10]

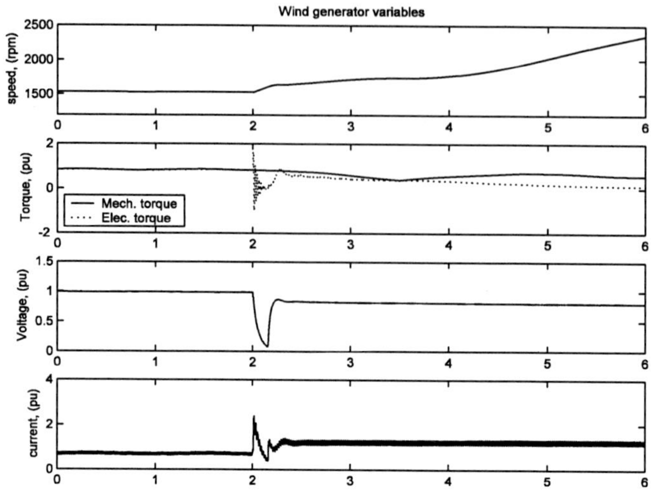

Figure 34. Simulation results with pitching but without rotor resistance control [10]

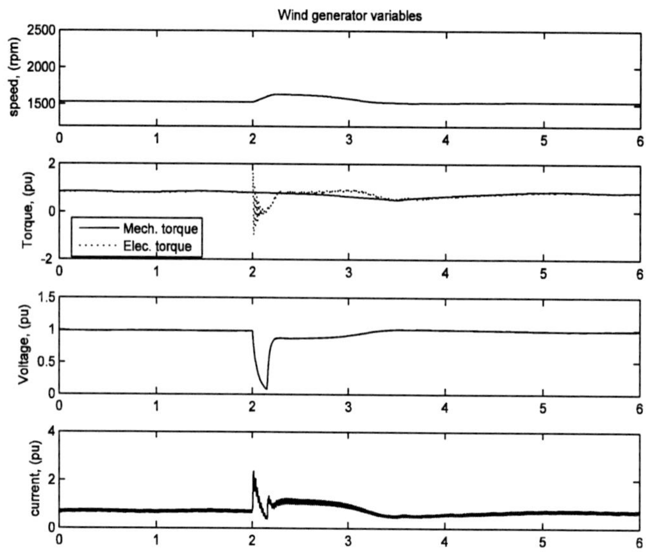

Figure 35. Simulation results with pitching and rotor resistance control [10]

Table 3. Comparison of four wind farm topologies

Comparison of Wind Farm				
Farm configurations (Fig.5.16)	**A**	**B**	**C**	**D**
Individual speed control	Yes	No	Yes	No
Control active power electronically	Yes	No	Yes	Yes
Control reactive power	Yes	Centralized	Yes	Yes
Short circuit (active)	Partly	Partly	Yes	Yes
Short circuit power	Contribute	Contribute	No	No
Control bandwidth	10-100 ms	200ms - 2s	10 -100 ms	10 ms – 10 s
Standby-function	Yes	No	Yes	Yes
Softstarter needed	No	Yes	No	No
Rolling capacity on grid	Yes	Partly	Yes	Yes
Redundancy	Yes	Yes	No	No
Investment	+	++	+	+
Maintenance	+	++	+	+

10. CONCLUSION

This chapter has discussed the electrical aspects of wind turbine systems in general. Various wind turbine systems with different generators and power electronic converters have been described. Different types of wind turbine systems will have quite different performances and controllability, which have been discussed with examples. The electrical topologies of windfarms with different wind turbines are outlined. It is shown that the wind farms need different configurations for best use of their particular type of turbines. The main grid connection requirements, which are closely related to the electrical characteristics of the turbine systems, have been discussed, and methods of improving wind turbine performance in power systems are illustrated.

REFERENCES

1. Blaabjerg, F., Chen, Z., Kjaer, S. B.,: "Power Electronics as Efficient Interface in Dispersed Power Generation Systems", IEEE Transactions on Power Electronics, Volume: 19, Issue: 5, Year: Sept. 2004, pp.1184- 1194.
2. Chen, Z., Blaabjerg, F.: "Wind Turbines - A Cost Effective Power Source", Przeglad Elektrotechniczny R. 80 NR 5/2004 pp 464-469 (ISSN 0033-2097).
3. B.J. Baliga, "Power IC's in the saddle", IEEE Spectrum, July 1995, pp. 34-49.
4. Chen, Z. Spooner, E : "Current Source Thyristor Inverter And Its Active Compensation System", IEE Proc. –Generation, Transmission and Distributions, Vol. 150, No. 4, July 2003, pp. 447-454.
5. Chen, Z: "Compensation Schemes for A SCR Converter in Variable Speed Wind Power Systems", IEEE Transactions on Power Delivery, Vol. 19, No 2, April 2004, pp. 813-821.
6. M.P. Kazmierkowski, R. Krishnan, F. Blaabjerg. *Control in Power Electronics-Selected problems*. Academic Press, 2002.
7. Chen, Z., Spooner, E. : "Voltage Source Inverters for High-Power, Variable-Voltage DC Power Sources", IEE Proc. –Generation, Transmission and Distributions, Vol. 148, No. 5, September 2001, pp. 439-447.
8. L.H. Hansen, P.H. Madsen, F. Blaabjerg, H.C. Christensen, U. Lindhard, K. Eskildsen, "Generators and power electronics technology for wind turbines", Proc. of IECON '01, Vol. 3, 2001, pp. 2000 – 2005.
9. A.K. Wallace, J.A. Oliver, "Variable-Speed Generation Controlled by Passive Elements", Proc. of ICEM '98, 1998.
10. Chen, Z., "Characteristics of Induction Generators and Power System Stability", Proc. of the Eighth International Conference of Electrical Machines and Systems (ICEMS 2005 Invited paper)
11. F. Iov, Z. Chen, F. Blaabjerg, A. Hansen, P. Sorensen, "A New Simulation Platform to Model, Optimize and Design Wind Turbine", Proc. of IECON '02, Vol. 1, pp. 561-566.

12. Florin Iov, Anca Daniela Hansen, Clemens Jauch, Poul Sørensen, Frede Blaabjerg, "Advanced Tools for Modeling, Design and Optimization of Wind Turbine Systems", Journal of Power Electronics, 2005.
13. Chen, Z., Spooner, E. :"Grid Interface Options for Variable-Speed, Permanent-Magnet Generators", IEE Proc. -Electr. Power Applications, Vol. 145, No. 4, July 1998, pp. 273-283.
14. Chen, Z., Gómez Arnalte S., McCormick, M: "A Fuzzy Logic Controlled Power Electronic System for Variable Speed Wind Energy Conversion Systems", 8th IEE International Conference PEVD'2000, London, September, 2000, (IEE Conf. Publ. No. 475) Page(s): 114 -119
15. Sun, Tao, Chen, Z., Blaabjerg, F. "Voltage Recovery of Grid-Connected Wind Turbines After a Short-Circuit Fault", Proc. of the 29th Annual Conference of the IEEE Industrial Electronics Society, IECON 2003, Roanoke, Virginia, USA, 2003. pp. 2723-2728 (ISBN: 0-7803-7907-1)
16. R. Pena, J.C. Clare, G.M. Asher, "Doubly-fed induction generator using back-to-back PWM converters and its application to variable speed wind-energy generation". IEE proceedings on Electronic Power application, 1996, pp. 231-241.
17. BTM Consults Aps. "International Wind Energy Department Word Market Update 2002", Forecast 2003-2007, 2003.
18. A.D. Hansen, C. Jauch, P. Soerensen, F. Iov, F. Blaabjerg. "Dynamic Wind Turbine Models in Power System Simulation Tool DigSilent", Report Risoe-R-1400 (EN), Dec. 2003, ISBN 87-550-3198-6 (80 pages).
19. Chen, Z., Spooner, E. "A Modular, Permanent-Magnet Generator for Variable speed Wind Turbines", IEE International Conference EMD'95, Conference Publication No. 412, 1995, pp 453-457.
20. Chen, Z., Spooner, E. : "Grid Power Quality with Variable-Speed Wind Turbines", IEEE Transactions on Energy Conversion, Vol. 16, No.2, June 2001, pp. 148-154.
21. IEC 61400-21: Power quality requirements for wind whines. (2001).
22. DEFU Committee reports 111-E (2nd edition): Connection of wind turbines to low and medium voltage networks 1998.
23. IEC 61400-12: Wind turbine generator systems. Power performance measurement techniques.
24. IEC 61000-4-15, "Electromagnetic Compatibility (EMC) —- Part 4: Testing and measurement techniques —- Section 15: Flickermeter —- Functional and design specifications," Bureau Central Commission Electrotech. Int., Geneva, Switzerland, Nov. 1997.
25. English version of Technical Regulations TF 3.2.6, "Wind turbines connected to grids with voltage below 100 kV –Technical regulations for the properties and the control of wind turbines", Eltra and Ekraft systems, 2004.
26. Chen, Z., "Issues of Connecting Windfarms into Power Systems", Proc. of 2005 IEEE/PES Transmission and Distribution Conference & Exhibition: Asia and Pacific. (Invited paper panel presentation paper)
27. Chen, Z., Blaabjerg, F., Sun, Tao "Voltage Quality of Grid Connected Wind Turbines", Proc. of the Workshop of Techniques And Equipments For Quality Ad Reliability Of Electrical Power, Bucharest, Romania, pp. 11-16, Printech, April 2004. (ISBN 973-652-961-4).
28. Blaabjerg, F., Chen, Z.,: "Wind Power – A Power Source Enabled by Power Electronics", Proc. of 2004 CPES Power Electronics Seminar, pp. I3-I14. April 2004..
29. Sun, Tao, Chen, Z., Blaabjerg, F,: "Flicker Study on Variable Speed Wind Turbines with Doubly-fed Induction Generators", Accepted for IEEE Transactions on Energy Conversion.
30. Sun, Tao, Chen, Z., Blaabjerg, F, "Transient Stability of DFIG Wind Turbines at an External Short-Circuit Fault", *Wind Energy*, 2005, 8:345-360.
31. Z. Saad-Saoud, N. Jenkins, "The application of advanced static VAr compensators to windfarms", Power Electronics for Renewable Energy, 1997, pp. 6/1 - 6/5.
32. Petru, T.; Thiringer, T.; "Modelling of wind turbines for power system studies", IEEE Transactions on Power Systems, Volume: 17 , Issue: 4 , Nov. 2002 Pages:1132 – 1139.
33. J.G.Slotweeg, H.Polinder, W.L.Kling, "Initialization of Wind Turbine models in power System Dynamics Simulations". IEEE Porto Power Tech, portugal, 10-13.September 2001, 6 p.
34. R. Fløglo, M. Gustafsson, R. Fredheim, T. Gjengedal, "Dynamic Simulation of Power Systems with Wind Turbines, A Case Study from Norway", NWPC '00, Trondheim, March 13-14, 2000, 6 p.

35. Akhmatov, Knudsen, "Modelling of windmill induction generators in dynamic simulation programs" (1999), Proc. IEEE Power Tech'99, 6 p.
36. Chen, Z., Hu, Y.: "Dynamics Performance Improvement of A Power Electronic Interfaced Wind Power Conversion System", Proceedings of 4th International Power Electronics and Motion Control Conference, IPEMC 2004, August 2004, Xi'an, China, pp.1641-1646.
37. Chen, Z., Hu, Y.: "Power System Dynamics Influenced by A Power Electronic Interface for Variable Speed Wind Energy Conversion Systems", Proceedings of 39th International Universities Power Engineering Conference UPEC 2004, pp.659-663.
38. J. Wiik, J. O. Gjerde, T. Gjengedal, and M. Gustafsson, "Steady state power system issues when planning large windfarms," IEEE PES 2002 Winter Meeting, New York, 2002.
39. Chen, Z., Blaabjerg, F. and Hu, Y. "Voltage Recovery of Dynamic Slip Control Wind Turbines with a STATCOM", Proc. of the 2005 International Power Electronics Conference IPEC 2005, April 2005, pp.1093-1100.
40. S.K. Salman, A.L.J. Teo, "Windmill modelling consideration and factors influencing the stability of a grid-connected wind power-based embedded generator", IEEE Transactions on Power Systems, Volume: 18 , Issue: 2 , May 2003 Pages:793 – 802.
41. J. Wiik, J. O.Gjerde, M. Gustafsson, and T. Gjengedal, "Dynamic simulations of wind power turbines in weak power systems," Wind power for the 21st century, Kassel, 2000.

Chapter 6:

Windfarm Power Connection

Thomas Ackermann

ABBREVIATIONS

AC	alternating current
DC	direct current
CIGRE	international council on large electric systems
HV	high-voltage
IGBT	insulated gate bipolar transistors
LCC	line commutated converter
PWM	pulse width modulation
VAr	(voltage x current) reactive power
VSC	voltage source converter

1. INTRODUCTION

This chapter provides an overview of the conceptual design options for:

(i) the electrical collector system for an offshore windfarm, and
(ii) the transmission system between an offshore windfarm to a grid connection point onshore.

The chapter presents and discusses methods for the evaluation of the key factors that govern the cost effective design and operation of the electrical collector systems and transmission system to shore. The main focus is on loss evaluation and reliability/redundancy assessment of different technical solutions. The methods can be used as part of an economic assessment of conceptual designs and can provide generic guidelines for optimal electrical system design in electrical system configuration. This chapter, however, will not provide a detailed economic assessment of the technical solutions, due to the rapid technical development in this area and the specific nature of individual offshore wind farms.

Initial offshore windfarms have a relatively small installed rated capacity (rated up to ~160 MW) and are placed comparatively close to the shore (up to ~20 km). As overall economic performance of offshore windfarms tends to increase with windfarm size, future projects will be significantly larger (up to ~1000 MW per windfarm) and most likely at a longer distance off shore (up to ~200 km). Therefore, this chapter will emphasize the design of the electrical collector system and the transmission system to shore for wind power plants with up to 1,000 MW of installed capacity. Due to its size, and also due to the fact that grid codes require windfarms to fulfil the same requirements in regards to grid support as conventional power plants, such installations should actually not be called offshore windfarms but offshore wind power plants.

1.1. SYSTEM OVERVIEW

Figure 1 shows a schematic overview of the layout of an offshore wind power plant collector and transmission system.

The collector system interconnects the wind turbines in the windfarm and connects them to the collecting point. For small wind power plants, without an offshore substation, the collecting point lies in the basement of a wind turbine. For larger wind power plants, the collecting point can be part of an offshore substation and very large offshore wind power plants might even have more than one collecting point/offshore substation.

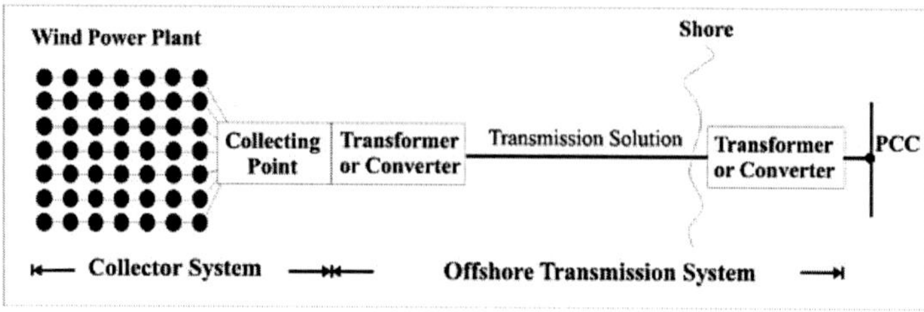

Figure 1. Schematic representation of an offshore wind power plant system. Collecting point can be an offshore substation. PCC = Point of Common Coupling (in this case to the onshore grid).

The offshore transmission system to shore usually begins at the transformer that raises the collector voltage level to the transmission voltage level. For small wind farms and a short distance to shore, the voltage in the collector system (typically around 30 kV) and to shore can be the same, which means a transformer is not required. In other cases, a transformer is usually necessary to increase the voltage. Such a transformer usually requires a separately installed offshore substation. If the offshore transmission system uses high-voltage DC, the transmission system begins at the converter stations which convert AC to DC.

Onshore, either a transformer or converter station is used to connect the offshore wind power plant to the onshore grid system. If the offshore transmission system operates at the same voltage as the onshore network, only switchgear is required at the point of common coupling.

2. THE OFFSHORE COLLECTOR SYSTEM

The standard design for the offshore collector system uses an AC network which collects the power production from each wind turbine. As most wind turbine generators operate at a generator voltage of 690 V to 1000 V, transformers are installed either in the nacelle of the wind turbine or in the basement of each wind turbine in order to increase the generator voltage to the voltage level in the collector system. The collector voltage used within onshore windfarms lies usually in the range of 33-36 kV; see also Table 1.

The mentioned voltage range is mainly used because the market offers competitive standardized equipment for this voltage range. Furthermore, higher voltage levels may require larger transformers and costs for transformers will increase accordingly. These transformers will be placed in the nacelle, in the tower or in a container next to the wind turbine and consequently, the size of the transformer might become an issue. In addition, switchgear, which is usually also installed in the basement of the wind turbines, increases in size with voltage levels. Switchgear design for voltages above 36 kV might not fit into the base of a wind turbine tower anymore. Nevertheless, research by Lundberg indicates that for small offshore wind power plants (around 60 MW) with a maximum distance to shore of 80 km, the most cost effective voltage is 45 kV (Lundberg 2003).

In practice, existing smaller offshore wind farms that are reasonably close to shore use comparatively low-voltage interconnection to shore, because the reduction in (load) losses is not sufficiently large to justify the additional costs for the equipment required for higher-voltages. The same applies to smaller offshore wind farms that are currently being planned. The seven turbines of the Swedish Utgrunden offshore windfarm (10 MW), for instance, are interconnected at a voltage of 20 kV. The cable to shore has a length of 8 km and is operated at 20 kV. Onshore the voltage is raised to 50 kV, see also Table 1.

Table 1. Overview of electrical parameters for offshore wind power plants built between 2000 and 2007 and planned selected wind power plants. [1] OSS = Offshore Substation; [2] 19 km offshore, 33 km onshore. [3] 27 km offshore, 3 km onshore. [4] 7 km offshore (3 phase cable), 2 km onshore (3 single phase cable), see Larsson (2008) for details, [5] 45 km offshore, 5 km onshore, [6] 128 km offshore, 75 km onshore, PCC = Point of Common Coupling.

Name, Country	Start	Turbines Rotor/ kW	Total Capacity	Distance to PCC [km]	Collector Voltage	OSS1	Transmission Voltage
Utgrunden, Sweden	2000	7*Enron Wind 70/1500	10.5 MW	8	AC, 20 kV	No	AC, 20 kV
Blyth, United Kingdom	2000	2*Vestas 66/2.000	4 MW	2	AC, 11 kV	No	AC, 11 kV
Middelgrunden, Denmark	2001	20*Bonus 76/2.000	40 MW	4	AC, 30 kV	No	AC, 30 kV
Yttre Stengrund, Sweden	2001	5*NEG-Micon 72/2000	10 MW	6	AC, 20 kV	No	AC, 20 kV
Horns Rev, Denmark	2002	80*Vestas 80/2.000	160 MW	19+33[2]	AC, 36 kV	Yes	AC, 150 kV
Palludan Flak, Samso, Denmark	2003	10*82,4/ 2300 Bonus	23 MW	5	AC, 30 kV	No	AC, 30 kV
Nysted, Lolland-Denmark	2003	72*Bonus 82/2300	165.6 MW	10	AC, 33 kV	Yes	AC, 132 kV
Arklow Bank, Irish Sea-Ireland	2003	7*GEW 104/3600	25 MW	10	AC, 38 kV	No	AC, 38 kV
North Hoyle, UK	2003	30*Vestas 80/2000	60 MW	10	AC, 33kV	No	AC, 33 kV
Scroby Sands, Norfolk-UK	2004	30*Vestas 80/2000	60 MW	4	AC, 33 kV	No	AC, 33 kV
Kentish Flat, UK	2005	30*Vestas 90/3000	90 MW	9	AC, 33 kV	No	AC, 33 kV
Barrow, UK	2006	30*Vestas 90/3000	90 MW	27+3[3]	AC, 33 kV	Yes	AC, 132 kV
Burbo Bank, UK	2007	25*Siemens 107/3600	90 MW	6.5	AC, 36 kV	No	AC, 36 kV
Lillgrund, Sweden	2007	48xSiemens 93/2300	110 MW	7+2[4]	AC, 33 kV	Yes	AC, 138 kV
Q7, NL	2008	60*Vestas 80/2000	120 MW	23	AC, 22 kV	Yes	AC, 150 kV
Horns Rev II, Denmark	est. 2009	91xSiemens 93/2300	215 MW	45+5[5]	AC, 36 kV	Yes	AC, 170 kV
Test field Alpha Ventus, Germany	est. 2009	6x5000 Repower, 6x5000 Multibrid	60 MW	66	AC, 30 kV	Yes	AC, 110 kV
Bard Offshore 1, Germany	est. 2010	80*Bard 122/5000	400 MW	128+75[6]	AC, 33 kV	Yes	33 kV transformed to 154 kV AC, than ±150 kV DC for 203 km to PCC

For the 40 MW Danish Middelgrunden offshore windfarm, a voltage of 30 kV was chosen for the connection between the 20 turbines and the approximately 3 km long cable to shore (Middelgrunden 2004). The Danish Horns Rev offshore wind power plant, which was constructed during 2002, has a capacity of 160 MW, and uses a voltage of 36 kV within the windfarm. Horns Rev was the first offshore wind power plant to use an offshore substation. At the offshore substation, the voltage is raised to 150 kV to feed the AC cable to shore (Horns Rev 2004). The 165 MW Danish Nysted offshore wind power plant uses 33 kV as the collection voltage. It also includes an offshore transformer station where the voltage is raised to 132 kV for the approximately 10 km AC connection to shore (Beck 2002).

Note that the overall network design of an offshore wind power plant does not necessarily correspond to the most energy-efficient network layout, i.e. lowest losses. The reason is that offshore substations are complex and require large support structures. Offshore substations are therefore very expensive and there is only limited experience regarding the reliability of offshore substations. Therefore, project developers tend to avoid using offshore substations unless they provide clear economic advantages.

2.1 Layout of the electrical collector system

The design of an overall windfarm layout aims at maximizing the overall wind production, taking into account the distributions of wind speed and direction as well as the turbulence between turbines, site ownership and planning permissions. The layout of onshore windfarms can therefore vary considerably, e.g. accounting for particular site conditions and complex terrain. The layout of offshore windfarms has less topographical and site-ownership constraints, but may be influenced by water depth and seabed conditions across its area. However, often the entire offshore site has approximately equal water depth, and so minimizing wake and turbulence effects between turbines becomes the main design criteria. Consequently, the distance between individual offshore turbines is generally larger than for similarly sized machines onshore.

In principle, the electrical collector system can be adapted to any windfarm layout although offshore windfarms tend to be larger in terms of total capacity and number of turbines, which increases the requirements on the design of the electrical collector system. However, installation costs and performance, e.g. losses, of the electrical collector system can vary substantially depending on the windfarm layout. The maximum number of turbines that can be connected to one feeder, for instance, will depend on the maximum cable rating of the cable connecting each feeder to the hub (Brakelmann 2003). Based on a collector voltage of 33 kV, the maximum capacity per cable is around 40 MVA. That means that each feeder can connect approximately 8 turbines with 5 MW rated capacity each.

Pierik et al. (2001), Gardner (1998) as well as Quinonez-Varela et al. (2006) have identified several possible electrical system layouts. Figure.3 provides a sample of possible electrical arrangements. Alternative A shows the most straightforward arrangement for a larger wind power plant, i.e. a radial layout of wind turbines. This configuration was adopted at both initial large offshore wind power plants in Denmark: Horns Rev and Danish Nysted. It also is included in the proposal for many other offshore wind power plant that are at a planning stage. The main advantage of this solution is the low cable costs achieved by tapered ratings between the wind turbines, see Figure 2. The average energy losses in radial configured collector systems are about 2 % of the annual power production of the wind power plant.

The major downside for the radial layout (option A of Fig 3) is the comparatively poor reliability, since cable or switchgear faults at the hub end of the radial cluster may prevent all downstream turbines from exporting for the duration of the fault.[1]

With some additional cabling and switchgear, ring layouts (Alternative B, C and D in Figure.3) can address the reliability of the simple radial layout by enabling alternative routes for exporting power during a cable fault. Standard practice would be for the ring to be operated in an open configuration

[1] King et al. 2008 shows some interesting results of energizing a wind farm using a radial layout. Similar studies looking into energizing issues as well as other switching transients must be studied for all layouts discussed in this chapter.

with automatic switching to reconfiguring the network after a fault incident. The additional security comes at the expense of increased cable costs for a given number of turbines, i.e. longer cables as well as increased cable ratings are required throughout the circuit and additional equipment to isolate the fault. In Alternative C, for example, both the cables in the first and in the second feeder must be sized for the total power output of all wind turbines connected to the feeder. This allows for the full output of the wind turbines in the upper feeder to be diverted through the second feeder by closing the cable section at the left end of the feeders. However, as wind turbines operate at full capacity for only a limited proportion of time, a compromise is to rate the back-up feeder at less than full capacity, and, during the time of a fault, to switch off some turbines in case all turbines were operate at full capacity.

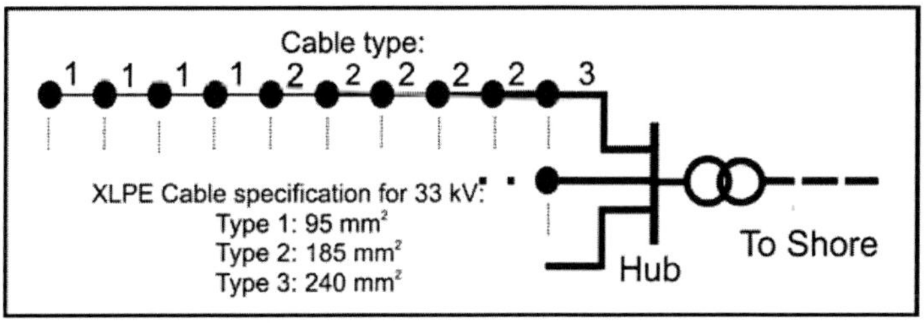

Figure 2. Example of possible cable rating for a radial network layout, based on Lillgrund Offshore wind power plant using 2.3 MW turbines, see Larsson 2008.

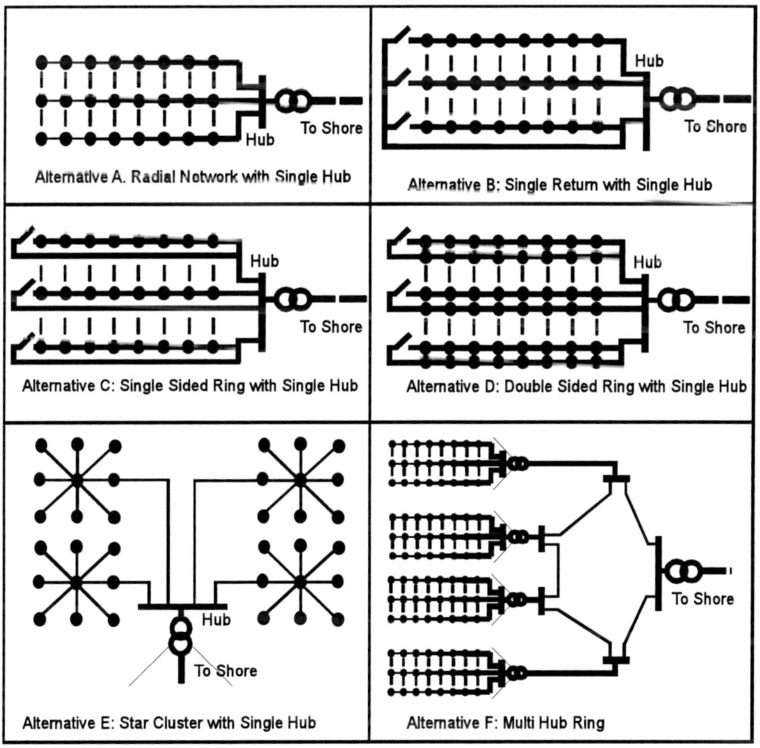

Figure 3. Schematic overview of different layouts for offshore collector systems, see also Ault et al. 2005.

Another suggested solution is the star-cluster layout (alternative E in Figure.3). This also requires only reduced cable ratings and provides a high level of security for the wind power plant as a whole, since one cable outage in general only affects one wind turbine. The drawback is that this layout requires longer 'diagonal' cables and some short sections of higher rated cabling, but this is not a major drawback.

Today, the likelihood of a fault and the associated costs are assumed to be less than the costs of the additional equipment. Therefore, ring or star configurations within the collector system are typically not considered. In order to reduce the likelihood of damage, cables are usually buried 1 to 2 meters into the seabed to protect them from ships' anchors, fishing equipment and strong sea currents.

Finally, in large offshore wind power plants or in case there are a number of offshore wind power plants in the same area, it might be advantageous to step-up the voltage at each turbine cluster (as shown in alternative F, Figure.3). System reliability issues associated with turbine feeders will depend on the electrical arrangement of the feeder (i.e. radial, ring or star). The main issue related to this layout is the cost of cables for different voltages and the cost of additional hubs that require a separate supporting offshore structure. The benefits of reduced losses and enhanced security (from multiple hubs) must be weighed against these additional expenses.

2.2 DC Collector Systems

The power output of generators used for offshore wind turbines is AC, however, variable speed turbines typically combine those generators with a back-to-back (AC/DC/AC) converter. Theoretically it is possible to separate the converter into an AC/DC converter installed at the wind turbine, followed by a DC collector system and a DC transmission to shore and a DC/AC converter close to the PCC. In other words, the DC bridge in the converter is replaced with an DC collector system and an high voltage DC transmission system. As the AC generator usually operates at 690 V, an additional DC/DC transformer (DC/DC switch mode converter/ buck booster) is usually required to reach the required voltage in the collector and transmission system. The disadvantage of this approach is that if all wind turbines are connected to the same DC/DC transformer, they will all work at the same operational speed. This operational speed can vary over time. Large offshore wind power plants, however, will cover such large areas that only a few turbines will be exposed to the same wind speed at any given time. An optimal aerodynamic efficiency cannot be reached because of the different operational speeds. An option would therefore be to connect wind turbines to the DC/DC transformer in clusters of approximately five turbines, see Figure 4. The five turbines of a cluster will operate at the same speed, which can vary over time. However, wind speed at each of the five turbines can vary, the overall aerodynamic efficiency of this option will still be less than if each turbine were to have an individual variable speed. The point of this option, however, is that the cost benefits of using clusters are larger than the drawbacks of the lower aerodynamic efficiency.

The general design concept allows for variations as discussed in more detail in Pierik et al. (2001), Courault (2001), Macken et al. (2001), Weixing and Boon-Teck (2002), Weixing and Boon-Teck (2003), Pierik et al. (2004), Martander (2002) and Lundberg (2003). These studies also include detailed economic analyses of this concept but the conclusions differ. Some companies have found this approach interesting and promising enough to study it further.

There are other, arrangements. For instance, the AC generator can be replaced with a DC generator or an AC generator with an AC/AC-AC/DC converter. In that case, a design option for a wind power plant would be similar to the design shown in Figure 4, but without AC/DC converters close to the wind turbines. Another option is to connect all wind turbines in series in order to obtain a voltage suitable for transmission, see Figure 5. This option, however, would require a DC/AC-AC/DC converter for a DC generator in order to allow each turbine to have an individual variable speed. The advantage of a series connection of DC wind turbines is that it does not require any offshore substations. For a more detailed discussion of this concept, see Lundberg (2003).

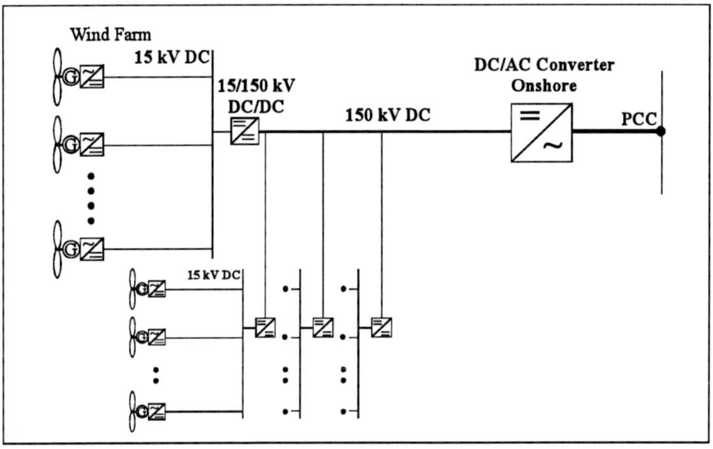

Figure 4. DC wind power plant design based on wind turbines with AC generators, based on Martander (2002). /Source: Wind Power in Power Systems, T. Ackermann [Editor], (c) _ John Wiley & Sons Limited. Reproduced with permission./

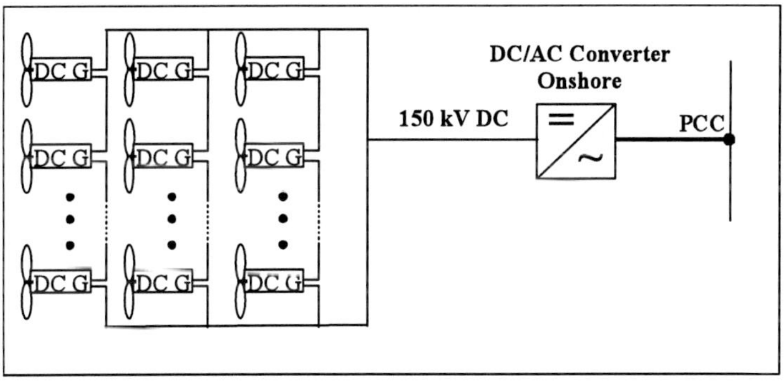

Figure 5. DC wind power plant design based on wind turbines with DC generators (DCG), based on Lundberg (2003). /Source: Wind Power in Power Systems, T. Ackermann [Editor], (c) John Wiley & Sons Limited. Reproduced with permission./

2.3 Variable Frequency AC Collector Systems

Another approach is based on a constant speed wind turbine using an AC permanent magnet induction generator. This turbine design has the drawback that pitch control cannot be used. This results in a less energy yield, as the rotation speed cannot be adapted to the wind speed, which leads to a smaller power coefficient. To overcome this drawback, a new control concept uses one central converter to control the frequency within the windfarm and thereby the rotational speed of the turbines, in order to gain a better power coefficient and hence greater energy yield. With this configuration, the collector system as a whole would be operated with a variable frequency, see Figure 6.

Similar to the DC collector system discussed in Section 0, this approach would result in the same operational speed for all turbines, hence optimal aerodynamic efficiency cannot be reached at all turbines. According to Troester (2008), such an approach combined with active stall controlled turbines would only cause a small reduction in energy yield (less than 4%) in comparison to the variable speed control turbines with individual converters. Due to the avoided costs of the individual converters at each turbine, such an approach might be economically interesting.

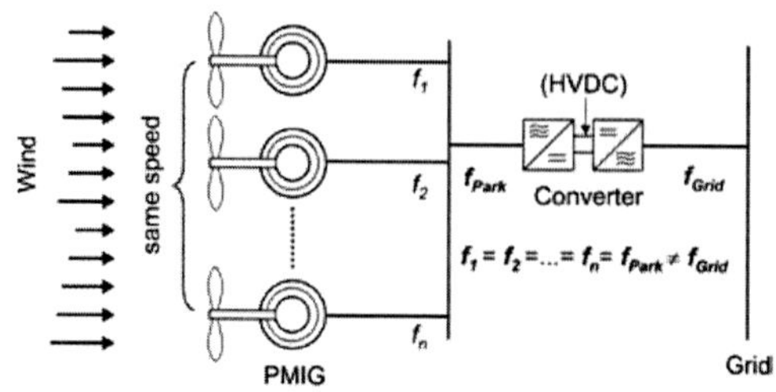

Figure 6. Constant speed turbines with central converter, see Troester (2008). PMIG = Permanent Magnet Induction Generator. Source: Troester (2008).

2.4 Collector System Optimization

To optimize the electrical collector system, i.e. to minimize total installation and operation cost, many factors should be included in the overall assessment of the various design options. Ault et al. (2005) lists the following parameters to determine the overall cost of a design option:

- Number of transformers at each voltage rating;
- Length of cables for each voltage rating and capacity;
- Number of cable terminations;
- Number of switchboard panel sections (including switchgear) for each voltage;
- Number of substation hubs required;
- Reactive compensation devices;
- Electrical flow losses;
- Production losses due to reliability issues in the collector system.

An optimization based on these parameters should result in the most economic solution. There is a large number of possible optimization methods that could be used, e.g. for optimizing distribution systems (Khator and Leung, 1997). Although, there are adequate methods, optimization is not an easy task mainly due to lack of sufficient cost data of equipment suitable for offshore applications. Furthermore, the optimization should not be limited to the offshore collector system, but should also include the transmission system to shore and possible upgrades of the power system onshore. The number of hubs, for instance, will affect the transmission system to shore. This also shows that the optimization is very case specific, at least for large (>> 100 MW) wind power plants. Nevertheless, the offshore wind power industry currently considers radial network solutions with no or one hub the most economic one for small wind power plants (< 200 MW) that are close to shore (< 30 km).

Figure 7. Left: Horns Rev, Right: Lillgrund Offshore substation.

3. OFFSHORE SUBSTATIONS

When constructed, the Horns Rev, Nysted and Barrow offshore substations were unique. Nothing similar had been built before, not even for the offshore gas and oil exploitation industry which usually only operates at voltage levels up to 13.8 kV. The existing offshore substations will provide important information about the reliability of offshore substations for future projects. The transformer of the Nysted offshore substation, for instance, experienced a significant failure in 2007, which led to a 41/2 month outage of the entire wind farm, see Andersen et. al. 2008. The experienced gained at the Nysted substation will influence the future design and possible redundancy of transformers on offshore substations.

The world's first offshore substation was built for the Horns Rev (Denmark) offshore wind power plant as a tripod construction as opposed to the Nysted (Denmark) and Barrow (UK) offshore substation which are based on a monopile construction, see Figure 6. The weight of the substation varies significantly for different design philosophies, see Table 2.

Table 2. Overview of Offshore Substations

Name	Wind Power Plant Capacity	Collector/Transmission Voltage [kV]	Design of Foundation	Weight of Substation
Horns Rev (DK)	160 MW	36/150	Tripod	1200 tons
Nysted (DK)	165.5 MW	33/132	Monopile, based on wind turbine monopile	670 tons
Barrow (UK)	90 MW	33/132	Monopile, based on wind turbine monopile	480 tons
Lillgrund (S)	110 MW	33/138	Gravity, based on wind turbine gravity foundation	520 tons

The Horns Rev substation includes a steel building with a surface area of approx. 20 x 28 metres which is placed about 14 metres above mean sea level. The platform accommodates, among others, the following technical installations (Horns Rev 2004, Larsson 2008):

- 36 kV switch gear
- 36/150 kV transformer
- 150 kV switch gear
- control and instrumentation system, as well as communication unit
- emergency diesel generator, incl. 2 x 50 tonnes of fuel
- sea water-based fire-extinguishing equipment
- staff and service facilities
- helicopter landing platform (helipad) crawler crane
- MOB (man over board) boat.

In comparison, the substation for the Barrow offshore wind power plant was designed to be much smaller, e.g. there is no helicopter landing platform; staff has to be lowered to the platform from a helicopter.

In the future, very large offshore wind power plants (>>250 MW) that are located far off the shore may require a number of offshore substations, see

Figure 7 The first step will be to connect each windfarm module with a maximum capacity of up to ~250 MW to a substation where the voltage is stepped up (or converted to HVDC) for transmission to the main substation. This main substation will provide the interconnection to shore,

Figure 7, alternative A. Another option is to connect each substation individually to shore, with an interconnection between the substations as redundancy, see also Figure 7, alternative B.

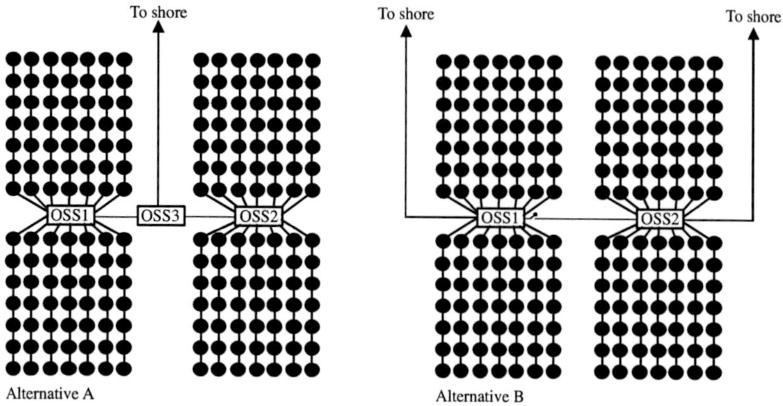

Figure 8. Possible layouts for a 980 MW (196 times 5 MW) offshore wind power plant. OSS = Offshore Substation / Source: Wind Power in Power Systems, T. Ackermann [Editor], (c) John Wiley & Sons Limited. Reproduced with permission./

4. TRANSMISSION TO SHORE

For the cabling to shore, either high voltage AC (HVAC) or high voltage DC (HVDC) connections can be used. For HVDC, there are two technical options:

(i) line commutated converter (LCC) based HVDC, and
(ii) voltage source converter (VSC) based HVDC technology.

All offshore wind power plants that are currently operating have adopted an AC alternative, see also Table 1. This is because these wind power plants are comparatively small and/or at a short distance from the shore. The situation may change as future wind power plants are expected to be larger and to be situated further off the shore. In the following, three different options for connecting offshore wind power plants to shore are discussed in more detail. Within the overall transmission system optimization, reliability issues and design features of offshore substations will be discussed.

4.1 High Voltage AC Transmission

An HVAC transmission system consists of the following main components: AC based collector system within the windfarm, an optional offshore transformer station including offshore reactive power compensation, three-core cross-linked polythene (XLPE) HVAC cable(s) to shore; and onshore an optional Static VAr Compensator (Eriksson et. al. 2003, Häusler and Owman 2002), see also Figure 9.

An example of HVAC transmission to shore is the Horns Rev 160 MW wind power plant in Denmark. As this wind power plant lies only 21 km offshore, compensation at the offshore transformer station was not necessary. The cable used was a 150 kV three core polyethylene insulation – XLPE-cable cable with 630 mm2 Cu conductor. The offshore power plant Nysted (Denmark) with a capacity of 165 MW uses a similar approach, i.e. a 29 km long XLPE cable to shore. Onshore an 80 Mvar capacitive and 65 Mvar inductive SVC for voltage control purposes is installed. The SVC is adequate sized to be able to compensate also the second offshore wind power plant to be connected to the same point of common coupling, see also Nielsen et. al. 2006.

For completeness it should be mentioned that also other possible AC transmission approaches for offshore wind power plants are discussed:

i) For large offshore wind power plants (>> 400 MW), the classical AC transmission solution will result in many parallel AC cables which is rather expensive and difficult to install.

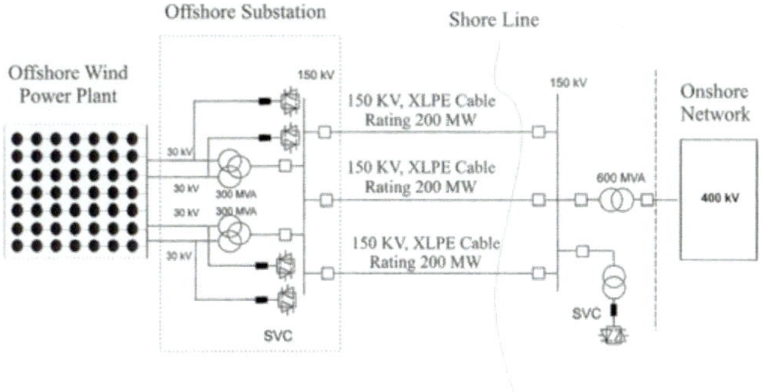

Figure 9. Single-line diagram for basic configuration of a 600 MW wind power plant with a HVAC solution, based on Eriksson et al. (2003) and Häusler and Owman (2002). /Source: Wind Power in Power Systems, T. Ackermann [Editor], (c) John Wiley & Sons Limited. Reproduced with permission./

Therefore a high capacity offshore transmission solution was proposed by Brakelmann based on a six phase system, bipolar AC system (Brakelmann et al. 2006 and Brakelmann et al. 2008). The proposed transmission system uses six single core cables connected to two conventional AC systems with opposite polarity. Those six-phase bipolar AC systems offer the opportunity to lay one pair of cables simultaneously in one trench. Because the sum of the currents in one pair of cables is zero the resulting magnetic field is negligible and, hence, the three pairs of cables are magnetically decoupled. This concept allows usage of maximum voltages and conductor cross sections. This results in substantially enhanced capacities for offshore AC connections. The losses of the proposed configuration are significantly lower than those associated with conventional AC submarine cables.

ii) Gas insulated cable (GIL) is a transmission system for high and extra high voltage applications. It is based on co-axial pipes of aluminum alloy, with the inner conductor at high voltage centred by solid insulators within an outer earthed enclosure. The space between the conductor and the enclosure is filled with a pressurized gas mixture to ensure the insulation. Because of its very high conductor cross-section the domain of GIL is high power transmission of electrical energy with currents exceeding 2000A. The first GIL installation onshore was realized in Germany in 1974. Today altogether some hundred kilometres phase lengths are in service all over the world. But as a result of the still very high installation costs compared to other transmission systems, all realized installations have only short lengths of up to 3 km. The advantage in regards to offshore wind power plants are higher capacity and, for the necessary length of offshore grid connections, comparatively low overall losses. In addition, no converter stations are necessary, see also Rathke (2008).

The above mentioned two approaches are currently not used for offshore wind power plants.

4.2 HVAC Transmission Capacity

The main factor limiting the transmission distance of HVAC transmission systems is the reactive power generated by the cable systems. Other limits, such as

i) voltage variation (< 10%) at the receiving end between no-load and full load, and
ii) phase variation (< 30o) are typically not critical constraints for the designs presented here (Brakelmann 2003a).
iii) Series resonance and resonance overvoltage can impact the design, but this is outside the scope of this chapter (see also Wiechowski et al. 2008).

Figure 10 provides a comparison of the transmission capacity of different cables operated at different voltages and different methods for compensation (only onshore or at both ends). The critical distance is achieved when half of the reactive current produced by the cable is equivalent to the nominal current at the end of one cable. For the considered cables, see Table 3, the critical distances are:

- $L_{max,132KV} = 370$ km
- $L_{max,220KV} = 281$ km
- $L_{max,400KV} = 202$ km

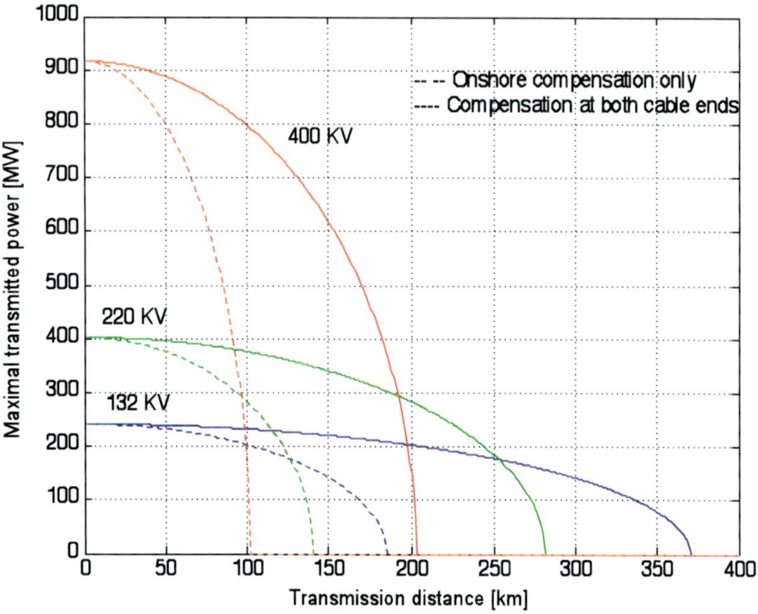

Figure 10. Transmission capacity versus transmission distance for different HVDC transmission cables for three voltages, 132 kV, 220 kV, and 400 kV. Cable based on data provided in Table 3

Table 3. Cable Parameters and Main Characteristics

Cable	132 kV	220 kV	400 kV
Resistance (W/m)	$48*10^{-6}$	$48*10^{-6}$	$45,5*10^{-6}$
Inductance (mH/km)	0.34	0.37	0.39
Capacitance (mF/km)	$0.23*10^{-3}$	$0.18*10^{-3}$	$0.18*10^{-3}$
Nominal current (A)	1055	1055	1323
Cable section[mm2]	1000	1000	1200
Max operating temperature [°C]	90	90	90

The results presented in Figure 10 take into account only reactive power compensation onshore or compensation at both sides of the cable. It can be seen that with an increasing distance to shore, reactive power compensation will be required at both ends of the cable, i.e. at the offshore substation, too. A 400 MW offshore wind power plant with two 150 kV, 120 km long cable systems, for instance, will require a 150 Mvar compensation both offshore and onshore (Eriksson et al. 2003). Furthermore, the maximum rating of AC cables is currently limited to about 200 MW per 3-phase cable, based on a voltage rating of 150 to 170 kV, compensation at both ends of the cable and maximum cable lengths of up to 200 km. Hence, a 1,000 MW wind power plant would need five cables for the AC connection to shore, excluding possible requirements for redundancy.

Theoretically, the transmission distance for a given capacity could be further extended by installing compensation equipment along the cable, i.e. in this case offshore. To date there is no practical experience of offshore compensation equipment, and the installation of additional platforms for compensation equipment will in many cases not be economically feasible. In some cases, however, it might be possible to use existing offshore platforms (from the oil industry or other offshore wind power plants, for instance) located close to the path of the HVAC transmission connection to shore.

Loss Calculation

According to Brakelmann (2003b), the cable losses per unit length can be calculated as followed

$$P' = \left(P'_{\max} - P'_D\right) \cdot \left(\frac{I}{I_N}\right)^2 \cdot \upsilon_\theta + P'_D \qquad (1)$$

where P'_{max} is the nominal total cable loss, P'_D is the dielectric loss, per core, I is the load current, I_N is the nominal current, υ_θ is the temperature correction coefficient that is:

$$\upsilon_\theta = \frac{c_\alpha}{c_\alpha + \alpha_T \cdot \Delta\theta_{\max} \cdot \left[1 - \left(\frac{I}{I_N}\right)^2\right]} \qquad (2)$$

where α_T is the temperature coefficient of the conductor resistivity [1/oK], c_α is a constant, i.e. $c_\alpha = 1 - \alpha_T((20^\circ C - \theta_{amb})$, $\Delta\theta_{max}$ is the maximal temperature rise, i.e. 90-15=75 °C, the ambient temperature is taken to be θ_{amb} =15°C.

As the cable current along the cable is not constant for a specific windfarm output, but depends on the position along a cable, i.e. $I=f(x)$, the following integral must be solved to obtain the cable losses per unit length:

$$P'_{lo} = \frac{P'_{\max}}{l_0 \cdot I_N^2} \cdot \int_{x=0}^{l_0} I^2(x) \cdot \upsilon_\theta(x) \cdot dx + P'_D \qquad (3)$$

Loss calculation considering the total transmission system should also include losses in the transformer as well as losses in the compensation units.

Results

Table 4 summarizes the cable loss for 500 and 1,000 MW wind power plant at several distances to shore and different voltages of the HVAC solution. The calculations are based on Brakelmann (2003b) and take into account the current distribution along the cable and the temperature dependence.

The proportional transmission system losses $l_\%$ are calculated as

$$l_\% = \frac{\left(\sum_i^N P_{lost,i} \cdot p_i\right) \cdot a}{\left(\sum_i^N P_{gen,i} \cdot p_i\right) \cdot a} \qquad (4)$$

where $P_{lost,i}$ represents the transmission losses at wind speed i, $P_{gen,i}$ is the power generated by the wind power plant at wind speed i, N is the number of wind speed classes considered for the model, pi is the probability of having a certain wind speed i, which is obtained by the Rayleigh distribution, a is the availability of the wind power plant. These calculations assume a technical availability of 100 %.

Table 4. AC transmission losses in percent of total wind power plant production for a 500 and 1,000 MW wind power plant with 9 m/s average wind speed. Cable characteristics are summarized in Table 3

%		500 MW	
Cable length	132 kV:3 cables	220 kV:2 cables	400 kV:1 cable
50 km	2.78	1.63	1.14
100 km	4.77	3.07	2.54
150 km	7.53	5.05	4.98
200 km	11.09	7.76	17.59

%		1000 MW	
Cable length	132 kV:5 cables	220 kV:4 cables	400 kV:2 cables
50 km	3.15	1.96	1.14
100 km	5.7	3.67	2.32
150 km	8.75	5.85	4.3
200 km	12.36	7.58	15.14

The shadowed cells in Table 4 represent the transmission options with the lowest losses. The table shows that the options using 220 kV and 400 kV result in the lowest losses, however, such cable designs are still under development (Rudolfsen 2003). Especially the 400 kV XLPE submarine cable can currently not be used for longer distances, as appropriate joints and splices are not available yet. For shorter distances, higher voltage ratings of up to 245 kV might be possible, which would increase the maximum rating to 350-400 MW (Rudolfsen 2001, Rudolfsen 2002). A three-core 1,000 mm^2 conductor cable operated at a system voltage of 230 kV with a maximum of +/- 10 % voltage variation between no-load and full-load can, for instance, transmit 350 MW over 100 km with losses of 4.3%, or 300 MW over 200 km with losses of 7.3 %, if the cable is compensated equally at both ends (Rudolfsen 2001). Most offshore wind power plants that are currently under development, however, will include only 132-150 kV AC transmission voltage to shore using three-core cable as a technically feasible solution.

Figure 11 shows the share that each transmission component contributes to the total transmission losses of a 500 MW wind power plant with a transmission distance of 100 km using a 132 kV cable. It can be seen that cable losses represent by far the largest portion of the total transmission losses. That means that in order to decrease total transmission losses, special attention should be given to the cable design.

For conventional applications, a cable is typically designed in such a way that no point in the cable is exposed to a temperature higher than 90 degrees Celsius (for XLPE cable) when loaded with maximum capacity for an infinitive time. The cable of a wind power plant with similar maximum capacity may theoretically be designed using a smaller conductor, since the windfarm output changes over time which allows the cable to "cool off" at wind speeds less than rated. That means that a wind power plant cable could be designed on the assumption that the cable will be exposed to the maximum rating for a limited number of days, e.g. 20 days continuously. This would allow the design with a smaller conductor size, however, it will result in increased losses. Therefore, the cable design constitutes an economic optimization between costs of the conductor, the reliability of the conductor and the economic value of the total losses in the cable over the lifetime of the project.

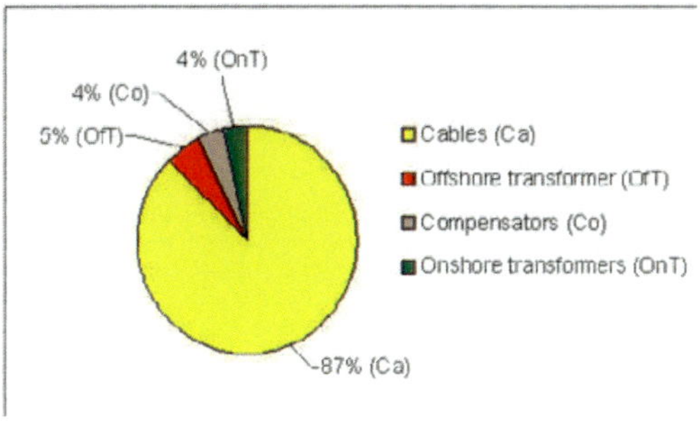

Figure 11. Contribution of each transmission component to total transmission losses for a 500 MW wind power plant, 9 m/s of average wind speed, at 100 km transmission distance, 3 three-core 132 kV submarine cables. Source: Todorovic (2004).

4.3 HVDC Transmission with Line Commutated Converter

The advantage of line commutated converter (LCC) based high voltage DC (HVDC) connections is certainly its proven track record. The first commercial LCC HVDC link was installed in 1954 between the island of Gotland and the Swedish mainland. The link was 96 km long, 20 MW rated and used a 100 kV submarine cable. Since then, LCC based HVDC technology has been installed in many locations in the world, primarily for bulk power transmission over long distances and for interconnecting power systems, e.g. the different island systems in Japan and New Zealand (Hammons et al. 2000). Other well-known examples using conventional HVDC technology are:

- the 1,354 km Pacific Intertie DC link with a rating of 3,100 MW at a DC voltage of ± 500 kV;
- the Itaipu link between Brazil and Paraguay, rated at 6,300 MW at a DC voltage of ± 600 kV (2 bipoles x 3,150 MW).

However, there is no experience of LCC based HVDC offshore substations or onshore in combination with wind power.

A thyristor based LCC HVDC transmission system for an offshore wind power plant would consists of the following main components:

- AC based collector system within the windfarm,
- an offshore substation with two three-phase two-winding converter transformers as well as filters
- either a Statcom or a diesel generator which would supply the necessary short-circuit capacity,
- DC cable(s)
- an onshore converter station with a single phase three-winding converter transformer as well as the relevant filters (Kirby et al. 2002, Cartwright et al. 2004), see also Figure 12.

The basic LCC HVDC transmission scheme uses a monopole, i.e. the sea is used for the return current, see Figure 13, alternative A. This approach reduced the cost of the interconnection since only one cable is necessary between the two converter stations. The sea return also keeps losses to a minimum since the return path has negligible resistance. However, the sea return current also results in several negative effects:

- Electrochemical corrosion of long buried metal objects in the vicinity, e.g.pipelines;
- Underwater earth-return electrodes in seawater may produce chlorine (Cl) or otherwise affect water chemistry;
- An unbalanced current path may result in a net magnetic field, which can affect magnetic navigational compasses for ships crossing over underwater cables.

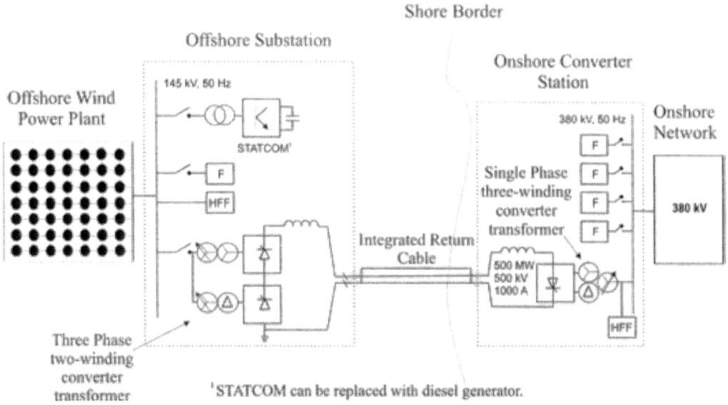

Figure 12. Basic configuration of a 500 MW wind power plant using a Line Commutated Converter HVDC system with a Statcom, based on Cartwright et al. (2004). For a configuration for a 1,100 MW wind power plant using LCC HVDC system with diesel generators on the offshore substation, see Kirby et al. (2002). F = Filter, HFF = High Frequency Filter. /Source: Wind Power in Power Systems, T. Ackermann [Editor], (c) 2005, John Wiley & Sons Limited. Reproduced with permission.

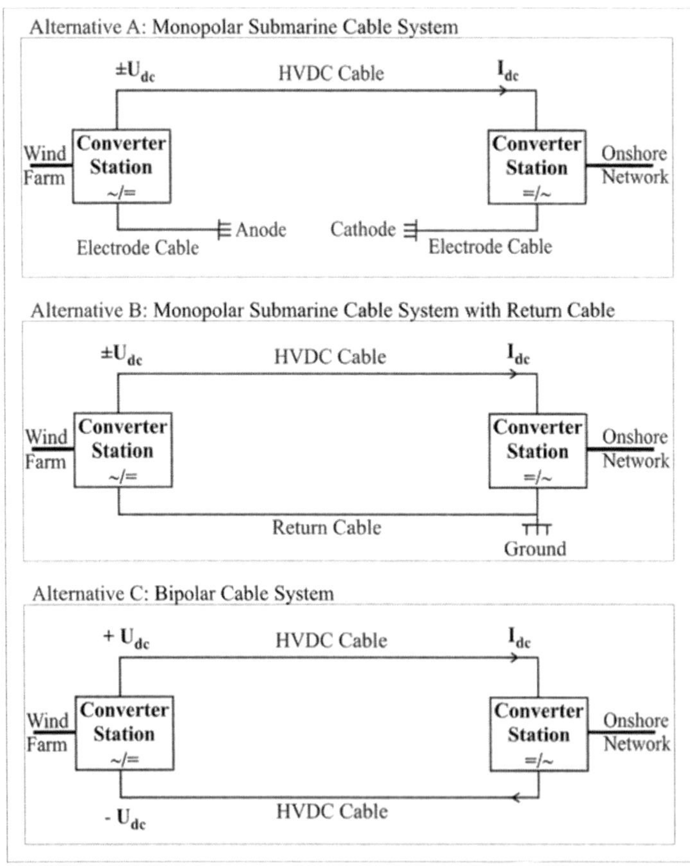

Figure 13. Overview of monopolar and bipolar cable configurations for HVDC systems.

Even though this LCC HVDC approach has been used for many years, the negative effects, mentioned above, probably means that today it would be almost impossible to obtain the relevant environmental permits. A solution to avoid the negative effects is a monopole scheme with a metallic return, see Figure 13, alternative B. The return current flows through a conductor in the form of a medium-voltage cable, which is either an independent cable or integrated into the HVDC cable.

The cables for HVDC LCC solutions are either XLPE cable for a voltage of up to 150 kV or paper-lapped cables for voltages of up to 500 kV. There are two designs of paper-lapped cables:

(i) the LPFF (Low Pressure Fluid-Filled) design, which typically uses oil as fluid, and
(ii) the MIND (Mass Impregnated Non-Draining) design, which is typically impregnated with oil.

XLPE cables are expected to be available soon for voltage levels of up to 300 kV.

This HVDC LCC technology requires converter stations both on- and offshore, as well as auxiliary service at the offshore converter station. The auxiliary service has to support the AC system at the offshore converter in order to enable the operation of the line-commutated converters even during periods with no or very little wind generated power. This auxiliary offshore service will be most likely supplied by diesel generator(s) (Kirby et. al. 2002). Another solution, also referred to as hybrid HVDC, comprises an LCC based HVDC converter and a Statcom, see Cartwright et al. (2004). The Statcom provides the necessary commutation voltage to the HVDC converter and reactive power compensation to the offshore network during steady state, dynamic and transient conditions.

HVDC LCC Loss Calculation

The loss calculation consists of two main parts: the transmission losses in the cable and the converter losses. Due to the lack of data for a detailed loss calculation of the converter station, losses can only be estimated by analysing data published on installed HVDC LCC converter stations, see also Siemens (2008) and ABB (2008). Converter stations have been built in sizes of 250 MW, 440 MW, 500 MW and 600 MW and published data about converter losses show a linear trend between 0,11% (no load) and 0,7% (rated power) of the rated power. Transmission losses can be calculated using the method suggested by Brakelmann (2003a), which takes into account variations in temperature in the cable in order to obtain more realistic results. Cable temperatures in buried subsea conditions may be significantly different than on land.

Results

For demonstration, the total system losses for different layouts based on a 500 and a 1,000 MW wind power plant have been calculated. The configurations and the corresponding system losses are presented in Table 4. The calculations include the assumptions that:

(i) If more than one converter station is used for the transmission, the total power is divided between the different converter stations depending on the configuration that gives the lowest total losses, and
(ii) Converter stations are shut down when the wind power plant's output is less than the system losses. Under these conditions, only the losses of protection and control devices are considered and these devices are supplied by the auxiliary power set e.g. a diesel generator.

The shadowed cells in Table.6 A represent the configuration with the lowest losses.

For some configurations, the share of each component in the total system losses is shown in Figure 14.

Table 5. A: Losses for different converter station layouts with 9 m/s of average wind speed in percentage of annual wind power production. (CS = Converter Station), B: Cable data calculated from Brakelmann (2003a), Siemens (2008), ABB (2008) and IEC (1994).

A

	"500 MW, 9 m/s"			"1000 MW, 9 m/s"			
Length Cable	**500 CS**	**2 x 250 CS**	**600 CS**	**2 x 500 CS**	**600 CS + 440 CS**	**500 CS + 600 CS**	**2 x 600 CS**
50 km	1.76	1.80	1.75	1.69	1.59	1.66	1.65
100 km	1.97	2.14	1.87	1.91	1.77	1.83	1.78
150 km	2.18	2.48	1.99	2.14	1.94	2.01	1.91
200 km	2.39	2.82	2.11	2.36	2.12	2.18	2.03

B

Rated power [MW]	**250**	**440**	**500**	**600**
Voltage level [kV]	250	350	400	450
Nominal current [kA]	1	1.257	1.25	1.333
Cable section [mm2]	1000	1400	1200	1600
Resistance [W/km] @ 20 °C	0.0177	0.0129	0.0151	0.0113
Max operating temperature [°C]	55	55	55	55

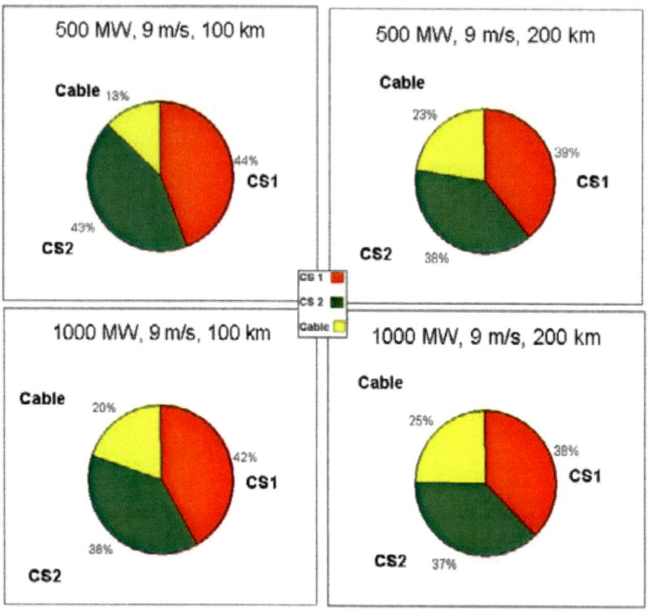

Figure 14. Contribution of each transmission component to the total transmission losses for a 500 MW and 1,000 MW wind power plant, 9 m/s of average wind speed, at 100 km and 200 km transmission distance, 3 three-core 132 kV submarine cables. Source: Todorovic, (2004). (CS = Converter Station)

The converter stations account for the highest share of the overall system losses; the contribution of the cable increases with cable lengths. It should be noted that the method used for calculating cable losses is much more detailed than the method for calculating losses in the converter station.

4.4 HVDC transmission with Voltage Source Converter

The voltage source converter (VSC)[2] based HVDC technology, is increasingly attracting attention. This comparatively new technology has become possible from progress in the high power electronics, namely Insulated Gate Bipolar Transistors (IGBTs). Consequently, Pulse Width Modulation (PWM) can be used for VSC converters, as opposed to thyristor based line-commutated converters used in the conventional HVDC technology.

The first commercial VSC-based HVDC link was installed by the ABB company on the Swedish island of Gotland in 1999. It is 70 km long, with 60 MVA at ± 80 kV. The link was mainly built in order to provide voltage support for the large amount of wind power installed in the South of Gotland. Between 1999 and 2000, a small demonstration project with 8 MVA at ± 9 kV was built in Tjæeborg, Denmark. This project is unique because the link is used to connect a wind power plant (3 wind turbines with a total capacity of 4.5 MW) to the Danish power system (Skytt et al. 2001). In 2000, a link – 65 km long, with 3 times 60 MVA at ± 80 kV – was built in Australia between the power grids of Queensland and New South Wales. A second link was installed in Australia in 2002. With a length of 180 km, this connection is the longest VSC-based HVDC link in the world. It has a capacity of 200 MVA at a DC voltage of ±150 kV. In the USA, a 40 km submarine HVDC link was installed between Connecticut and Long Island in 2002. The link operates at a DC voltage of ± 150 kV and has a capacity of 330 MVA. In 2005, ABB installed the first two 40 MW VSC offshore converter stations for the offshore gas platform Troll A in Norway. The link is 68 km long with a rating of 2 times 40 MVA at ± 60 kV (Eriksson et al. 2003). The first offshore application using an HVDC VSC system to connect a wind power plant to an onshore grid is scheduled to become operational in 2009. The 400 MW NORD E.ON 1 HVDC link will connect the offshore wind power plant Bard Offshore 1 (80 turbines with a rated capacity of 5 MW) to the German onshore grid. First, the collector AC voltage (33 kV) will be transformed to 154 kV AC, than a VSC based HVDC system with ±150 kV DC will be used for the connection to shore (total transmission distance 203 km.)

A VSC based HVDC transmission system consists of the following main components: AC based collector system within the wind power plant, an offshore substation with the relevant converter(s), DC cable pair(s) and an onshore converter station (Häusler and Owman 2002; Eriksson et al. 2003), see also Figure 16 and Figure 17. The VSC HVDC is typically operated as a bipolar system, see Figure 13, alternative C, and is usually not connected to ground, i.e. not earth-connected.

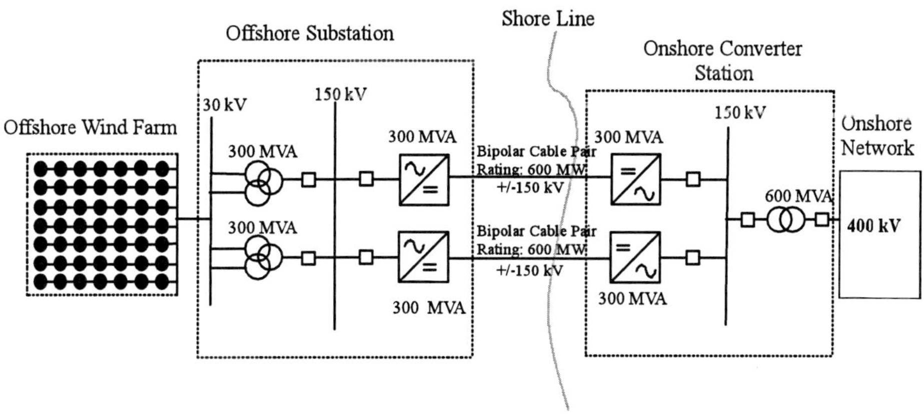

Figure 15. Single-line diagram for a 600 MW wind power plant using two Voltage Source Converter HVDC system, each converter station with a 300 MW rating. Based on Eriksson et al., 2003/Source: Wind Power in Power Systems, T. Ackermann [Editor], (c) 2005, John Wiley & Sons Limited. Reproduced with permission./

[2] Also known as forced commutated converter.

A VSC based HVDC link does not require a strong offshore or onshore AC network, it can even start up against a non-load network. This is possible because in a VSC the current can be switched off, which means that there is no need for an active commutation voltage. Furthermore, the active and reactive power supply can be controlled independently. An operation with multiterminal VSC schemes, e.g. interconnecting a number of offshore wind power plants, is also in principle possible, see also Hendriks et al. (2006).

VSC use IGBT semiconductor elements with a switching frequency of approx. 2 kHz. This design has resulted in comparatively large converter losses of 1,5 to 2 % per converter station. However, some manufacturers now claim that losses have been reduced to 1%.

The advantage of high frequency switching is low harmonic levels, i.e. a reduced need for filters. Also, rating per converter is presently limited to 400 to 550 MW, while the cable rating at +/-150 kV is 600 MW. Figure 17 shows the possible design for a 500 MW offshore wind power plant based on converter stations with 500 MW rating.

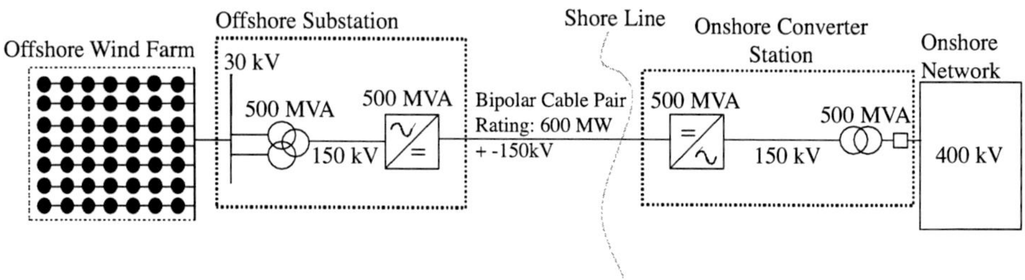

Figure 17. Single-line diagram for a 500 MW wind power plant using one Voltage Source Converter HVDC system based on a converter station with a 500 MW rating. /Source: Wind Power in Power Systems, T. Ackermann [Editor], (c) 2005, John Wiley & Sons Limited. Reproduced with permission./

HVDC VSC Loss Calculation

The loss calculation consists of two main parts: the transmission losses in the cable and the converter losses. Similar to the LCC technology, detailed data for the loss calculation of the VSC stations are difficult to obtain. Total system loss data, however, are available, for instance for the Cross Sound Cable (Railing 2004) and the Murray Link Project (Mattsson 2004). As the total system losses contain three components (2 stations + the cable) it is possible to determine the converter station losses if the cable losses can be calculated. These transmission losses can be calculated by using again Brakelmann's cable theory (Brakelmann 2003b), this method, therefore, enumerates the losses for the converter stations (350 MW and 220 MW) reasonably accurately, even without detailed knowledge of the converter design.

Results

For demonstration purposes, the total system losses for different layouts based on a 500 and a 1,000 MW wind power plant are calculated. Table 6 presents the configurations and the corresponding system losses. For these calculations, the converter stations are assumed to shut down if the power production in the wind power plant is less than transmission system losses.

The shadowed cells in Table 6 A represent the configuration with the least losses. For some configurations, the share of each component in the total system losses is shown in Figure 17.

Similar to LCC, VSC stations account for the largest share of the overall system losses; the proportion of the cable loss increases with cable lengths.

Table 6. A: Losses for different converter station layouts with 9 m/s of average wind speed in percentage of annual wind power production. (CS = Converter Station), B: Cable data [Brakelmann, (2003b), IEC (1978) and IEC(1994)].

A

	"500 MW, 9 m/s"			"1000 MW, 9 m/s"	
Length Cable	**350 + 220 CS**	**2 x 350 CS**	**500 CS**	**3 x 350 CS**	**2 x 500 CS**
50 km	4.0452	4.2115	4.428	4.0233	4.0893
100 km	4.4307	4.5758	4.8664	4.5228	4.5597
150 km	4.8168	4.9401	5.3064	5.0221	5.0317
200 km	5.203	5.3043	5.748	5.5215	5.505

B

Rated power [MW]	**220**	**350**	**500**
Nominal current [kA]	0.793	1.2	1.677
Cable section [mm2]	1300	1300	2000
Resistance [W/km] @ 20 °C	0.0138	0.0138	0.009
Max operating temperature [°C]	70	70	70

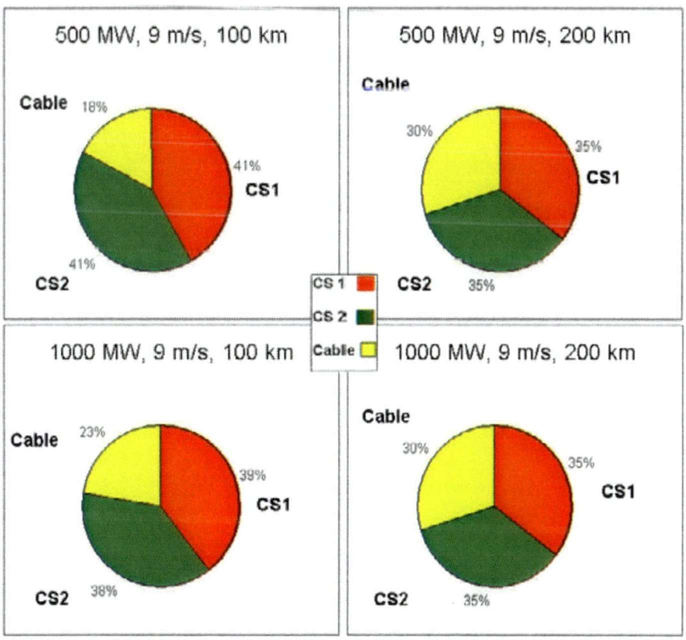

Figure 17. Contribution of each HVDC VSC transmission component to the total transmission losses for a 500 MW and 1,000 MW wind power plant, 9 m/s of average wind speed, at 100 km and 200 km transmission distance. Source: Barberis (2005). (CS = Converter Station).

4.5 Combination of Different Transmission Technologies

It may be beneficial to combine different transmission methods to improve some features of the overall system such as system reliability or power stability. For example, HVDC VSC transmission may be used to improve the stability of the onshore power system, e.g. during high wind power plant production and low onshore demand.

As an example consider a 1,000 MW wind power plant 50 km offshore connected via one 200 MW HVDC VSC transmission system and four HVAC cables with a capacity of 200 MW each. At times of high onshore demand, the HVAC systems are mainly used to reduce overall transmission losses. During times with low onshore loads, the HVDC VSC is predominantly used to support the onshore network.

A detailed analysis of the advantages and disadvantages of different combinations of HVAC, HVDC LCC and HVDC VSC will largely depend on the individual case and does not lie within the scope of this chapter.

4.6 Summary of Transmission Technologies

Table 7 compares the three standard transmission solutions. The technical capabilities of each system can probably be improved by adding additional equipment to the overall system solution.

Table 7. Comparison of Transmission Technologies

	HVAC	LCC HVDC	VSC HVDC
Maximum available capacity per system	• 200 MW at 150 kV; • 350 MW at 245 kV	• ~ 1200 MW	• 350 MW • 550 MW and 1100 MW announced
Voltage	• Up to 170 kV; 245 kV in the near future	• Up to ± 500 kV	• Up to ± 150 kV; ± 300 kV announced
Transmission capacity depending on distance?	• Yes	• No	• No
Total system losses	• Distance dependent	• 2 – 3 % (plus requirements for ancillary services offshore)	• 4 – 6 %
Black-start capability	• (Yes)	• No	• Yes
Fault contribution?	• Yes	• Less than HVAC;	• Less than HVAC;
Technical capability for network support	• Limited	• Limited	• Large range of possibilities.
Offshore substations in operation?	• Yes	• No	• Yes
Space requirements for offshore substation.	• Small	• Dependent on capacity. Converter larger than for VSC.	• Dependent on capacity. Converter smaller than LCC, but larger than HVAC substation.

5. RELIABILITY ASSESSMENT

Technical equipment can fail due to a number of reasons. Offshore, the repair of equipment is expensive, with the loss of exported power adding to the cost.

Therefore, offshore transmission system design must include a detailed reliability assessment. This assessment should compare the impact of different technologies, e.g. HVAC versus HVDC technology, on the overall system reliability and the impact of different configurations. In many cases, the transmission technology influences the layout of the transmission system to shore, i.e. the number of cables. For instance, a 1,000 MW offshore wind power plant can be connected with a single bipolar HVDC LCC connection to shore, while an HVAC solution might require up to 5 cables. With HVDC LCC, a faulty cable will result in the power loss of the entire wind power plant. With HVAC a single faulty cable still allows, however, 800 MW be transmitted to shore. Therefore the choice of technology affects reliability, not only because different equipment (with different reliability) is used, but also due to different configurations. The general transmission design has therefore not only to take into account different technical transmission solutions, with different investment and operating costs, but also the overall reliability related to the different transmission technologies and the number of connections to shore.

The reliability assessment requires detailed information about typical failure rates, including average time for repair and the associated costs. Such data are very hard to obtain, since there will be few comparable offshore wind power plants worldwide. Furthermore, costs of a cable failure depend strongly on the availability and current position of cable repair ships, which affects repair time. Nevertheless, it is, very difficult to protect subsea cables from damage, particularly if the cable crosses major shipping and fishing routes. It has been reported that anchors of large ships may penetrate 13 metres into the seabed. It is not viable to bury a cable at such a depth.

In general, cables damage may be reduced by using different cable routes to shore or by using back-up cables that follow another route to shore, see Figure 7, alternative B. Besides the significant costs of a second connection to shore, it may be difficult to carry this out in practice. At many locations, environmental restrictions would make it very difficult to find a second cabling route. Moreover, a second connection point to the onshore network may not always be available, or permitted. Currently, no offshore wind power plant has any redundancy for its cabling to shore. Experience from the operation of the first large offshore wind power plants will show whether redundancy for onshore connections will become necessary in the future.

In the following, a general method for the reliability assessment of a transmission solution based on HVDC LCC transmission is briefly presented.[3] It uses HVDC LCC because there are detailed data from existing HVDC LCC installation around the world. The data refer, of course, to shore-to-shore installation. The figures do not include the reliability issues associated with an offshore substation. Useful data for other transmission configurations are even harder to find, because

i) HVAC submarine installation, e.g. as between islands, typically use voltages of 36 kV or less, and
ii) there are few HVDC VSC applications, with these operative for only a few years.

For reliability assessment, the per unit term "energy not supplied" (Un) is defined as the proportion of energy produced by a power generator, such as a wind power plant, that cannot be transmitted due to failures (forced outages) in the transmission system (see also Equation 4). For a wind power plant, this estimate has to include the likely generation at the time of the failures, allowing for the wind conditions; thus in Europe, a failure in winter is likely to lose more transmittable energy than in summer

$$Un = \frac{Energy \quad not \quad transmitted}{Energy \quad that \quad could \quad have \quad been \quad transmitted} \tag{5}$$

Maintenance (scheduled outages) is another factor that contributes to "energy not supplied". However, it is assumed that maintenance takes place during periods with low wind speeds and easy access to wind turbines and thus its contribution to the unavailability of the system is less significant.

[3] A comparison of different techniques for offshore wind power plant reliability assessment can be found in Barberis Negra et al. (2006).

5.1 HVDC LCC Transmission Systems Reliability

The major source for statistical data of HVDC LCC (Line Commuted Converter) performance are the CIGRE[4] reliability reports, see Christofersen et al. (1996), Vancers et al. (2000, 2002 and 2004). In these reports, the number of forced outages and the equivalent outage hours for all the components of an HVDC LCC transmission system are summarized. The equivalent outage hours represent the hours that the transmission system was not operating due to a fault in a specific component. In these reports the data on forced outages are classified into six categories, namely:

- AC and auxiliary equipment (AC-E)
- Valves (V)
- Control and Protection (C&P)
- DC equipment (DC-E)
- Transmission line or Cable (TLC)
- Other (O)

Based on the CIGRE reports, the fault probabilities for the above mentioned components can be calculated, see Table 8, where the data presented refer to both the transmitting and receiving substations.

Table 8. Failure probabilities for HVDC LCC transmission components. Based on data for 222 pole · years and 13300 km · years. Calculated from Christofersen et al. (1996), Vancerset al. (2000, 2002 and 2004).

Component	Failure Probability
Converter Transformer	0.009251
AC auxiliary equipment	0.001397
Thyristor Valves	0.001175
DC equipment	0.000647
Protection and control	0.000545
Other	0.000124
Submarine Cable (per 100 km)	0.004462

5.2 Method for Calculating the "Energy Not Supplied"

The following method is explained by using three cases, the first case is based on a single cable always fully loaded, the second case is a cable system with two parallel cables also always fully loaded and the final case is based on two parallel cables with varying loading from a wind power plant.

First Case: 'Energy not Supplied' for a System with One Cable and Full Loading, Generator at Full Capacity.

First, a HVDC LCC transmission scheme with a monopole cable and integrated return cable is assumed, see Figure 13, alternative B. For the unavailability study the system can be schematically represented as shown in Figure 18. All equipment is represented only once, because the data in Table 8 represent the failure probability of the onshore as well as offshore converter station.

The state in which the converter transformer (CT) is not operating is defined as OCT, and, according to Table 8, the probability of this state, is:

$$F(O_{CT})=0.009251$$

[4] International Council on Large Electric Systems

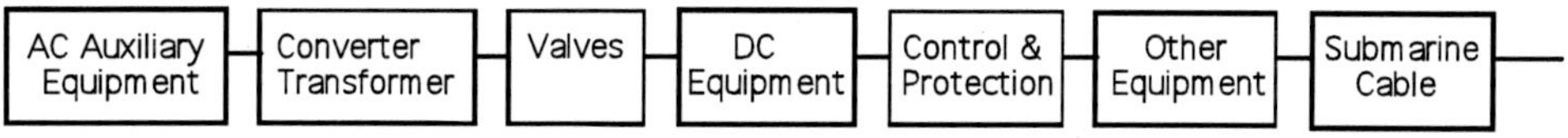

Figure 18. Schematic representation of a monopole HVDC LCC system.

In the same way, the states $O_{AC\text{-}E}$, O_V, $O_{DC\text{-}E}$, $O_{C\&P}$, O_{Cable}, O_o, are introduced for the AC auxiliary equipment, valves, DC equipment, protection and control, cable and other respectively. For the transmission scheme presented in Figure 19 the probability of a system fault can now be written as:

$$F(O_{Tot}) = F(O_{AC-E} \cup O_{Trans} \cup O_{Valves} \cup O_{DC-E} \cup O_{C\&P} \cup O_O \cup O_{Cable})$$

For a serial system, like the one in Figure 18, it is assumed that only faults at one component at a time can occur. In this case the above equation can be rewritten as:

$$F(O_{Tot}) = F(O_{AC-E}) + F(O_{Trans}) + F(O_{Valves}) + F(O_{DC-E}) + F(O_{C\&P}) + F(O_O) + F(O_{Cable})$$

It is now possible to calculate the energy not supplied due to failure in the transmission system depending on the cable lengths. For the calculation of the energy not supplied it is assumed that the theoretical available energy production is the same as the maximum transmission capacity, i.e. the fluctuating wind power plant output is not considered so far.

Table 9. Energy not supplied in percentage of theoretical energy supply for the system described in Figure 19 for different transmission lengths, see also Lazaridis (2005).

Transmission distance (km)	50	100	150	200	250	300
Energy not supplied in % of theoretical energy supply	1.5370	1.7601	1.9832	2.2063	2.4294	2.6525

Second Case: "Energy not Supplied" for a System with two Parallel Cables and Full Loading, Generator at Full Capacity.

It is now assumed that the transmission system consists of an HVDC LCC system with two parallel monopole cables. The schematic representation of this system is shown in Figure 19.

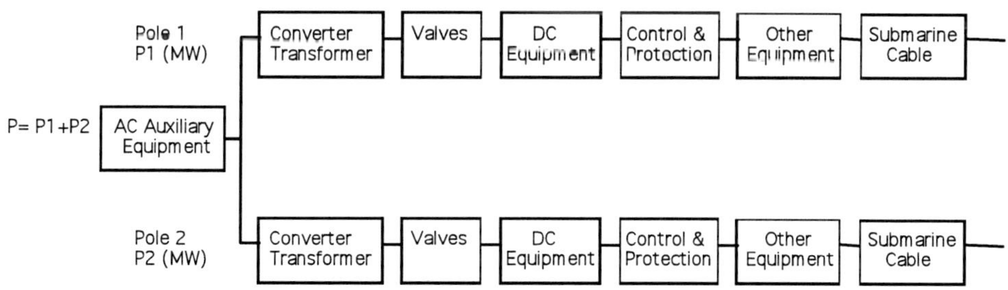

Figure 19. Schematic representation of an HVDC LCC system with two cable systems, which share the auxiliary equipment.

In the system described in Figure 20 the total transmission capacity is P (MW), with Pole 1 rated at P1 (MW) and Pole 2 rated at P2 (MW). The probability that the first pole will not be operating due to a fault in one of its components (common auxiliary equipment is not included) is given by:

$$F(O_{P1}) = F(O_{Trans}) + F(O_{Valves}) + F(O_{DC-E}) + F(O_{C\&P}) + F(O_O) + F(O_{Cable})$$

The above equation also describes the probability of failure in the second parallel pole. The failures in the parallel poles are assumed to be independent of each other, hence the probability of a fault in both parallel poles is given by:

$$F(O_{P1\&P2}) = F(O_{P1} \cap O_{P2})$$

or

$$F(O_{P1\&P2}) = F(O_{P1}) \cdot F(O_{P2})$$

The two parallel poles can now operate in the four following modes:

- Pole 1 in operation ("ON"), Pole 2 in operation ("ON")
- Pole 1 in operation ("ON"), Pole 2 out of operation ("OFF")
- Pole 1 out of operation ("OFF"), Pole 2 in operation ("ON")
- Pole 1 out of operation ("OFF"), Pole 2 out of operation ("OFF")

Each mode results in a specific probability and results in a specific transmission capacity. Table 10 provides an overview of the different operation modes and their corresponding probabilities as well as the related transmission capabilities.

Table 10. Operation modes and corresponding probabilities of a parallel HVDC LCC system, see also Lazaridis (2005).

		Pole 1	Pole 2	Total
Mode 1	State	ON	ON	
	Transmission capability	P1 (MW)	P2 (MW)	(P1+P2)(MW)
	Probability	$1\text{-}F(O_{P1})$	$1\text{-}F(O_{P2})$	$[1\text{-}F(O_{P1})]*[1\text{-}F(OP_2)]$
Mode 2	State	ON	OFF	
	Transmission capability	P1 (MW)	0 (MW)	P1 (MW)
	Probability	$1\text{-}F(O_{P1})$	$F(O_{P2})$	$[1\text{-}F(O_{P1})]*F(OP_2)$
Mode 3	State	OFF	ON	
	Transmission capability	0 (MW)	P2 (MW)	P2 (MW)
	Probability	$F(O_{P1})$	$1\text{-}F(O_{P2})$	$F(O_{P1})*[1\text{-}F(OP_2)]$
Mode 4	State	OFF	OFF	
	Transmission capability	0 (MW)	0 (MW)	0 (MW)
	Probability	$F(O_{P1})$	$F(O_{P2})$	$F(O_{P1})*F(OP_2)$

It is now possible to calculate the energy not supplied for the transmission system depending on the cable lengths. It is assumed that the transmission capacity P1+P2 is actually larger than the available energy production. Variable wind power plant production is still not considered.

Table 11. Energy not supplied of the system described in Figure 20 for different transmission lengths. Continuous production of 700 MW, transmission capacity Pole 1 = 500 MW, Pole 2 = 250 MW, see also Lazaridis (2005).

Transmission distance (km)	50	100	150	200	250	300
Energy not supplied in % of theoretical energy supply	1.44	1.64	1.85	2.06	2.27	2.48

Third Case: "Energy not Supplied" for a System with two Parallel Cables and a Wind Power Plant
This case assumes the same transmission system as in the second case, but the continuous power production of 700 MW from the previous case is replaced with a wind power plant rated at 700 MW. The average wind power produced by the wind power plant can be calculated as:

$$P_{avg} = \int_{w_{cin}}^{w_{cout}} P(w) \cdot R(w)\, dw$$

Where:

w_{cin} and w_{cont}: the cut in and cut out speed of the windfarm respectively,
$P(w)$: the power produced by the wind power plant for wind speed w based on an aggregated power curve for a wind power plant, which can be calculated according to Holttinen et al. 2004.
$R(w)$: the probability of a wind speed w occurring, according to the Rayleigh distribution

For the average wind speed of 8 m/s and a wind power plant of 700 MW installed capacity the average power produced is:

$$P_{avg} = 333.9\,MW$$

For the four modes presented in Table 10 the following can now be said:

Mode 1: All of the produced power is being transmitted, therefore the power not transmitted is:

$$P_{non\_tr\_mode1} = 0$$

Mode 2: For a power production up to P_1 MW all of the produced power is being transmitted. If the produced power is assumed to be y, where y is greater than P_1, the non-transmitted power will be (y-P_1). According to the Rayleigh distribution and the aggregated model of the windfarm, there is a very specific probability that y will be produced by the windfarm. This probability is named *F(y)*. If the transmission system operated continuously in this mode, the average value of the non-transmitted power, using data for the season of the failure (e.g. summer or winter) would be:

$$P_{non\_tr\_mode2} = \int_{P1}^{P} (y - P_1) F(y)\, dy$$

where P is the maximum power that the wind power plant can produce.

Mode 3: Similarly to mode 2, the non-transmitted power is:

$$P_{non\_tr\_mode3} = \int_{P2}^{P} (y - P_2) F(y)\, dy \ \text{[MW]}$$

Mode 4: None of the P_{avg} produced by the windfarm is transmitted, therefore the power not transmitted is:

$$P_{non\_tr\_mode4} = P_{avg}$$

where P_{avg} is the average power produced by the windfarm.

It is now possible to calculate the "energy not supplied" through the transmission system depending on the cable lengths, see Table 12, and the time period considered (e.g. one year, 6 months).

Table 12. "Energy not supplied" for the system described in Figure 20 for different transmission lengths. Wind power plant rated at 700 MW as power input, 8m/s yearly average wind speed.

Transmission distance (km)	50	100	150	200	250	300
Energy not supplied in % of theoretical energy supply	1.0345	1.1787	1.3232	1.4681	1.6134	1.7590

5.3 Results

Table 13 presents the "energy not supplied" for different HVDC LCC configurations for two different wind power plants. The transmission losses are not considered and the technical availability of the wind power plant is assumed to be 100%.

Table 13. Energy not supplied for different HVDC LCC transmission configurations

(%)	500 MW windfarm			1000 MW windfarm			
Transmission Distance	500 CS	2x250 CS	600 CS	2x500 CS	600 CS+ 440 CS	500 CS+ 600 CS	2x600 CS
50 km	1.53	1.16	1.53	1.15	1.11	1.03	0.90
100 km	1.76	1.33	1.76	1.31	1.27	1.17	1.03
150 km	1.98	1.49	1.98	1.48	1.42	1.31	1.15
200 km	2.20	1.66	2.20	1.64	1.58	1.46	1.28
250 km	2.42	1.82	2.42	1.80	1.74	1.60	1.40
300 km	2.65	1.99	2.65	1.97	1.89	1.75	1.53

The results in Table 13 show that the energy transmitted increases with two transmission parallel poles (i.e. there is less 'energy not supplied') and that oversizing the transmission system, i.e. the transmission system has a larger rating due than the wind farm, also leads to more energy transmitted.

6. ECONOMIC EVALUATION OF THE TRANSMISSION SYSTEM

In addition to wind speed, transmission losses and transmission reliability, total investment costs for the different transmission systems must be included in an overall economic evaluation of different transmission options. Investment costs, however, are very case specific because many factors affect the overall investment costs such as water depth, ocean currents, copper price, steel price, installation costs etc.

Therefore, the investment costs for the different system components cannot be evaluated here, and so a detailed economic analysis cannot be presented. Different studies, however, indicate that, independent of wind power plant size, HVAC solutions are the most competitive ones for a transmission length of max. 60 to 80 km. For transmission distances exceeding 80 km, HVDC solutions appear to be more competitive than HVAC. However, typically' neither takes into account the installation of offshore compensation equipment for the HVAC transmission solution nor include upgrade costs in the onshore power system, which the transmission technology may require.

Finally, economic evaluation of HVDC LCC versus HVDC VSC transmission solutions currently faces significant restrictions, due to the limited data available about the technical performance and the investment costs.

7. CONCLUSIONS

This chapter has discussed the conceptual design for the electrical collector system of an offshore wind power plant, with an overview of three different transmission technologies (HVAC, HVDC LCC and HVDC VSC). The transmission losses and the reliability of different transmission technologies and designs are also analyzed.

A generic overall economic analysis of the different transmission technologies is very difficult, because investment costs are very case-specific. However, independent of the size of the wind power plant, HVAC solutions can be considered the most competitive technology for transmission lengths less than 60 to 80 km.

ACKNOWLEDGEMENTS

The author thanks Nicolai Barberis Negra, Lazaros Lazaridis, Jovan Todorovic, Stefan G Johansson, Flemming Krogh, Per-Anders Löf, Nis Martensen and Eckehard Tröster for valuable comments. Also thanked for their contributions, are all the presenters and participants in the Stockholm (2000, 2001, 2002), Billund (2003), Glasgow (2005), Delft (2006) and Madrid (2008) workshops on Large-Scale Integration of Wind Power into Power Systems as well as on Transmission Networks for Offshore Wind Farms for their contribution (see also http://www.windintegrationworkshop.org/).

REFERENCES

ABB (2008), List of projects found at www.abb.com (last visit January 2008).

Ackermann, T. (2005), *Transmission Systems for Offshore Wind Farms, in Wind Power in Power Systems*, Editor T. Ackermann, Wiley, 2005.

Andersen, N., Marcussen, J., Jacobsen, E., Nielsen, S. B., Experience gained by a Major Transformer Failure at the Offshore Platform of the Nysted Offshore Wind Farm, *Proceedings of 7th International Workshop on Large-Scale Integration of Wind Power into Power Systems as well as on Transmission Networks for Offshore Wind Farms*, Editor: Betancourt U., Ackermann T., published by Energynautics, Langen, Germany, May 2008.

Ault, G., Gair, S. and McDonald, J. R. (2005), Electrical System Designs for the Proposed 1GW Beatrice Offshore Wind farm, *Proceedings of Fifth International Workshop on Large-Scale Integration of Wind Power and Transmission Networks for Offshore Wind Farms*, Editors: Matevosyan, J. and Ackermann, T., published by the Royal Institute of Technology, Stockholm, Sweden, April 2005.

Barberis Negra N. (2005), *Losses Evaluation of HVDC Solutions for Large Offshore Wind Farms*, Master Thesis, Royal Institute of Technology, Stockholm, Sweden, January 2005.

Barberis Negra N., Holmstrom O., Bak-Jensen B. and Sorensen P. (2006), Comparison of Different Techniques for Offshore Wind Farm Reliability Assessment, *Proceedings of 6th International Workshop on Large-Scale Integration of Wind Power and Transmission Networks for Offshore Wind Farms*, Editor: Hendriks, R. L., Bart, C. U., Ackermann T., published by Energynautics, Langen, Germany, October 2006.

Brakelmann, H. (2003), Aspects of Cabling in Offshore Wind farms, *Proceedings of Fourth International Workshop on Large-Scale Integration of Wind Power and Transmission Networks for Offshore Wind Farms*, Editors: Matevosyan, J. and Ackermann, T., published by the Royal Institute of Technology, Stockholm, Sweden, October 2003.

Brakelmann, H. (2003a), Efficiency of HVAC Power Transmission from Offshore Windmills to the Grid, *IEEE Bologna PowerTech Conference*, Bologna, Italy, June 23-26, 2003.

Brakelmann, H. (2003b), Loss determination for long three-phase high-voltage submarine cables, ETEP, 2003, pp 193-198.

Brakelmann, H., Burges, K., Jensen, M. and Schütte, T. (2006), Bipolar Transmission Systems with XLPE HVAC Submarine Cables, *Proceedings of 6th International Workshop on Large-Scale Integration of Wind Power and Transmission Networks for Offshore Wind Farms*, Editor: Hendriks, R. L., Bart, C. U., Ackermann T., published by Energynautics, Langen, Germany, October 2006.

Brakelmann, H., Brüggmann, J. and Stammen, J. (2008), Connection of Wind Energy to the Grid by an Optimized HVAC Cable Concept, *Proceedings of 7th International Workshop on Large-Scale Integration of Wind Power into Power Systems as well as on Transmission Networks for Offshore Wind Farms*, Editor: Betancourt U., Ackermann T., published by Energynautics, Langen, Germany, May 2008.

Cartwright, P., Xu, L. and Saase, C. (2004), Grid Integration of Large Offshore Wind Farms Using Hybrid HVDC Transmission, *Proceedings Nordic Wind Power Conference*, held at Chalmers University of Technology, Sweden, March 2004.

CA-OWEA (2001), Grid Integration, *Energy Supply & Finance, a CA-OWEA (Concerted Action on Offshore Wind Energy in Europe) report*, sponsored by the EU, available at: http://www.offshore-wind.de/media/article000320/CA-OWEE_Grid_Finance.pdf

Christofersen, D.J., Elahi, H., Bennett, M.G. (1996), A *Survey of the Reliability of HVDC Systems Throughout the World During 1993-1994*, CIGRE, Paris, 1996 Report 14-101.

Courault, J. (2001), Energy Collection on Large Offshore Wind Farms – DC Applications, P*roceedings of Second International Workshop on Transmission Networks for Offshore Wind Farms*, Editor: Ackermann, T., published by the Royal Institute of Technology, Stockholm, Sweden, March 2001.

Eriksson, E., Halvarsson, P., Wensky, D. and Hausler, M. (2003), System Approach on Designing an Offshore Windpower Grid Connection, P*roceedings of Fourth International Workshop on Large-Scale Integration of Wind Power and Transmission Networks for Offshore Wind Farms*, Editors: Matevosyan, J. and Ackermann, published by the Royal Institute of Technology, Stockholm, Sweden, October 2003.

Gardner, P., Craig, L., Smith, G., Electrical Systems for Offshore Wind Farms (1998), *20th BWEA Conference: 'Wind Energy – Switch on to Wind Power'*, 1998.

Germanischer Lloyd (1998), *Offshore Wind Turbines*, Report Executive Summary, Germanischer Lloyd, Wind Energy Department, Hamburg, Germany.

Häusler, M. and Owman, F. (2002), AC or DC for Connecting Offshore Wind Farms to the Transmission Grid?, *Proceedings of Third International Workshop on Transmission Networks for Offshore Wind Farms*, Editor: Ackermann, T., published by the Royal Institute of Technology, Stockholm, Sweden, April 2002.

Hammons, T.J., Woodford, D., Loughtan, J, Chamia, M., Donahoe, J., Povh, D., Bisewski, B., Long, W. (2000), *Role of HVDC Transmission in Future Energy Development*, IEEE Power Engineering Review, February 2000, pp. 10-25.

Hendriks, R. L., Paap, G. C. And Kling, W. L. (2006), Development of a Multiterminal VSC Transmission Scheme for Use with Offshore Wind Power, *Proceedings of 6th International Workshop on Large-Scale Integration of Wind Power and Transmission Networks for Offshore Wind Farms*, Editor: Hendriks, R. L., Bart, C. U., Ackermann T., published by Energynautics, Langen, Germany, October 2006.

Holttinen H. and Norgaard P (2004)., A Multi-Machine Power Curve Approach, *Nordic Wind Power Conference 1-2 March 2004,* Chalmers University of Technology, Göteborg, 2004.

IEC Standards 60228 (1978), IEC Conductors on Insulated Cables,1978

IEC Standards 60287 (1994), IEC Electric Cables, 1994

Khator, S.K., Leung, L.C. (1997), Power Distribution Planning: A Review of Models and Issues, *IEEE Transactions on Power Systems* vol. 12, No. 3, pp. 1151-1159, 1997.

King R., and Jenkins N. (2008), Switching Transients in Large Offshore Wind Farms, *Proceedings of 7th International Workshop on Large-Scale Integration of Wind Power into Power Systems as well as on Transmission Networks for Offshore Wind Farms*, Editor: Betancourt U., Ackermann T., published by Energynautics, Langen, Germany, May 2008.

Kirby, N. M., Xu, L., Luckett, M. and Siepman, W. (2002), HVDC Transmission for Large Offshore Wind Farms, *IEE Power Engineering Journal*, Vol. 16, Issue 3, June 2002, pp. 135-141.

Larsson, A. (2008), Practical Experiences gained at Lillgrund Offshore Wind Farm, *Proceedings of 7th International Workshop on Large-Scale Integration of Wind Power into Power Systems as well as on Transmission Networks for Offshore Wind Farms*, Editor: Betancourt U., Ackermann T., published by Energynautics, Langen, Germany, May 2008.

Lazaridis L. (2005), *Economic Comparison of HVAC and HVDC Solutions for Large Offshore Windfarms under Special Consideration of Reliability*, Master Thesis, Royal Institute of Technology, Stockholm, Sweden, February 2005.

Lundberg, S. (2003), *Configuration Study of Large Wind Parks, Licentiate thesis, Chalmers University of Technology, School of Electrical and Computer Engineering*, Technical Report No. 474L, Göteborg, Sweden, October 2003.

Macken, K. J. P., Driesen, L. J. and Belmans, R. J. M. (2001), A DC Bus System for Connecting Offshore Wind turbines with the Utility System, *Proceedings European Wind Energy Conference 2001*, Copenhagen, Denmark, July 2001, pp. 1030-1035.

Martander, O. (2002), *DC Grids for Wind Farms*, Licentiate thesis, Chalmers University of Technology, School of Electrical and Computer Engineering, Technical Report No. 443L, Göteborg, Sweden, March 2002.

Mattsson I., Ericsson A., Railing B.D., Miller J.J., Williams B., Moreau G., Clarke C.D. (2004), *Murray Link – The longest underground HVDC cable in the world*, Paper B4-103 presented at the Cigré conference, Paris, France, Aug. 29 – Sept. 03, 2004.

Middelgrunden (2004), http://www.middelgrunden.dk/.

Nielsen, S. B., Jensen, C. O., Andersen, N. Megos, J. And Claus, M. (2006), Dynamic Voltage Control of Offshore Wind Turbines by using a SVC, *Proceedings of 6th International Workshop on Large-Scale Integration of Wind Power and Transmission Networks for Offshore Wind Farms*, Editor: Hendriks, R. L., Bart, C. U., Ackermann T., published by Energynautics, Langen, Germany, October 2006.

PB Power (2002), *Concept Study – Western Offshore Transmission Grid*, by PB Power for DTI, UK.

Pierik, J., Damen, M.E.C., Bauer, P. and de Haan, S.W.H. (2001), *ERAO Project Report: Electrical and Control Aspects of Offshore Wind Farms, Phase 1: Steady State Electrical Design, Power Performance and Economic Modeling, Volume 1: Project Results*, Technical Report ECN-CX-01-083, ECN Wind Energy, June 2001.

Pierik, J., Morren, J. de Haan, SW.H., van Engelen, T., Wiggelinkhuizen, E. and Bozelie, J. (2004), Dynamic models of wind farms for grid-integration studies, *Proceedings Nordic Wind Power Conference*, held at Chalmers University of Technology, Sweden, March 2004.

Quinonez-Varela, G., Ault, G. W. And McDonald, J. R. (2006), Steady-state Performance Analysis of Collector System Designs for Large-scale Offshore Wind Farms, *Proceedings of 6th International Workshop on Large-Scale Integration of Wind Power and Transmission Networks for Offshore Wind Farms*, Editor: Hendriks, R. L., Bart, C. U., Ackermann T., published by Energynautics, Langen, Germany, October 2006.

Railing B.D., Miller J.J., Steckley P., Moreau G., Bard P., Ronström L., Lindberg J. (2004), Cross Sound cable project – Second generation VSC technology for HVDC, *Paper presented at the Cigré conference*, Paris, France, Aug. 29 – Sept. 03, 2004.

Rathke, C., Siebert, M. and Hofmann, L. (2008), Grid Integration of Offshore Wind Farms using Gas – Insulated Transmission Lines, *Proceedings of 7th International Workshop on Large-Scale Integration of Wind Power into Power Systems as well as on Transmission Networks for Offshore Wind Farms*, Editor: Betancourt U., Ackermann T., published by Energynautics, Langen, Germany, May 2008.

Rudolfsen, F. (2001), *Power supply to offshore plattforms via long AC submarine cables*, in Norwegian, (Strømforsyning til offshoreplattformer over lange vekselstrøm sjøkabelforbindelser), Prosjektoppgave Høsten 2001, NTNU, Institute for Electrical Power Technology, Norway.

Rudolfsen, F. (2002), Power Transmission Over Long Three Core Submarine AC Cables, *Proceedings of Third International Workshop on Transmission Networks for Offshore Wind Farms*, Editor: Ackermann, T., published by the Royal Institute of Technology, Stockholm, Sweden, 2002.

Rudolfsen F., Balog G.E., Evenset G. (2003), Energy Transmission on Long Three Core/Three Foil XLPE Power Cables, *JICABLE – International Conference on Insulated Power Cables*, 2003.

Siemens (2008), 'Losses of converter station', www.siemens.com (last visit January 2008).

Steen Beck, N. (2002), Danish Offshore Wind Farm in the Baltic Sea, *Proceedings of Third International Workshop on Transmission Networks for Offshore Wind Farms*, Editor: Ackermann, T., published by Royal Institute of Technology, Stockholm, Sweden, April 2002.

Todorovic, J. (2004), *Losses Evaluation of HVAC Connection of Large Offshore Wind Farms*, Master Thesis, Royal Institute of Technology, Stockholm, Sweden, December 2004.

Troester, E. (2008), Constant Speed Turbines on a Grid with Variable Frequency – A Comparison in Terms of Energy Capture, *Proceedings of 7th International Workshop on Large-Scale Integration of Wind Power into Power Systems as well as on Transmission Networks for Offshore Wind Farms*, Editor: Betancourt U., Ackermann T., published by Energynautics, Langen, Germany, May 2008.

Vancers I., Christofersen D.J., Bennett M.G., Elahi H. (2000), *A Survey of the Reliability of HVDC Systems Throughout the World During 1997-1998*, CIGRE, Paris, 2000 Report 14-101.

Vancers I., Christofersen D.J., Leirbukt A., Bennett M.G. (2002), *A Survey of the Reliability of HVDC Systems Throughout the World During 1999-2000*, CIGRE, Paris, 2002 Report 14-101.

Vancers I., Christofersen D.J., Leirbukt A., Bennett M.G (2004), *A Survey of the Reliability of HVDC Systems Throughout the World During 2000-2002*, CIGRE, Paris, 2004, Report 14-101.

Weixing, L. and Boon-Teck, O. (2002), Multiterminal LVDC System for Optimal Acquisition of Power in Wind-farm using Induction Generators, *IEEE Transactions on Power Electronics*, Volume: 17, Issue: 4, July 2002, pp. 558 – 563.

Weixing, L. and Boon-Teck, O. (2003), Optimal Acquisition and Aggregation of Offshore Wind Power by Multiterminal Voltage-Source HVDC, *IEEE Transaction on Power Delivery*, Vol. 18, Issue 1, January 2003, pp. 201-206.

Wiechowski W., Hygebjerg J. C. and Borre Eriksen, P. (2008), Higher Frequency Performance of AC Cable Connections of Offshore Wind Farms – Studies of the Danish TSO, *Proceedings of 7th International Workshop on Large-Scale Integration of Wind Power into Power Systems as well as on Transmission Networks for Offshore Wind Farms*, Editor: Betancourt U., Ackermann T., published by Energynautics, Langen, Germany, May 2008.

Chapter 7:

The regulatory framework: grid integration and market incentives

Carlo Degli Esposti

1. INTRODUCTION

This chapter first considers the regulatory framework in European for the generation of electricity from wind, and then focuses on offshore windfarms, especially considering grid integration and market incentives. These regulations are driven by the significant growth of wind energy across Europe and the need to regulate the environmental impacts. Regulations are needed so wind energy has fair and transparent access to the electricity market with respect to other generation sources. Particular regulations relate to the technical management of the grid by the Transmission System Operator (TSO), which requires new approaches so large and amounts of electricity can be injected into the grid at varying power and from a range of windfarm locations. Under European Union direction, these new technical practices are set out in unified "Grid Codes", which TSOs should publish with the appropriate Regulatory Authorities. In this chapter, therefore, the technical aspects will be discussed within the regulatory perspective, stressing the key elements for successful integration of wind power according to the grid codes.

2. REGULATION OF RENEWABLE ENERGY

The European Commission (EC) has increasingly supported legal instruments for the promotion of Renewable Energy in Europe. For instance in 1996, it authorised the first Directive to open electricity markets for prioritised acceptance of Renewable Energy Sources (RES) by Transmission System Operators [1]. In November 1997, the EC adopted Communication COM (97) 599, called the 'White Paper on Renewable Energy Sources'. Accordingly, governments should promote renewable energy supplies within overall energy policy objectives, with the aim of strengthening security of supply, competitiveness, environmental protection and sustainable development. To reach these goals, the White Paper proposed the doubling the proportion of RES within the EU's gross inland energy consumption establishing, with an indicative Community objective of 12% by 2010. The White Paper also contains a comprehensive Strategy and Action Plan to reach this objective. In particular, the White Paper stressed the "*enormous potential for Offshore Windfarms*". In the EC view, although the present costs of installation are clearly a drawback, the exploitation of offshore sites requires a specific promotional campaign; only by the exploitation of these remote sites can the target for feasible penetration of Renewable Energy Supplies for Electricity (RESE) be reached.

The EC White Paper was followed, in 2001, by 'Directive 2001/77'. This had the specific purpose of promoting and increasing the contribution of RESE in the EU's electricity markets, so creating the basis for the growth of RESE in the Community. Directive 77 became a most important milestone for EU regulated market incentives for renewables. The Directive set EU and National indicative targets to increase RESE in conformity with national indicative targets. In particular, for the combined 'EU 15' countries by 2010, renewables should be 12% of the total primary energy supply and RESE 22%

of the overall electricity supply. This latter target was subdivided into different National targets.

For all the Member States (EU 25), similar targets were negotiated and added to the Accession Treaty by May 2004, so the renewables consumption target for the EU 25 became 21%. Overall, these actions met the targets of the Kyoto Protocol and strengthened security of supply with respect to limited energy resources available in Europe.

Also in 2001, the EC adopted a set of Community Guidelines to assess if aid given by member states for environmental protection is compatible with the internal market. These guidelines were valid to 2007. In particular, point 5 stated that the beneficial effects of such measures on the environment must outweigh the distorting effects on competition. There was specific reference to the promotion of renewable energy by tax exemptions or by the refunding of eligible costs, (i.e. extra investment costs necessary to meet environmental targets). Moreover, state aid based on avoided external costs should be less than 0.05 €/MWh. Every 10 years, national governments must notify the Commission about such governmental aids, so the Commission can decide if there has been any breach of Community legislation.

An important decision regarding market distortion was made by the European Court of Justice concerning the German Feed-In-Tariffs (FIT). The verdict was given that this Feed-in-tariff does not involve any transfer of state resources to any electricity market player, since it is not a form of state aid within the meaning of Art 87(1) of the EU Treaty.

3. SUPPORT SCHEMES FOR ELECTRICITY PRODUCTION

3.1 Market Integration

One of the main questions is how new renewable generators should be integrated into the framework of a European electricity market. Different national legislation regarding support schemes and other rules might lead to a distortion in this market. In accordance with the principle of subsidiarity, the choice of which support schemes to adopt should be made by each Member State. However, under the Directive 2001/77, the EC was required to publish by 27 October 2005 a report evaluating the coexistence of the different mechanisms adopted by each Member State to promote renewable energy supplies, along with their cost-effectiveness. In accordance with the Article 4 of the RES Directive 2001/77, "*this report shall, if necessary, be accompanied by a proposal for a Community framework with regard to support schemes...*"

Among the requirements for fulfilment by a potential Community-wide framework, is that it should "be compatible with the principals of the Internal Electricity Market". Currently, most adopted promotion mechanisms are based on one or more of: (i) direct price support schemes (known as feed-in systems), (ii) obligated quotas, often with certificated trading, (iii) investment aid, and (iv) tax exemptions or reductions. However, it is generally agreed that the continued use of some of these solutions may induce significant distortions in the electricity market.

3.2 Current support schemes

Presently, every Member State and other European countries have implemented their own strategies regarding support schemes for electricity generation from renewable sources, which usually is a combination of two or more mechanisms. The main strategies currently adopted by EU countries, plus Switzerland and Norway, are presented below. Each country has legislated one or more of the following schemes.

3.2.1 Feed-in tariffs

The feed-in model is the most widespread instrument to stimulate electricity generation using renewable energy sources (RESE). This mechanism obligates utilities to buy RESE from any renewables generators, including microgenerators, at a previously fixed unit price or premium. This simple mechanism ensures a long-term stable price structure, which overcomes an important market risk of otherwise unstable prices. The fixed price can be either (i) obligated for both the electricity itself and the renewable value together; or (ii) obligated only for the renewable value, with the electricity sold in the open market.

3.2.2 Tendering

Tendering is based on the principle of "giving all players an equal opportunity but provide a subsidy only to a limited number of investors", in order to drive down costs. The renewable generators must compete in a series of capacity or energy-bidding rounds, from which only the cheapest offers are selected.

Theoretically, the system uses competition to emphasis economic efficiency and, in operation (in the UK in the past, and presently of the Republic of Ireland), this method should be very efficient in driving down the overall costs of supporting RESE. In practice, however, there have been significant difficulties for fair implementation. In particular, there is no long-term financial stability for manufacturers and developers. Such stability is a key issue to provide confidence to investors. Tendering processes with competitive bidding are inherently uncertain regarding tariffs and hence income.

3.2.3 Quota obligation or Renewables Portfolio Standards

In this mechanism, suppliers or consumers are obligated by national law to provide a fixed percentage of the energy produced and/or consumed as renewable energy. If the percentile quota is not fulfilled, penalties are applied. This system plays an important role in the definition of the demand - and therefore, the market - for renewable electricity generation. The mechanism is relatively recent and has attracted attention both in and outside Europe. Within Europe, obligated quotas are applied by Belgium, Italy, Sweden and United Kingdom.

3.2.4 Certificates

Associated with obligated quotas, certificates have been introduced to add flexibility to the quota obligations. Such '*Tradable Renewable Certificates*' (TRCs) are not a support scheme by themselves, but, they can be considered as a tool to implement a transparent renewable support strategy. TRCs facilitate the implementation of support schemes by labelling, certifying and monitoring renewable electricity generation. The principle of TRCs is based on the allocation certificates to generators, based on the meter readings of renewable electricity generation (e.g. one certificate per MWh of RESE). The TRCs are issued to renewables generators by a central authority, so certifying that a certain (renewable) amount of electricity has been produced, no matter where it has been fed into the grid. The generators then sell the certificates to the obligated suppliers or consumers on an open market. Eventually, the certificates are *redeemed* by the central authority (i.e. retired from circulation, or, in other words, definitively associated to a given amount of RESE). Fines from non-compliance may be recirculated as a bonus to those suppliers or consumers that have met their obligations. The TRC system establishes a market for renewable values[1], as distinct from the market in actual electricity.

3.2.5 Fiscal measures

Other market instruments supporting RES can be grouped as 'fiscal measures', which may be applied independently or in association with other methods. Examples are grants to manufacturers, developers and/or purchasers, tax reductions (e.g. VAT reduction) and, in effect, penalties associated with fossil-fuel consumption (e.g. regarding pollutant emissions).

3.2.6 Green Pricing

If a large proportion of consumers are willing to pay an extra price for Renewables, a support scheme may be less necessary. This situation is called Green Pricing. It is an option provided by electricity

[1]The certificate trade may be understood as a form of carbon trading (Ed)

providers (utilities, suppliers and stand-alone producers) that allows their customers to both support investment in renewable energy technologies and decrease their 'carbon footprint'. Through green pricing, participating customers pay a premium per unit of electricity, which allows extra payment to renewable energy generators. In practice, most consumers are not willing to pay such extra costs, so that in most of the EU Member States, another support scheme is necessary to achieve the targets set by EC Directives and national governments. It needs to be stressed, that for individual consumers, Green Pricing is the easiest, and perhaps the only, way to have 100% RESE.

4. INTEGRATING RESE INTO THE ELECTRICITY MARKET

Within an open European electricity market, competition is the main driver for the allocation of resources. The competition will certainly be financial, but may also involve other factors, such as minimising environmental impact. Therefore, it is the electricity price and its environmental values that ideally drive investments.

The liberalisation of the European electricity markets affects the implementation of RESE policy. Liberalisation may limit the instruments used, but does not by itself set any limits for policy goals for RESE volumes. The criteria are that such policies must not distort the electricity market unfairly in favour of particular players. Environmental policy may seek to enhance investment in specific renewable technologies, in order to compensate for economic disadvantages, e.g. the normal market not including the environmental (external) costs of greenhouse gases from fossil fuels.

4.1 Prices and costs in an electricity market

The main cost elements for the production of RESE are investment costs, operational costs, balancing costs and grid costs. In traditional, non-liberalised power systems, the private RESE generator is accountable only for the investment cost, the operational cost and perhaps part of the costs of connecting the device to the grid. Balancing costs (which are particularly significant only for large proportions[2] of wind energy) are usually borne by "the system". The costs borne by the RESE owner may be been assisted with a subsidy and, in countries with a 'feed-law', a fixed electricity price, independent of the market value of physical electricity. Therefore, for such systems, the RESE generator does not have to consider the normal retail price of the electricity at the given location.

In an open liberalised electricity market, a RESE generator, or the supplier to whom the generator sells, should be treated by the TSOs on equal terms with other producers. Feed-law and other market incentives may apply, but the charging of balancing and grid costs should be in line with the principles of an open electricity market. Consequently, the financial flow becomes transparent; and it becomes clear what support RESE needs to be profitable. If the support given by society corresponds with the actual environmental value that the RESE contributes (e.g. the abatement of external costs from substituted fossil fuel) an optimal allocation of resources can be maintained.

Other aspects arise when comparing the EU IEM Directive and the RES Directive 2001/77, requiring some matters still to be clarified.

4.2 Security of supply

No form of electricity generation can be guaranteed to operate at any particular time, therefore all forms of generation need 'back-up' plant. The failure of large, centralized plant may be rapid and unexpected, so methods to maintain grid supply in such unlikely events (security of supply) are normal. Most supplies of RESE are of a naturally varying nature and of relatively small scale, yet for these also the grid has to balance other generators, frequency and voltage to give security of supply. Thus for all plant, security of supply has a cost.

The EU Internal Electricity Market Directive 1997 (the IEM Directive) states that security of supply must be guaranteed, while the RES Directive 2001/77 does not prescribe any kind of support for

[2] i.e. greater than about 20% (Ed)

ancillary services or additional reserve needed to keep, at least, the same level of security of supply. These are important fundamental conditions in an open European electricity market. The purpose of implementing a market approach is that decentralised individual decisions, made by competing commercial market players, lead to optimal allocation and usage of resources. However, this can only occur if the market players are forced to make decisions that account for *all* fundamental conditions, including the conditions and the value added from a security of supply perspective. The key instruments to give investors the necessary feedback are locative signals to avoid the concentration of all wind plant in one region and the concentration of all conventional power plants elsewhere, as well as remuneration by the market of balancing reserve and other ancillary services. If this is not so, then the design of the electricity market is flawed.

4.3 Priority dispatch

The IEM Directive requires non-discriminatory procedures among the market players, while the RES Directive asks for priority when dispatching generation from renewable sources - although all generators produce the same kind of electricity. However, priority dispatch of RESE directly influences the open electricity market. The RES Directive states effectively that RESE production should be given *priority dispatch*. In an electricity market, central dispatch is replaced by a price signal. If the renewable energy plant is not forced to relate to the price signal in the operation, the interaction between renewable energy and traditional production technologies is random. As an example, the bottleneck capacity at the German-Danish border is reduced because of the priority dispatch for German windfarms next to the border. This means that German wind energy is prioritized with respect of Danish renewable energy that is imported to Germany.

4.4 Cross border transfer of renewable value

To avoid cross border distortions in the European electricity market, we need harmonized rules and support schemes. The "flow" of environmental benefit from one country to another should not be related to the physical transmission capacity. Part of the value of RESE is given by the local electricity market price, although environmental factors are both local and global. The market price depends on the transmission capacity from this area. Ideally, payment for the environmental benefit should be a purely financial transfer of money from the beneficiary of the benefit, to the producer of the RES.

If, for instance, renewable generation must be physically imported, this can lead to some congestion problems, causing increasing prices and then distortions, which can and should be avoided. For example in 2002, several periods of congestion appeared at the German-Dutch border because supported RESE in The Netherlands had to be physically imported. Separating the price for the actual electricity from the price of the environmental benefit, e.g. by using Tradable Renewable Certificates, is a solution for this problem.

In conclusion, RESE policy should be implemented so not to distort the electricity market. Market tools are generally more effective for this than regulatory schemes; however, externalities have to be internalised to obtain the full picture. In the open electricity market, the RESE generators should be treated by the TSOs on equal terms with other producers. The balancing and grid costs should be in line with the principles of an open electricity market. Comparing the IEM Directive and the RES Directive there are some matters which need to be clarified, especially Security of Supply and Priority Dispatch. To avoid cross border distortions in the European electricity market we need harmonized rules and support schemes, with the inclusion of external costs.

5. SYSTEM COSTS

Wind power penetration has increased significantly during recent years. Its optimal capacity and distribution for each power system remains unsolved, with the constraints both economic and technical. Historically, European power systems were designed and developed for conventional sources of centralised electrical power (hydro, gas-, coal- and oil-fired, nuclear). Now with dispersed

RESE, often of small scale (microgeneration), the integration of much accumulated capacity of varying supply, requires a change in the system itself and/or in its operation. This affects the cost of energy and of the related services (e.g. price for reserve capacity) along the entire chain of supply.

5.1 Network connection costs

The RESE generator always has to pay for the connection of the actual plant up to an agreed 'grid connection point', which at the nearest is at a substation on the site. The responsibility for the connection from that substation to the established grid network varies according to the custom and regulations of the particular country. The amount charged to the generator for this, i.e. 'the network connection costs', vary considerably across the European Union. In some Member States, the specific national Regulatory regime issues grid connection charges which are unrelated to the actual connection costs for the new power plant. In other cases, when connection costs are specified, inconsistencies emerge when comparing the different connection/upgrade assets included in these costs. Spreading the costs over all consumers is called 'socialising' of the costs.

Three different categories of connection costs has been proposed [2]. In particular, it is possible to distinguish between 'shallow', 'intermediate ('shallowish') and 'deep connection' charges.

A shallow network connection charging system consists in socializing all the grid installations/upgrade to connect the new windfarm from the contractually agreed connection (usually an on-site substation) point to the transmission or distribution system. These costs are initially born by the TSO; the generator is responsible on his side for all costs concerning the wind installations and connection equipment to the substation connection point. Such a tariff for connection is characteristic of all those countries having a rapid development of wind generation, e.g. Germany and Denmark. This approach has alleviated several uncertainties related to grid reinforcing and other system upgrade costs, by sharing them between all market participants, who can equally benefit in their turn by the increased sustainable energy supply. Nevertheless, shallow connection charges can prevent optimum economic-efficient grid development of the total ('deep') system, since the evaluation of broader network extension planning is neglected, which would favour the best distribution of new wind generation into the grid.

An intermediate connection charges are a combination of a shallow approach and a deep contribution by the wind producer to the total grid reinforcement needed for a particular windfarm development, which normally increases proportionally to the new capacity connected. Such deeper effects are certainly likely for the relatively large capacities of offshore windfarms. This charging system has been adopted in UK, where from 2005 any new wind generation site connected to distribution grid did not pay for deep connection charges but, only a contribution to the network reinforcement proportional to the new generation capacity.

This approach should be also foreseen for other situations where the political decision for rapid RESE development is planned. This is the case in Germany, (ref DENA , Deutsche Energie-Agentur GmbH, German Energy Agency) for new onshore and offshore wind generation.

Eventually, the choice between these approaches depends on the particular national regulatory regime action plan. It remains an open question whether or not the in-feed characteristics of offshore windfarms have to be considered differently from onshore installations, which are usually of less extent and less capacity. This could be furthermore justified by the requirements for isolation from grid disturbances and fault ride-through for offshore plants.

5.2 System Operation costs

One of the major technical challenges for the grid and market integration of new wind generation is the variable nature of wind power, which is similar in effect to the variation of demand. The accuracy of wind forecasts depends on the 'look ahead' time, and this inherent uncertainty is compounded by the cubic relationship between wind speed and wind power at the most likely wind speeds. Consequently, grid operators handling large amounts of wind power have to assess the costs of balancing total generation with consumption in their control areas. Depending on the contracts, these costs may be passed to consumers or charged to the wind generators.

The capacity of Primary Control held in UCTE[3] is fixed at 3000 MW and is based upon the "worst case" of the sudden disconnection of 2 large sets (2 x 1500 MW). Primary control holding is allocated between UCTE members based on their annual production volumes. In Nordel[4], Primary Control holding is fixed at 600 MW (Frequency Regulation reserve) and 1000 MW (Momentary Disturbance reserve). However, in England and Wales (E&W) the amount of Primary Control held is optimised on a half hourly resolution and is dependant on total system demand, and the largest single in-feed to the system. For the instantaneous loss of 1320 MW generation, this results in a requirement for a Primary frequency-response reserve range of 400-1500 MW and a Secondary frequency response reserve range of 900-1300 MW.

There are also differences in the frequency criteria for activation of Primary Control. In UCTE a frequency deviation of –0.2 Hz will activate the entire Primary Control reserve. However, in Nordel the amount of Primary Control activated increases with frequency deviation, from –0.1 Hz to –0.5 Hz when the entire reserve is fully activated. In E&W, the Primary Control reserve is fully activated for a frequency deviation of -0.8 Hz.

Varying approaches are also taken to time error correction. UCTE can use Secondary Control, by simultaneously altering the target frequency of all participating generators. This generally occurs when the time error reaches ~30 seconds. In E&W, the target frequency of individual generators held on Primary Control can be altered to correct time within the limits ~_10 seconds. Some similarities also exist in the approach to Primary Control services. There are two common ways of achieving service, either commercially or by mandatory obligation. In Germany, Poland, Denmark and Sweden, all frequency control services are procured on a commercial basis and there is no obligation on generators to provide services. In other countries, e.g. France, E&W, Norway, Austria, Spain, Italy, Slovenia and Switzerland, frequency control services are mandatory and all large generators are obliged to provide them. The remuneration of these services is discussed later.

In E&W mandatory primary control services are complimented by additional 'commercial' services. These are either enhanced services beyond a generator's mandatory requirement (which are based upon a 4% droop characteristic), or commercial demand reduction (the demand is on low frequency relays, which are activated for deviations of around 0.3 Hz).

The technical requirements of Primary Control are very similar across Europe, with full response power being delivered immediately, i.e. full delivery within 30 seconds after a disturbance. Secondary control (UCTE only) is also used after a system disturbance, whereby system frequency and power interchanges must begin to return to their set point values after 30 seconds, and must be completed within 15 minutes of the incident. Given this time horizon, it is evident that the wind-induced imbalanced have to be tackled by the means of the secondary control.

6. TECHNICAL INTEGRATION OF RENEWABLES

At European and national levels, present and future plans will increase the share of electricity produced by renewable energy sources (RESE) significantly. They shall contribute to meeting the targets of the Kyoto protocol and support the security of supply with respect to limited energy resources in Europe. This was indicated by the European Commission White Paper on Renewable Sources of Energy promotion in the EU and the Directive 2001/77/EC issued soon after.

At European level, such Directives set the framework for the national targets and support mechanisms. The exact situation in Europe differs from country to country. Circumstances may also differ as between synchronous interconnected systems and 'island' systems. The capacity targets and

[3]"Union for the Co-ordination of Transmission of Electricity" (UCTE) is the association of transmission system operators in continental Europe

[4]Nordel, Organisation for the Nordic Transmission System Operators

the future portfolio of RESE depend on the national situations. Nevertheless the biggest growth potential is for wind energy, especially offshore-wind energy, due to the larger wind speed at offshore sites as compared with adjacent inshore sites [2]. The expectations of the European Wind Energy Association for total wind capacity in Europe increased from 28.5 GW in 2003 to 180 GW in 2020. Due to different support-schemes, restrictions in licensing and a limited number of suitable locations, the RESE capacity tends to focus on just a few regions in Europe. The effective scarcity of inshore sites and the variable in-feed characteristic of wind energy, are the key factors for integrating the wind power.

6.1 Grid extension

One of the major issues concerning new RESE capacity, is the need for new grid infrastructure. New offshore windfarms will normally be built far away from the main load centres and their capacity requires connection with high-voltage transmissions lines. New overhead lines will therefore be necessary to transport the electricity to the nearest connection point. Such investments are exclusive to or at least mainly driven by, new RESE generation sites. The variable contributions from wind power must be balanced with other generation capacity located elsewhere. This adds to the requirements for grid reinforcements. In Germany for example, about 850 km of new 380 kV lines will be needed to support the integration of future windfarms.

The time for licensing procedures for new lines requires several years, perhaps as much as ten years. Any delays in grid extension cause delays in RESE investments because windfarms cannot earn an adequate return on investment without an adequate grid connection. New lines are therefore critical for the success of new RESE, but there is no European-wide harmonised rule yet for who should pay for them. It is therefore advisable that licensing procedures for new RESE sites and for grid infrastructure go on hand in hand. The legal framework and administrative procedures have to be set properly to speed up the licensing of grid infrastructure. To have further incentives for this, the legal and regulatory framework needs to be properly set for the investments in new infrastructure.

6.2 System stability

Historically, windfarms behave differently from conventional plants with respect to grid stability. If faults in the grid causing a voltage drop, conventional power stations with large synchronous generation normally continue to generate and therefore support system stability. However, the turbines on most windfarms to date tend to shut down and so do not support stability. Simulations in Germany show that a generation outage of more than 3000 MW of such turbines is possible in the coming years which might result in a blackout scenario. This situation would occur in any other country if the wind capacity becomes large with an increasing share of wind energy, unsuitable grid codes and turbines without the pitch control needed for interruptible generation.

As counter measures, suitable European-wide harmonised grid codes for new windfarms and other RES defining their electrical behaviour in critical grid situations are needed. Consequently, the best available technology for windfarms should be used. Furthermore it might be necessary that existing windfarms not fulfilling the actual grid code requirements must be upgraded or replaced (i.e. electrical behaviour of wind turbines during grid faults). Finally, sufficient capacity of non-varying synchronous generation has to be in the system at any time to keep it stable. An unlimited priority of wind in-feed as laid down in some national legislation is therefore impossible with respect to the detrimental effects on the security of supply. Balance management

6.3 Cost of Imbalance management of windfarms

The hour-to-hour and day-to-day forecast of electricity production over a distributed network from wind energy are reasonably accurate. To balance the difference between generation and demand, balancing power is needed. In Germany by 2015, the predicted installed wind capacity of 36,000 MW may require about 7,000 MW of balancing-power, i.e. a controllable capacity equal to 20% of the wind capacity. This amount would have to increase significantly with further wind capacity installation.

Depending on the different national rules, different parties are responsible for providing the balancing-power and reserve capacity. If the TSO has to contract for it, the costs will become part of the consumer tariff, paid by all customers, and so there is no incentive for the windfarms to reduce the costs of grid integration. Yet if the market is to solve the problem of imbalance, the wind generators should be made responsible for any imbalances from their contracts , as already occurs in some countries. Moreover, such actions will encourage the use of energy storage to smooth otherwise variable supply.

6.4 Impact on cross-border electricity transits

In situations of strong wind regions with a high density of wind generation and low load , excess 'export' electricity will flow to neighbouring interconnected grid regions. For instance, generation from windfarms in northern Germany has produced significant negative operational impacts in the connected Dutch and Polish systems. The physical flows from Germany to the Netherlands have occasionally been more than twice as large as otherwise expected, so causing critical grid situations. The effect of such flow patterns may reduce the capacity for cross-border trade. This situation is even promoted by the priority dispatch for RESE, since additional RESE generation causes flows bottlenecks that force the TSO to reduce the tradable capacity. If this effect is increased, the more wind generation will be concentrated into particular regions, such as particular offshore areas. Consequently, non-harmonised support schemes for RESE in Europe may well cause additional congestion in the system.

REFERENCES

[1] Directive 96/92/EC of the European Parliament and of the Council, 19 December 1996 concerning common rules for the internal market in electricity.

[2] Thomas Ackermann, *"Wind Power in Power System"*, p. 479, Wiley, 2005

Chapter 8:

Offshore Turbines: Dynamics and Fatigue

Jan Van der Temple

1. INTRODUCTION

This chapter starts with basic explanations concerning the dynamics of wind turbines and moves to their location at offshore sites. The natural elements of wind, wave, tidal range and tidal current, all activate the dynamics of the machines. The unified dynamic-design for the very complex machines evolves step by step.

2. GENERAL TERMINOLOGY

An overview of the terminology is shown in Figure 1.

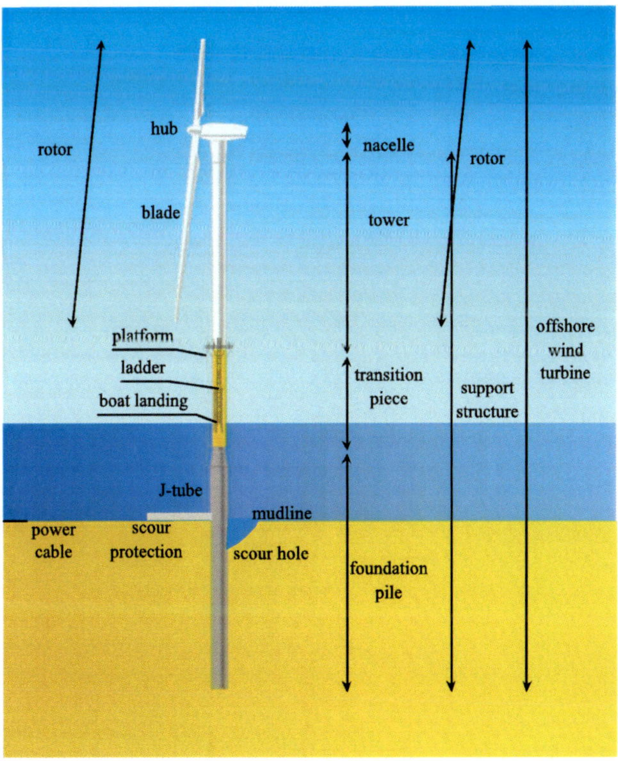

Figure 1. Overview of offshore wind turbine terminology

3. STOCHASTIC RANDOM PROCESSES

Loads and responses of offshore wind turbines vary in time. For time-series data analysis, i.e. in the 'time domain', the most important aspects are maximum, minimum, mean, strange peaks and slow variations. To make the data more useful for dynamic and fatigue analysis, the time series can be transformed into the frequency domain. The data can then be plotted as a spectrum. For example, Fig. 2a shows a measured time series of the mudline bending stress of an offshore wind turbine, and Fig. 2b shows the corresponding spectrum created from these data. In the spectrum, several characteristic responses, as indicated on Fig 2b, can be distinguished far more easily than in the time series.

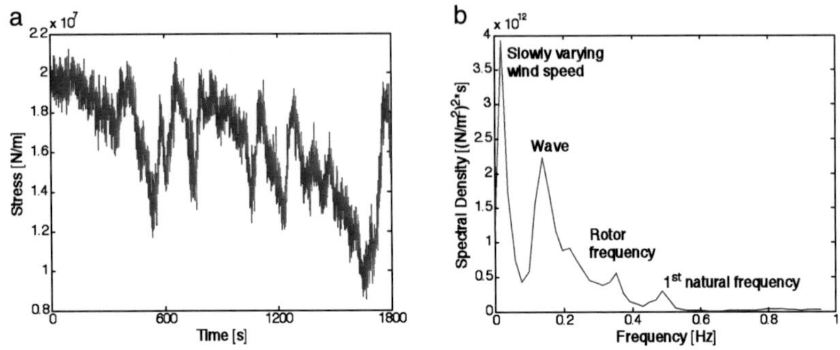

Figure 2. (a) Time domain record of measured mudline bending stress variation (b) The frequency domain spectrum of the same time trace

4. WAVES AND CURRENTS

4.1 Sea surface description

The main driver of sea waves is the wind. Only absolute absence of wind may allow a mirror-like sea surface, every slightest gust of wind will cause ripples and strong winds create waves with amplitudes to 30 m in Atlantic deep-water locations. Normal sea surfaces result from numerous waves travelling in different directions and so mathematical models must replicate these pseudo-random processes. If we reduce the sea surface to a single point and measure the surface elevation in time, a seemingly random sea, e.g. of Fig. 3a, produces a seemingly random time series, e.g. Fig. 3b. This time varying signal can be Fourier transformed to a so-called 'energy density' or 'wave' spectrum, as shown in Fig. 3c.

Some characteristic parameters can be defined from both the frequency spectrum and from the time series. The *significant* wave height of the spectrum, H_s, is defined as the mean of 1/3 of the waves which have the highest amplitudes in the time series, which, in practice, equals about 4 times the standard deviation σ of the time series. The standard deviation is also the square root of the zeroth order moment of the spectrum, m_0. A noteworthy feature of this parameter is that a visual estimate by experienced sailors, H_v, compares rather well with the statistical H_s value.

From the time series, the mean zero crossing period T_z can be found by dividing the measurement time by the number of up-crossings of the zero elevation. This period can also be found from the spectrum as the square root of the zeroth moment over the second moment. A summary of the relations between time domain and spectral parameters is given in table 1, [1][2].

The shape of a wave spectrum may be fitted to various curves. A frequently used spectral shape is the Pierson-Moskowitz wave spectrum [3]. The shape was fitted to measurements taken in the Atlantic Ocean during long periods of constant environmental conditions. The spectrum is therefore based on the input of one single forcing parameter of the average wind speed and so describes the sea surface elevation due to the wind speed for a fully developed sea at infinite fetch. The Pierson-Moskowitz spectrum description can be adjusted to have Hs and Tz as input, as in (1):

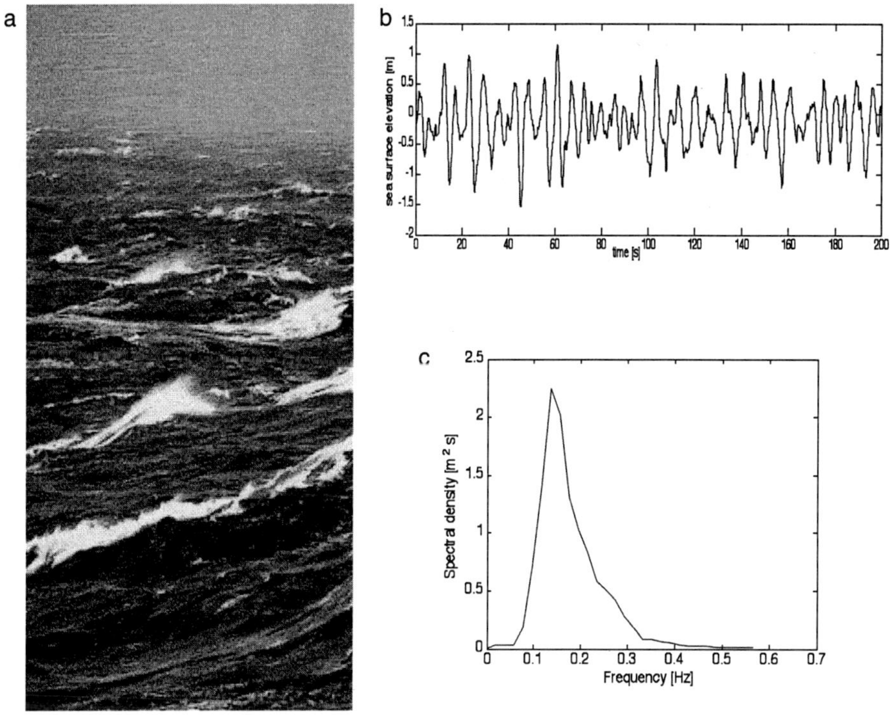

Figure 3. (a) Actual sea surface elevation. (b) Single point time recording of sea surface elevation. (c) Wave spectrum of measured time recording of sea surface elevation.

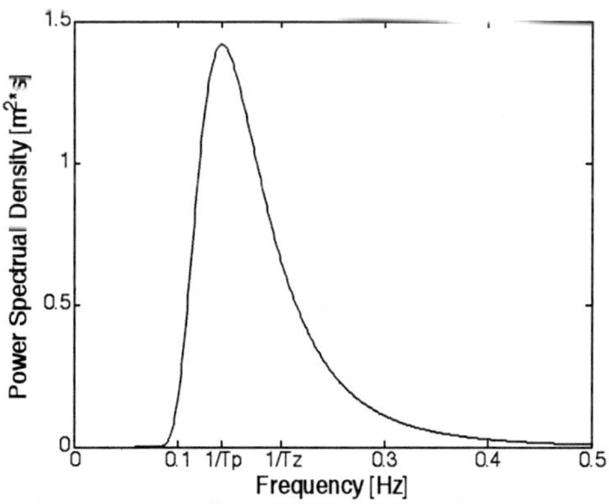

Figure 4. Pierson Moskowitz spectrum for H_s = 1.5 m and T_z = 5 s with mean zero-crossing frequency at $f_z = 1/T_z = 0.2$ Hz and peak spectral frequency at $f_p = 0.14$ Hz

$$S_{PM}(f) = \frac{H_s^2}{4\pi T_z^4 f^5} \exp\left(-\frac{1}{\pi} \left(f T_z \right)^{-4} \right) \quad (1)$$

Fig. 4 shows a plot of the Pierson-Moskowitz spectrum for the parameters H_s and T_z from the wave signal in Fig 3.

Table 1. Relations between time series and spectral parameters for waves

Description	Relation
spectral moments (n = 0,1,2...)	$m_n = \int_0^\infty f^n S(f)\,df$
variance or mean square	$\sigma^2 = m_0$
standard deviation or root-mean-square (RMS)	$\sigma = \sqrt{m_0}$
significant wave height	$H_s \approx 4\sigma$
visual estimate of the wave height	$H_v \approx H_s$
mean zero crossing period	$T_z = \sqrt{\frac{m_0}{m_2}}$
mean period of the spectrum	$T_m = \frac{m_0}{m_1}$
mean crest period	$T_c = \sqrt{\frac{m_2}{m_4}}$
estimate of the most probable maximum wave height in a sea state for 1000 waves (appr. 3 hours) for a Pierson-Moskowitz spectrum	$H_{max} = 1.86 H_s$

The mean of the zero-crossing frequency $f_z = 1/T_z$ is shown in the plot. A visually more characteristic parameter is the frequency at which the peak occurs, f_p, and its inverse T_p. Note that this peak refers to the peak of the spectrum and has nothing to do with the mean crest period T_c referring to wave crests. The relation between T_p and T_z for the PM spectrum is T_p = 1.41 Tz. Equation (2) can be re-written for T_p:

$$S_{PM}(f) = \frac{5}{16}\frac{H_s^2}{T_p^4 f^5}\exp\left(-\frac{5}{4}\left(fT_p\right)^{-4}\right) \tag{2}$$

The peak period, T_p, is important for development of the Pierson-Moskowitz spectrum using further measurements of wave spectra of the Joint North Sea Wave Project from which the JONSWAP spectrum originated [4]. This spectrum represents sea states that are not fully developed under a certain wind condition. The wave spectrum shape is therefore much more peaked. The JONSWAP spectrum is actually an extended version of the Pierson-Moskowitz spectrum, incorporating a peak enhancement factor, which is controlled by a peak shape parameter γ_{JS}. When the shape parameter is taken as γ_{JS} = 1, the JONSWAP spectrum is equal to the Pierson-Moskowitz spectrum. A typical value for not fully developed seas is γ_{JS} = 3.3. The shape of the slopes before and after T_p, are controlled by slope factors σ_a and σ_b, equation (3).

Because the peak enhancement within the JONSWAP spectrum increases the total area under the spectrum, a normalising factor F_n must be introduced to ensure that the area (the zeroth order moment) under the JONSWAP spectra still represents the real energy density of the sea state. Several derivations exist for this normalising factor; equation (3) shows one of the formulations from [2].

$$S_{JS}(f) = F_n \cdot S_{PM}(f) \cdot \gamma_{JS} \exp\left(\frac{-(f-f_p)^2}{2\sigma_{JS}{}^2 f_p^2} \right)$$

$$\sigma_{JS} = \begin{cases} \sigma_a \text{ for } f \le f_p \text{ (typically: 0.07)} \\ \sigma_b \text{ for } f > f_p \text{ (typically: 0.09)} \end{cases} \tag{3}$$

$$F_n = \left[5\left(0.065\gamma_{JS}^{0.803} + 0.135 \right) \right]^{-1} \text{ for } 1 \le \_JS \le 10$$

Furthermore, the relation between T_p and T_z for a JONSWAP spectrum differs from that for a Pierson-Moskowitz spectrum. An approximation is shown in equation (4) [5]:

$$T_p / T_z \approx 0.327 e^{-0.315\gamma_{JS}} + 1.17 \tag{4}$$

Using either the zero-crossing period or the peak period as basis for the wave spectrum, can result in different spectra. Fig 5 shows Pierson-Moskowitz and JONSWAP spectra for these two cases.

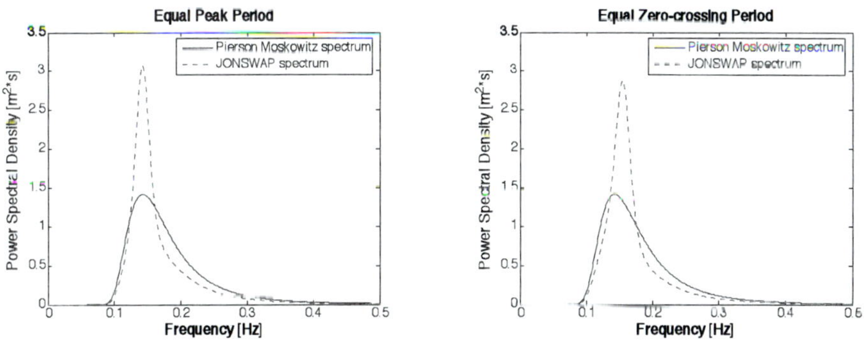

Figure 5. (a) Left-hand side: Pierson-Moskowitz (continuous trace)) and JONSWAP (dashed trace, peaky) spectra for H_s = 1.5 m and T_z = 5 s, and equal peak period. (b) Right-hand side: same spectra, but with equal zero-crossing period

Although Pierson-Moskowitz and JONSWAP are the most commonly used spectra, other descriptions exist. Remember that all these "standard" spectra try to reproduce the actual measured spectra at a certain location under certain circumstances. When detailed, long-term measurements are available for a site, more finely tuned spectra can be used. In all other cases, the designer can only select an appropriate shape based on the parameters available for a site.

4.2 Representation of a random sea

With a suitable wave spectrum, calculation starts by converting the spectrum back into sinusoid waves (IFFT). The sinusoids have amplitude and frequency that can be derived from the energy density given by the spectrum. The phase angle is appointed randomly to each sinusoid. The sum of all these harmonic waves is the sea surface elevation at each point in time.

For every harmonic wave, the water particle motion can be described by linear wave theory according to Airy [2]. In deep water, the elementary water 'particles' move in circles in accordance with the harmonic wave. The diameter of the circle decays with depth below the surface. When the water depth becomes shallow relative to the wave-length, $\lambda_{wave} < 0.5d$, the effect of the seabed transforms the circular motion into an elliptic motion. These particle motions are shown in Fig 6.

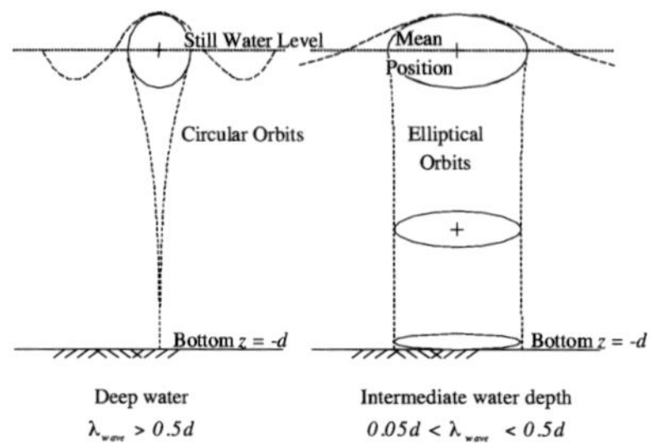

Figure 6. Particle orbits according to the Airy theory

The horizontal water particle kinematics are described by equation (5), with the z-axis pointing upwards from the free water surface ($-d \le z \le 0$) and position x horizontally in the wave direction; t is time:

$$u(x,z;t) = \hat{\xi}\, 2\pi f \frac{\cosh k_{wave}(z+d)}{\sinh k_{wave} d} \cos(k_{wave}x - 2\pi ft)$$
$$\dot{u}(x,z;t) = \hat{\xi}\,(2\pi f)^2 \frac{\cosh k_{wave}(z+d)}{\sinh k_{wave} d} \sin(k_{wave}x - 2\pi ft) \quad (5)$$

with:

ξ	wave amplitude (= 0.5 H)	[m]
k_{wave}	wave number = $2\pi\lambda_{wave}$	[m^{-1}]
f	wave frequency	[Hz]
λ_{wave}	wavelength	[m]
d	water depth	[m]

Because Airy linear wave-theory is only valid up to the 'still water' level (i.e. the average level), it does not describe the kinematics in the wave profile itself (see Fig 7). Several methods have been suggested to correct this. One of the most commonly used methods is Wheeler stretching [6] [2]. The wave kinematics profile over the water depth is determined for every time step. This profile is then stretched to the instantaneous wave surface elevation. The method is shown in Fig 7.

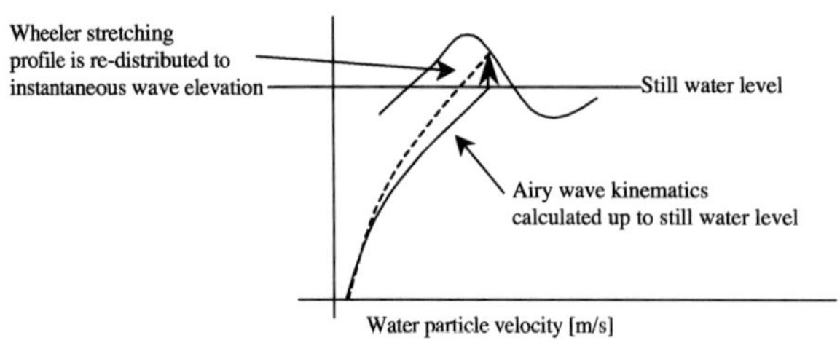

Figure 7. Normal Airy wave kinematics up to the still water level and Wheeler stretching to the instantaneous wave elevation.

4.3 Wave loads on structures

Loads on vertical towers

The wave particle kinematics can now be used to calculate the loads on a structure with the Morison Equation (6). The relative velocity of the structure can also be incorporated but is ignored here, because its magnitude is very small compared to the water particle velocities. The Morison equation is an empirical formula to calculate the hydrodynamic loads on slender members per unit length:

$$\begin{aligned} f_{Morison}(x,z,t) &= f_d(x,z,t) + f_i(x,z,t) \\ f_d(x,z,t) &= C_d \cdot \tfrac{1}{2}\rho_{water} D \cdot \left|u(x,z,t)\right| u(x,z,t) \\ f_i(x,z,t) &= C_m \cdot \frac{\rho_{water}\pi D^2}{4} \cdot \dot{u}(x,z,t) \end{aligned} \tag{6}$$

with:

$f_{Morison}$	hydrodynamic load	[N/m]
f_d	hydrodynamic drag load	[N/m]
f_i	hydrodynamic inertia load	[N/m]
C_d	hydrodynamic drag coefficient	[-]
C_m	hydrodynamic inertia coefficient	[-]
ρ_{water}	density of water	[kg/m^3]
u	water particle velocity	[m/s]
$\dot{u}$	water particle acceleration	[m/s^2]
D	diameter of cylinder section	[m]

Fig 8 shows the representation of a slender vertical member under hydrodynamic loads. The sum of drag and inertia load is the total hydrodynamic load on the cylinder. Note that velocity and acceleration have a 90° phase difference, so inertia and drag loads will also be out of phase. This means that, in general, the vector sum for the maximum load is not equal to the magnitude either of the maximum drag or of the maximum inertia load.

The coefficients C_m and C_d are determined by (i) the specific load case under consideration (extreme or fatigue), (ii) the shape of the structure, (iii) the presence of marine growth and (iv) other factors. Typical values for fatigue calculations on monopile structures are C_m = 2.0 and C_d = 0.7, but the designer has to make sure the values are correct for each specific load case [2].

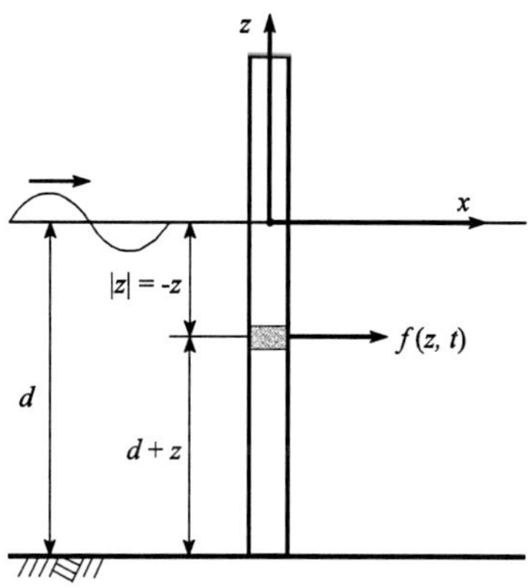

Figure 8. Slender vertical tower with hydrodynamic loads

Horizontal and inclined members

When members of a structure are inclined or horizontal, the Morison equation can still be used, but the amount of preparation increases significantly. Not only do the horizontal velocities and accelerations need to be calculated per structural interval (with the Airy theory), but the velocity and acceleration vectors (in both the horizontal and vertical directions) must be transposed to those perpendicular to the member under consideration. The overall structural loading can then be found by combining all individual section loads in their local grid in relation to the global grid.

Next to the bookkeeping required for the different inclinations of structural members, also a phase shift occurs as a wave top first reaches one leg of a structure and some time later the next. Although the calculation required for more complex structures increases significantly when compared to a single vertical tower, however computer programs handle all these calculations very effectively.

Adding sea current

When a structure is subjected to wave particle velocity and acceleration, and to a marine current, the velocity of the current must be incorporated into the Morrison calculation of the total hydrodynamic load. As the velocity is only a parameter in the drag term, the combined wave and current drag load equation will be:

$$f_d = C_d \cdot \tfrac{1}{2}\rho_{water}\, D \cdot \left(\left| (u + U_c) \right| (u + U_c) \right) \tag{7}$$

with:

u	wave induced particle velocity	[m/s]
U_c	current velocity	[m/s]

Diffraction

The basic assumption of the Morison equation is that the submerged members for which the wave loads are calculated do not affect the waves. As long as the cylinder diameter is relatively small compared to the wave length, this assumption is valid. For large diameter structures, like the monopile support structures for offshore wind turbines, placed in relatively shallow water with consequently reduced wavelength, the validity of the Morison equation can be compromised due to diffraction of the waves.

To incorporate diffraction into the Morison equation, the MacCamy-Fuchs correction is introduced [7]. This correction reduces the magnitude of the inertia coefficient. The correction factor is dependent on the ratio of diameter to the wavelength: D/λ_{wave}. Fig 9 shows the reduction of C_m for increasing ratio.

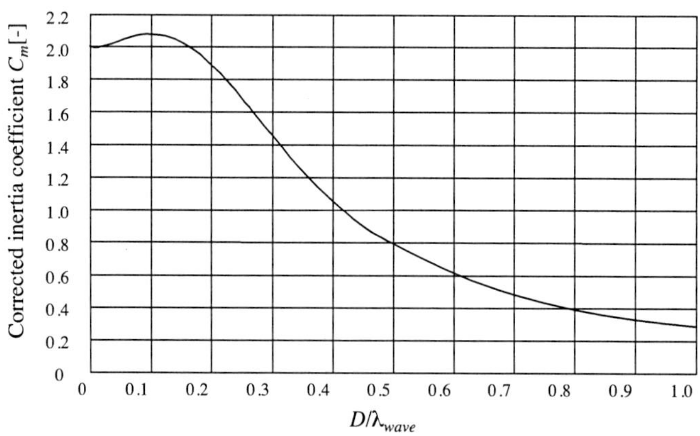

Figure 9. MacCamy-Fuchs diffraction correction of the inertia coefficient C_m in the Morison equation for cylinder diameter over wave length

4.4 Extreme waves, non-linear wave theories and breaking waves

To determine the extreme wave loads on a structure, the expected maximum amplitude wave for the site must be known. When long-term measurements of wave heights are available, a distribution curve can be fitted through the maximum values and extrapolated to find the 50 and 100 year return-period maximum wave heights. It is also possible that the maximum wave height is limited by to the breaking-wave limit due to reduced water depth at the site or in the vicinity (sand banks).

To calculate the wave forces on the structure, the linear Airy theory does not always completely describe the more non-linear features of these extreme waves. Fig. 10 shows which theory to use in which situation, based on the relation between wave height, period and water depth. As the non-linear methods can only be used to model deterministic waves, they are not suitable to model stochastic wave fields.

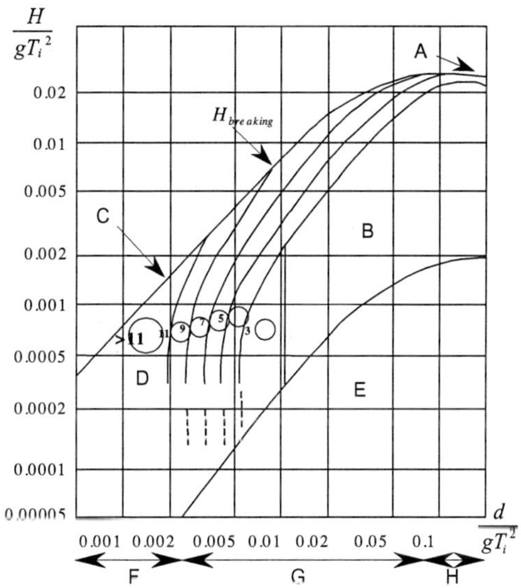

A Deep water breaking limit $H/\lambda_{wave} = 0.14$
B Stokes' 5th order, New Wave or 3rd order Stream Function
C Shallow water breaking limit H/d = 0,78
D Stream function (showing order number)
E Linear / Airy or 3rd order Stream Function
F Shallow water
G Intermediate depth
H Deep water

Figure 10. Regions of applicability of different wave theories [2]

The area of applicable theories is enveloped by the breaking wave limit. Waves are assumed to break when $H/d > 0.78$. The probability of breaking waves at a specific offshore wind farm site needs to be assessed. At sites with a steep sea floor slope, at sandbanks or rocky outcrops, the probability of breaking waves is high. This is for instance the case at Blyth (east coast of Northumbria, England). Because waves breaking directly onto the pile will induce large loads on the structure, analysis that is more detailed may be required in such situations. Although many experiments have been conducted to find a suitable empirical formulation [8] [9] for the loads from breaking waves on structures, much uncertainty remains. DNV [10] prescribes equation (8) to calculate the breaking wave load:

$$F = \tfrac{1}{2}\rho_{water} C_S A u^2 \tag{8}$$

with:

C_s	slamming coefficient (range: 2 – 6.3)	[-]
ρ_{water}	water density	[kg/m^3]
A	area exposed to breaking wave	[m/s]
u	water particle velocity in breaking wave crest	[m]

Fortunately, the probability of breaking waves is relatively small on a nearly horizontal seabed without abnormal bathymetrical characteristics; most waves just break at or near the coast and not within the offshore wind farm.

4.5 Long-term wave description

All wave calculations discussed in the previous sections deal with dynamic sea conditions that are assumed 'mathematically stationary'. The wave conditions during these periods are called 'sea-states' and they typically last 3 hours. When long-term effects of the wave climate at a certain site are studied to find the probability of extremes or to calculate the lifetime fatigue damage, a large number of data are required. These data can be taken by measurements from buoys, platforms and satellites, or can be estimated from large meteorological and oceanographic models fed and calibrated with measurements from a large number of stations. The resulting sea-states are binned into a scatter diagram. This diagram gives the probability of the occurrence of each combination of H_s and T_z for that location. Combinations of H_s and T_p are also common. Scatter diagrams based on different wave periods should not be confused, see previously. Every bin represents the probability of occurrence of wave conditions having that specific combination of parameter values, with which the wave spectrum according to a selected spectral model can be determined. The size of the bins can be selected arbitrarily; larger bins will remove much detail but reduce the total number of sea-states. A typical bin size for H_s is steps of 0.5 m and for T_z (or T_p) steps of 2 s.

When looking at the scattered occurrences in a wave scatter diagram, the visual impression is that there is a linear relationship hidden in the scatter plot, with trend increasing from bottom-left to top-right. The general direction of this plot is of course obvious; large waves have large periods (or they become too steep and break) and small waves have smaller periods. However, to fix a linear relationship would considerably reduce the variability reflected in the diagram and discard valuable information. Because scatter diagrams are very site-specific, it is even more dangerous to superimpose the assumed linear relationships of one location onto the next. Fig 11 shows the scatter diagram for the OWEZ or "NL1" location [11].

T_z [s]	0-1	1-2	2-3	3-4	4-5	5-6	6-7	7-8	Sum:
H_s [m]									
6.5-7.0									0.0
6.0-6.5								0.1	0.1
5.5-6.0							0.1	0.1	0.2
5.0-5.5							0.1	0.1	0.2
4.5-5.0							1		1.0
4.0-4.5							4		4.0
3.5-4.0						4	5		9.0
3.0-3.5						19	0.1		19.1
2.5-3.0					0.1	38			38.1
2.0-2.5					27	43			70.0
1.5-2.0				0.1	115	5			120.1
1.0-1.5				6	220	1			227.0
0.5-1.0				236	145	1			382.0
0.0-0.5	1		1	113	14	0.1			129.2
Sum:	1.0	0.0	1.0	355.1	521.1	111.1	10.4	0.3	1000

Figure 11. Wave scatter diagram for Hs and Tz with occurrence in parts per thousand for the OWEZ location [11]

4.6 Sea level elevation: tides

The mean sea level elevation varies in time. The main driver of these variations is the tide, but also storm surges may temporarily raise or lower the mean elevation. Tides are driven by the gravitational pull of the Moon and the Sun. This pull creates two bulges of water on both sides of the globe, underneath which the earth rotates. This results in 2 high and 2 low tides every 24 hours for most locations. These high and low tides are further influenced by the local bathymetry. The tidal wave becomes trapped within basin-like areas like the North Sea, and consequently rotates within the basin, so creating an amphidromic system. The tidal wave rotating within the basin is driven by the Coriolis force due to the Earth's rotation, giving it a counter clock-wise rotation in the Northern Hemisphere. The tidal wave rotates around amphidromic points where the tidal range is zero. Away from these points, the tidal range increases. Fig. 12 shows the tidal ranges in the North Sea with amphidromic points; medium tidal ranges of nearly 2 m occur at the Dutch coast and more than 4 m along the east coast of Britain. Outside the main basin, large ranges of up to 12 m occur in the English Channel and in the Severn Estuary, south of Wales. The latter tidal range is also driven by the funnelling effect of the local bathymetry, and, mainly, by the resonance of water movement up and down the estuary in tune with the tides.

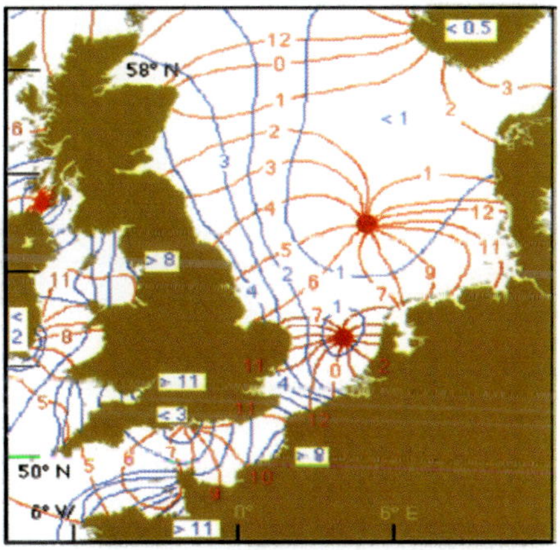

Figure 12. Tidal ranges in the North Sea. Blue lines: co-range line (locations with the same tidal range, metre). Red lines: co-tidal lines (locations that experience the tide at the same time) [12]

The tidal range also has monthly fluctuations. When the Sun and Moon are aligned with each other, at full moon and new moon, the tidal bulges become extra large, causing a 'Spring Tide' with higher high-water and lower low-water. When Sun and Moon to Earth-directions are oriented at 90o angle of each other, the tidal range is smaller than usual, a Neap Tide. For design considerations, the Spring Tides are usually critical for the height of boat landing, and they can affect the extreme load calculations. For installation of turbine foundations, the range of the neap tide is important; installation of gravity-based structures is easier at Neap Tide because the longer period of near constant water level and thus minimum currents.

Sea map water-depths are usually related to the mean of the lowest astronomical tide (LAT). This means that sailors always know the minimum water depth at locations on the map.

4.8. Currents

Sea currents are mostly driven by the tides and ocean circulations, although the outflow of rivers, differences in temperature or salinity and storm surges may cause extra local currents. The currents are usually largest where large tidal differences occur, or where local bathymetry influences the tidal currents drastically. An example of the latter is Horns Rev, where the tidal wave of the North Sea is pushed around the corner of Blåvands Huk. This regularly causes local surface current speeds of 2 m/s.

The sea-current velocity variation is of a time scale much larger than the time scales for load variations associated with the design of offshore wind turbines. It is therefore common to assume the instantaneous surface current velocity and direction to be constant in design calculations.

5. WIND

5.1 Wind speed

Measuring the intensity of the wind first originated from marine applications. In 1838, Admiral Beaufort of the British Royal Navy created a look-up table to relate wind force to ship behaviour. Dividing the wind force in 13 classes (0-12) proved an effective method to log wind behaviour throughout the British fleet.

Though the Beaufort scale is still used in meteorological forecasts and by sailors, it is common for wind speed to use units of metres per second. Because the wind speed is constantly changing, the main feature of wind speed is its mean. Either over short intervals called gusts (3-10 s), or as 10 minute means, daily means, monthly means or yearly means.

When taking a longer measurement period, the time varying character of the wind can be captured in a wind spectrum, covering frequency ranges from years to seconds. This spectrum, after Van der Hoven [13], is shown in Fig 13.

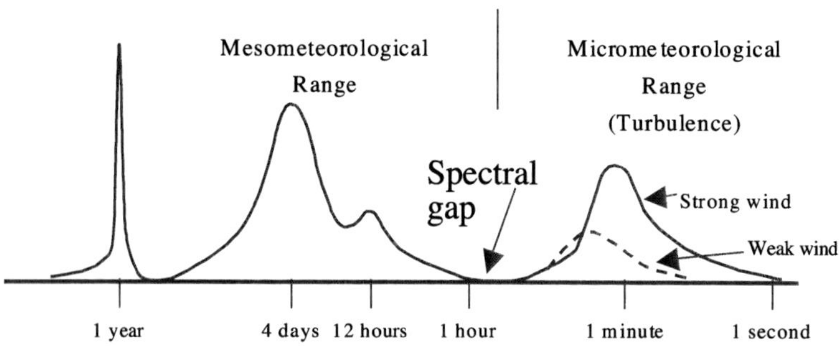

Figure 13. Wind speed spectrum over a broad range of frequencies [13]. Turbulence region indicates strong and weak wind variaition.

The frequency content on the left hand side represents the yearly changes, pressure systems and diurnal changes. On the right hand side, the turbulence is visible. The solid line represents strong turbulence during a period of high-speed wind, the dotted line shows reduced turbulence at lower-speed wind. The details of turbulence will be described in the next section. Note Van der Hoven's gap around the 1-hour period, known as the "spectral gap", which separates the slowly changing and the rapidly changing ranges. In this latter frequency range, the mean wind-speed over a period of 10 minutes to 1 hour can be considered constant.

Recent research on the spectral gap revealed that the perceived gap was a coincidental feature of Van der Hoven's analysis for plotting short term and long-term measurements [14]. Although this is probably true, the assumption that in the 10 minute to 1-hour range the mean wind speed is constant has proved to be an effective model for wind turbine design in the past and will therefore be adhered to here.

5.2 Wind shear and turbulence

Introduction

The wind velocity measured in the field shows variations in space, time and direction. A momentary representation of a typical wind speed distribution is shown in Fig. 14.

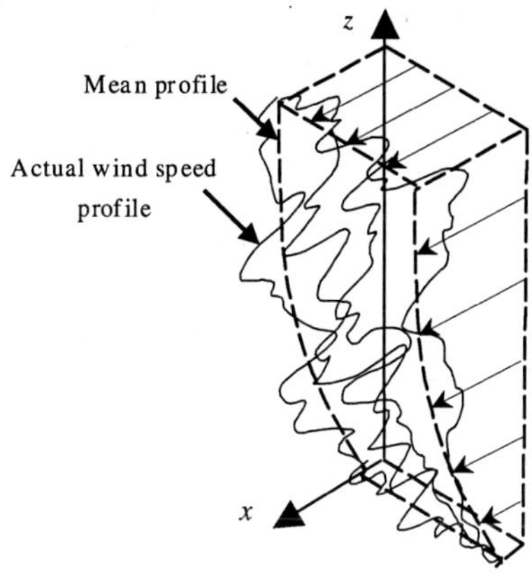

Figure 14. Actual wind speed profile

From the figure, it is clear that the mean wind speed increases with height; a phenomenon called wind shear. Furthermore, the actual wind speed at any location varies in time and direction around its mean value, due to turbulence.

Wind shear

In the lower 2 km of the Earth's atmosphere, the atmospheric boundary layer, the wind speed is affected by friction with the Earth's surface. This effect reduces the wind speed from its undisturbed value at 2 km height, to nearly zero at the surface. To describe the effect, two models are commonly used: the logarithmic profile and the power law profile. Both profiles are fitted curves to measured wind shear. The logarithmic profile and the power law profile are described by equations (9) a, b:

$$\text{(a)}\ V_w(z) = V_{w,r} \cdot \frac{\ln\left(\frac{z}{z_0}\right)}{\ln\left(\frac{z_r}{z_0}\right)} \qquad \text{(b)}\ V_w(z) = V_{w,r}\left(\frac{z}{z_r}\right)^{\alpha_{shear}} \tag{9}$$

with:

$V_w(z)$	mean wind speed at height z	[m/s]
$V_{w,r}$	mean wind speed at the reference height z_r	[m/s]
z_r	reference height	[m]
z_0	surface roughness-length	[m]
α_{shear}	power law coefficient	[-]

Typical values for the surface roughness-length for different types of terrain are given in Table 2. Det Norske Veritas (DNV) mentions the use of $z_0 = 0.05$ m for offshore [10], unless stated differently. Germanischer Lloyd (GL) advises in their offshore wind regulations to use $z_0 = 0.002$ m [15].

Table 2. Roughness length and power law factor for different types of terrain [16]

Type of terrain	z_0 [m]	α_{shear}
Flat desert, rough sea	0.001	0.12
Calm sea	0.0002	

Turbulence

When wind is measured in the field, a time varying wind speed can be found as shown in Fig 15. From this time series the statistical parameters can be calculated: the mean wind speed and the standard deviation.

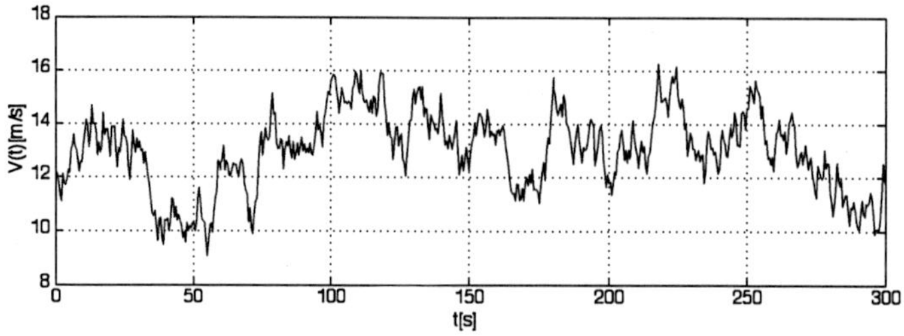

Figure 15. Time varying wind speed measurement

The turbulence intensity is defined as the standard deviation of the time varying wind speed divided by the mean wind speed (usually presented as percentage):

$$I_t = \frac{\sigma}{V_w} \tag{10}$$

The turbulence intensity is dependent on height and the roughness of the terrain. Rougher terrain and lower altitude give larger turbulence intensities. Design standards give descriptions of the turbulence intensity based on these roughness and altitude parameters. For design, fixed turbulence levels can be selected for specific sites. Fig. 16 shows the recommended turbulence intensities as a function of wind speed for different standards [15] [17] [18].

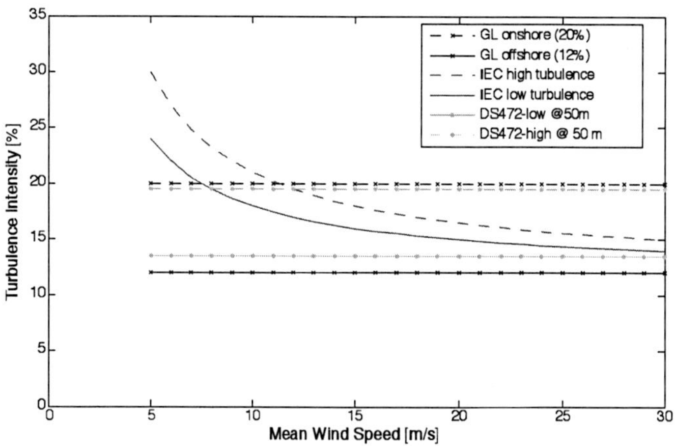

Figure 16. Turbulence intensity as a function of wind speed from different standards [15] [17] [18]

5.3 Extreme wind speeds and gusts

Two basic extreme wind load cases can be defined: (i) the extreme hourly-mean wind-speed with given return period (1, 50, 100 years), and (ii) the extreme incident wind-speed within a given short interval.

The extreme hourly-mean wind-speed is found by taking a large number of measured means and selecting the extremes above a chosen threshold. These extremes are then plotted on log-linear scale over the number of occurrences per time interval (month, year). By fitting a distribution curve through these measured values, an interpolation can be made of the expected maximum in periods beyond the measurement period. One of the most commonly used is the Gumbel distribution, as shown in Fig. 17.

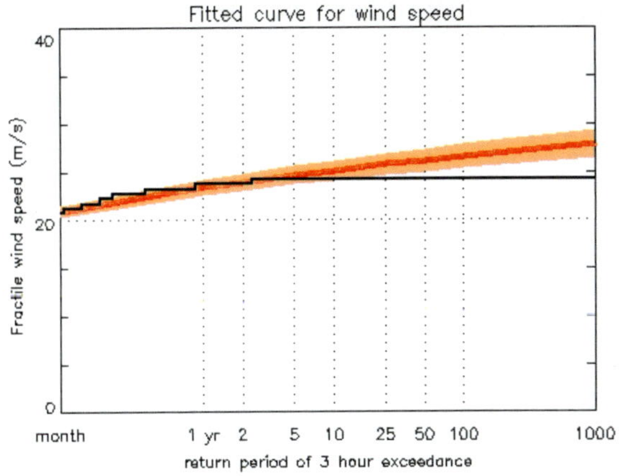

Figure 17. Gumbel fit (red line) for data recordings of mean wind speed (steps in black line) [19]

In onshore wind turbine design, the design standards give a prescribed extreme wind speed for different classes. Table 3 shows the values given by the IEC [17]. The classes are used as reference for the design of the turbine; they are not directly linked to specific sites. Offshore sites are classed as "S" (special), when they are yet undefined and require site-specific analysis.

Table 3. Location classes according to IEC with yearly hourly mean wind speed ($V_{w,\ ext\ 1h}$) and extreme 10-minute mean wind speed ($V_{w,\ ext\ 10min}$) in [m/s] at hub height

Class	I	II	III	IV
$V_{w,\ ext\ 1h}$ [m/s]	10	8.5	7.5	6.0
$V_{w,\ ext\ 10min}$ [m/s]	50	42.5	37.5	30

For offshore wind turbines, the site-specific analysis can be based on extreme measured values and a Gumbel distribution, although, for a quick result, the standard classes may give a first order estimate.

To determine extreme incident wind-speed, or gust within a certain time interval, the tail of the turbulence spectrum can be used; an approximate formulation was derived by Wieringa [20], which shows good agreement with theoretical results. The gust factor $G(t)$ is defined as:

$$G(t) = 1 + 0.42 I_t \ln \frac{3600\,\mathrm{s}}{t} \tag{11}$$

The gust factor is used to convert the hourly mean wind speed to the gust wind speed for duration t:

$$V_{gust}(t) = G(t) \cdot V_w \tag{12}$$

Fig.19 shows the gust factors for different durations for the turbulence intensity for onshore (20%) and offshore (12%) prescribed by Germanischer Lloyd [15].

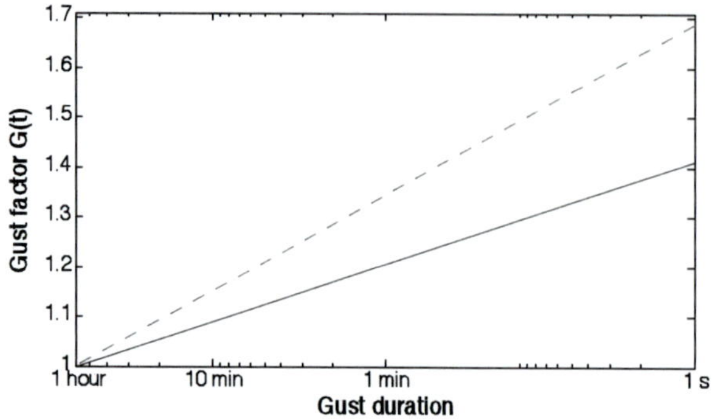

Figure 19. Gust factors for different durations and onshore (20%) (dotted) and offshore (12%) turbulence intensity as specified by GL [15].

5.4 Long-term distribution of wind speed

In addition to short-term load calculations, the wind climate at the site of an offshore wind farm needs to be known to calculate potential power production and structural fatigue. The first characteristic value in this respect is the yearly average wind speed. Fig 20 shows the average annual wind speed for the North Sea from [21].

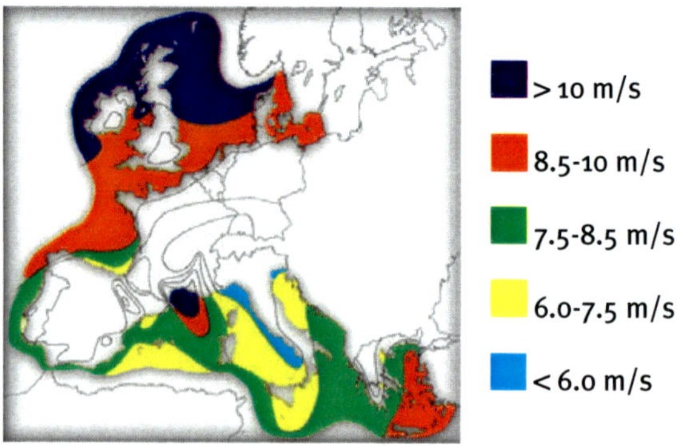

Figure 20. Yearly average wind speed at 100 m height for the European Seas [21]

For most offshore locations in the northerly half of the North Sea, the average annual wind speed at a height of 100 m ranges from 8 m/s to 10 m/s. For onshore sites, the yearly average is significantly less. Coastal sites may match the 8 m/s but further inland the value drops rapidly to a typical value of 4 m/s for wooded southern Germany. A common way to describe the annual wind speed distribution is by use of the Weibull distribution. The curve represents a fit of the probability density per wind speed at a given location, as shown in equation (13) for Horns Rev (10 km off the west coast of Jutland). It is based on two parameters: a scale parameter c_w, related to the annual mean wind $V_{w,\, year}$ speed and a shape parameter kw describing the variability about the mean. Equation (13) represents the Weibull probability density function [22]:

$$f(V_w) = \frac{k_w}{c_w}\left(\frac{V_w}{c_w}\right)^{k_w - 1} e^{-\left(\frac{V_w}{c_w}\right)^{k_w}} \tag{13}$$

From this expression, the yearly mean can be reproduced analytically. A faster method is the empirical formulation by Lysen [23]:

$$V_{w,year} = c_w(0.568 + 0.433 / k_w)^{\frac{1}{k_w}} \tag{14}$$

Typical values for k_w are: 1.75 for inland areas, 2.0 on the coast and 2.2 offshore. A larger k_w value (e.g. 2.2) means that the peak becomes more probable (lower on the figure) and the tail flatter, meaning that the distribution contains less extreme high or low values of wind speed. For offshore locations, this is true because diurnal patterns (heating of the surface in the daytime, cooling at night) are much less significant than on land. Fig 21 shows the Weibull distributions for typical onshore, coastal and offshore conditions.

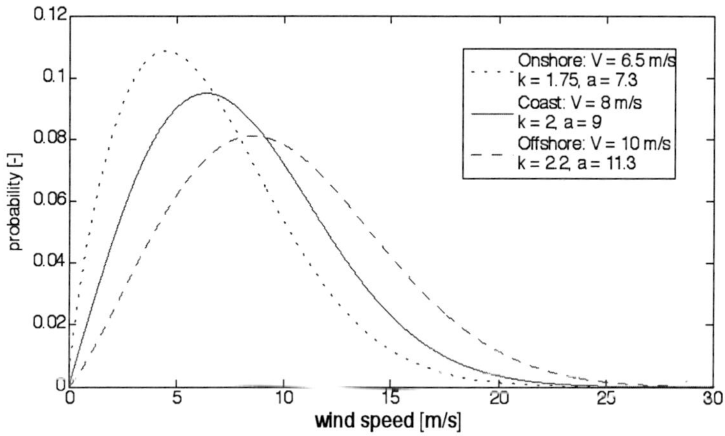

Figure 21. Weibull annual wind distributions for typical onshore, coast and offshore conditions

6. THE TURBINE

6.1 Introduction

Wind turbines come in many different types and shapes. This chapter only deals with the most commonly used type in present-day wind energy production: the upwind, 3-bladed wind turbine.

To calculate the loads on the turbine, blade-element momentum theory is used. The momentum theory assumes a stream tube as shown in Fig. 22. The turbine can be seen as an actuator disk, slowing the flow of air, creating a wake with reduced airflow velocity. The load that the actuator disk exerts on the flow to achieve this is the axial load F_{ax}.

By defining the induction factor a, equation (15), the axial force of the actuator disk on the flow of air can be found using the momentum theory assuming incompressible, homogeneous and horizontal flow:

$$a = \frac{V_0 - V_{disk}}{V_0} \tag{15}$$

Using the Bernoulli to calculate the resulting load on the actuator disk, we find [16]:

$$F_{ax} = \tfrac{1}{2} A_{disk} \rho_{air} V_0^2 \cdot 4a(1 - a) \tag{16}$$

with:

a	induction factor	[-]
V_0	undisturbed wind velocity	[m/s]
V_{disk}	wind velocity at the actuator disk	[m/s]
A_{disk}	area of the rotor disk	[m^2]

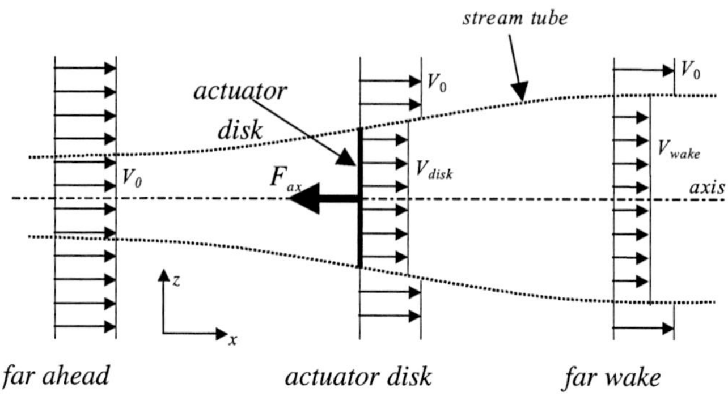

Figure 22. Actuator disk with stream tube

Because neither the axial load nor the induction factor is known, blade element theory is used to solve the equation. The blade is considered to consist of blade elements, each with their own aerodynamic features. The blades are shaped according to a series of airfoil profiles to create the desired aerodynamic performance. The airfoils near the root have a large thickness to accommodate the internal structures that transfer the blade loads to the rotor axis. Near the tip, where most torque is generated, the airfoils are optimised for performance. The airfoil variation is shown in Fig. 23. The airfoils originate from aero-elastic design programs; they are validated in test projects that describe their shape and aerodynamic performance as tested in wind tunnels.

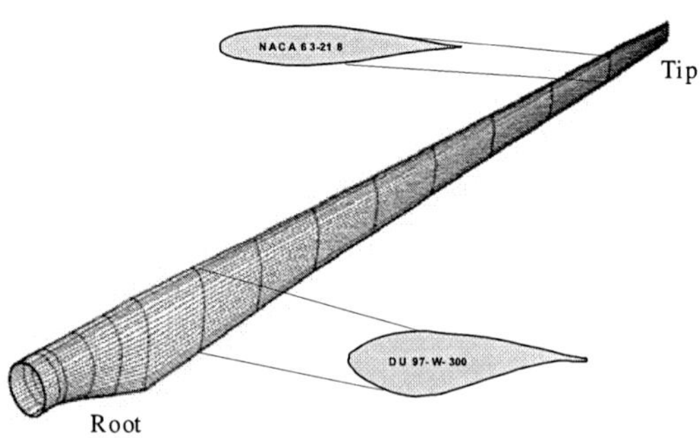

Figure 23. Rotor blade with thick airfoils at the root, thin airfoils at the tip

The wind load on a structure can be calculated with an equation similar to that for the calculation of hydrodynamic loads. The density of the air and the airflow velocity are used together with an aerodynamic coefficient that is dependent on the surface roughness, the shape of the structure and the airflow velocity. For a single section of a tubular tower its value is typically $C_{aero} = 0.7$, which is the drag coefficient. The wind load on this section is given by:

$$F_{aero} = \frac{1}{2} C_{aero} \rho_{air} A V_{section}^2 \tag{17}$$

with:

F_{aero}	wind load	[N]
C_{aero}	aerodynamic coefficient (shape, surface dependent)	[-]
ρ_{air}	density of air (1.225)	[kg/m^3]
A	exposed area of the section	[m]
$V_{section}$	wind velocity at the centre of the section	[m/s]

Airfoils experience not only a drag force, but also a lift force, which is calculated in the same manner. The lift and drag coefficients for airfoils are dependent on the angle of attack as shown in Fig. 24.

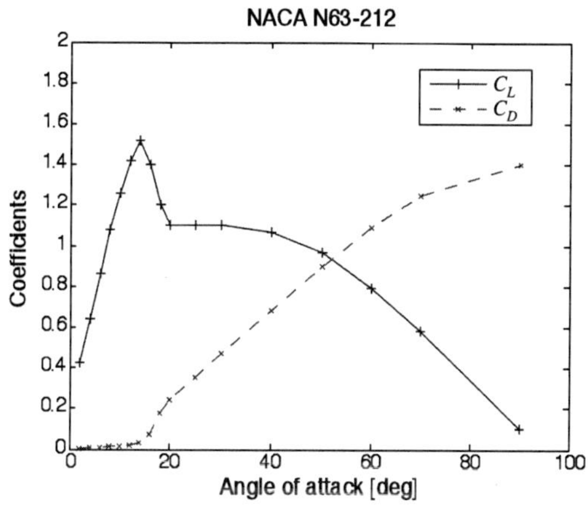

Figure 24. Lift coefficient (C_L) and drag coefficient (C_D) curve per angle of attack for the NACA N63-212 airfoil

To calculate the aerodynamic loads on the blade, each blade element is considered separately with their own airfoil characteristics. At this stage, each element is considered infinitely long, so assuming no boundary effects and no aerodynamic interaction between elements. The loads on the blade elements are generated by lift and drag, which are induced by the wind speed and the rotation speed, combined in the relative wind speed over the blade:

$$\text{a} \quad V_{rel} = \sqrt{V_{disk}^2 + V_{rot}^2} \tag{18}$$

$$\text{b} \quad V_{disk} = V_0(1-a) \text{ and } V_{rot} = \Omega \cdot r$$

with:

V_{rel}	relative wind speed at a blade section	[m/s]
V_{disk}	wind velocity at airfoil	[m/s]
V_{rot}	linear rotation speed at a blade section	[m/s]
Ω	angular rotation speed	[rad/s]
r	distance of blade element to axis of rotation	[m]

Due to this relative wind velocity, lift and drag loads are exerted on the blade element as shown in Fig. 25. Note that the drag load is not drawn on the same scale as the lift load. Would that be the case, the drag load arrow (FD) would be shorter than the size of its tip, making it invisible in the plot. The lift and drag loads are calculated with equations (19).

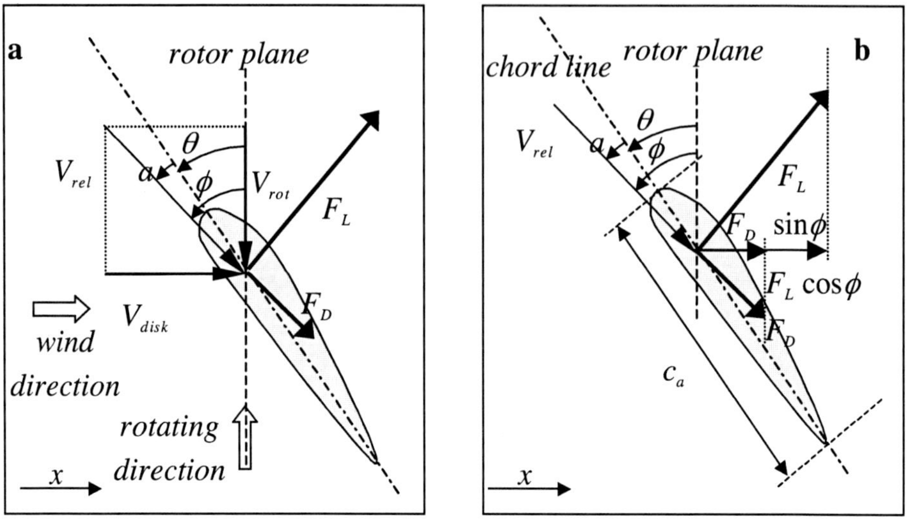

Figure 25. a) Lift and drag load on the blade element b) Resulting loads in the x direction

$$F_L = \tfrac{1}{2} C_L(\alpha) \rho_{air} V_{rel}^2 c_a \Delta r \qquad F_D = \tfrac{1}{2} C_D(\alpha) \rho_{air} V_{rel}^2 c_a \Delta r \tag{19}$$

with:

F_L	aerodynamic lift	[N]
F_D	aerodynamic drag	[N]
$C_L(\alpha)$	aerodynamic lift coefficient	[-]
$C_D(\alpha)$	aerodynamic drag coefficient	[-]
ρ_{air}	mass density of air	[kg/m^3]
c_a	airfoil chord length	[m]
r	radial length of blade element	[m]
α	angle of attack	[deg]
θ	pitch angle	[deg]
φ	angle of inflow	[deg]

The load in x direction per blade element is:

$$F_x = F_L \cos\phi + F_D \sin\phi \tag{20}$$

Making the total axial load:

$$F_{ax} = N_b \sum_{r=root}^{r=tip} F_{x,r} \tag{21}$$

When the axial load is solved iteratively by using blade element theory (21) and momentum theory (17), the aerodynamic loads on the structure are known. From this, the power production can be calculated:

$$P = F_{ax} V_{disk} = 2\rho_{air} A_{disk} V_0^3 a(1-a)^2 \tag{22}$$

By defining the power coefficient C_p as the power P divided by the total power in the air flow through the rotor plane, equation (23) is found:

$$C_P = \frac{P}{\frac{1}{2}\rho_{air} V_0^3 A_{disk}} = 4a(1-a)^2 \tag{23}$$

The maximum of the power coefficient can be determined to be $C_P = 16/27 = 0.593$ at $a = 1/3$ which is known as the Lanchester-Betz limit [24]. The full blade-element momentum theory is far more refined than the basics described here, incorporating, among others, tip and root effects. These effects are described well in [16] and are incorporated in the GarradHassan software 'Bladed', which has been used for turbine modelling for the work of this chapter.

6.2 Managing power capture

Although the theoretical maximum power extraction from the wind is limited to the Lanchester-Betz limit, several parameters influence the actual power capture. If a blade configuration is optimised for aerodynamic power capture so the angle of attack is optimal per airfoil, the power will still be dependent on the relation between rotor speed and wind speed, combined in the relative wind speed V_{rel} in equation (24). We define λ as the tip-speed ratio: the ratio between the speed of the blade tip and the wind speed:

$$\lambda = \frac{V_{tip}}{V_0} = \frac{\Omega R}{V_0} \tag{24}$$

The relation between the tip-speed ratio and the power coefficient is plotted in Fig. 26. The curve shows a peak, for this particular blade, at $\lambda = 8$.

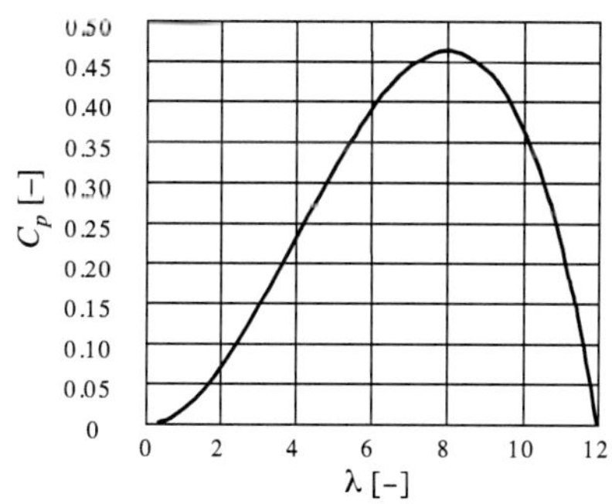

Figure 26. Typical C_p-λ curve

To capture maximum power at every wind speed, the rotation speed should be changed to keep the C_P-λ curve at its maximum Fig. 26 shows the power capture of a variable speed turbine tuning its rotation speed to maximum power. Because the generator is connected to the electricity grid, the grid demands the power output to be of constant frequency. Because of advances in power electronics, the rotation speed and the grid frequency can nowadays be fully uncoupled without loss of power quality. However, until about 2006 this option was not always available, so turbines were fitted with constant speed generators. The power capture could therefore only be optimal at one wind speed, which is also shown in Fig. 27.

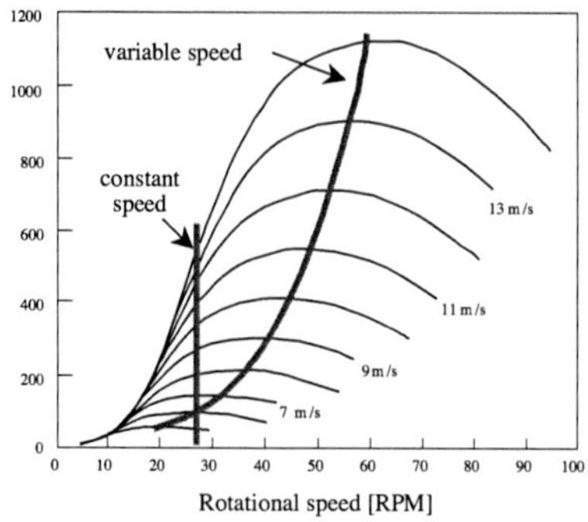

Figure 27. Power output as function of rotational speed for different wind speed classes

In Fig 27, the power capture of the variable speed turbine continuously increases with larger wind speeds. For actual turbines, the maximum power capture is controlled to be constant at the rated capacity of the generator. The optimum rotor-generator configuration is the result of economic optimisation. The wind speed at which the generator reaches maximum power is called the rated wind speed, V_{rated}. To limit the power capture to match the generator, two basic options exist.

1) *Passive and active stall regulation at large angle of attack.* If the blade pitch is constant, as wind speed increases from zero, both the angle of attack and the lift coefficient increase. However, the lift coefficient reaches a maximum value, as shown in Fig 24. When the wind speed increases further, the lift coefficient decreases, so the airflow over the blade becomes detached in a somewhat erratic manner and the blade stalls (right side of Fig 24). Consequently, the power extracted from the wind is limited.

   For fixed pitch blades, the stall at large angle of attack is called *'passive stall'*.
   For controlled pitch blades, the stall at large angle of attack is called *'activate stall'*.

2) *Pitch regulation at small to zero angle of attack.* The other, smoother, option to control the power output above rated wind speed is to pitch the blade to make the angle of attack smaller so the lift decreases smoothly, corresponding with the reduced lift coefficients on the left side of Fig 24. This is called *pitch control.*

The theoretical relation between wind speed and power production can now be described. A wind turbine needs a certain minimum wind speed to start operation, typically 3 m/s. This wind speed is called the cut-in speed, $V_{cut\text{-}in}$. From this wind speed, the turbine power follows the power curve until the rated power is reached, V_{rated}. Above rated wind speed the turbine continues at rated power. Finally, for very high wind speeds with excessive turbulence, the turbine is shut down at $V_{cut\text{-}out}$, the cutout wind speed. For present-day turbines $V_{cut\text{-}out}$ = 25 m/s typically, although some turbines continue to $V_{cut\text{-}out}$ = 36 m/s. This value is also the result of cost-component optimisation. Fig. 28 shows the power (or *P*-*V*) curve for the Vestas V80 turbine [25].

6.3 The turbine in turbulent flow

When a turbine is in operation, its rotor rotates through a three-dimensional turbulence field. This field can be described by the correlated turbulence effects described earlier. A graphic representation of such a turbulence field is shown in Fig. 29. This turbulence field was modelled in Bladed for the offshore turbine at Blyth, North East England; the field covered an area of 100 by 100 m at a wind speed of 10 m/s with 12% turbulence intensity. Areas of faster and slower wind speeds are clearly distinguishable.

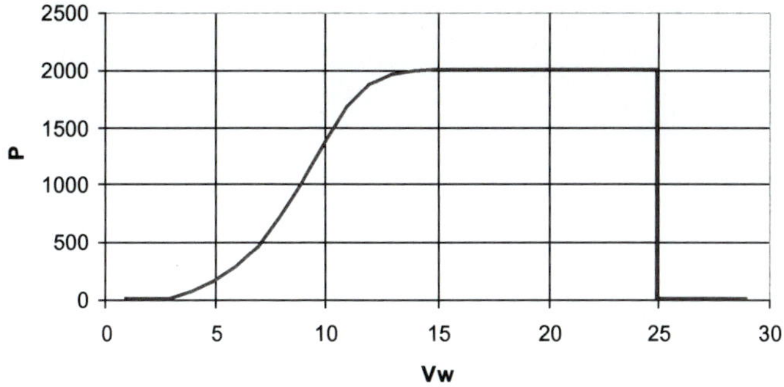

Figure 28. Power curve for Vestas V80 turbine [25]

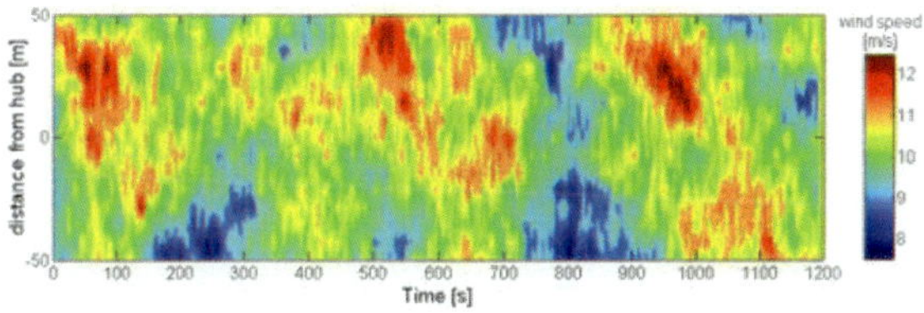

Figure 29. Turbulent eddies in a wind field

When a rotating blade passes through an eddy, it experiences a short period of larger or smaller wind speed. Fig. 30 zooms in at such an eddy occurring between 500 s and 550s in Fig. 29. Because the length of the eddy is such that a blade will pass through the eddy several times in successive rotations, the turbine structure will experience a load peak at the rotation frequency of the rotor, called 1P; upper figure. This *eddy slicing* will create not only a load peak at the frequency of 1P but also at the frequency of all the blades passing at N_bP = 3P for a three-bladed turbine, as shown in the lower figure.

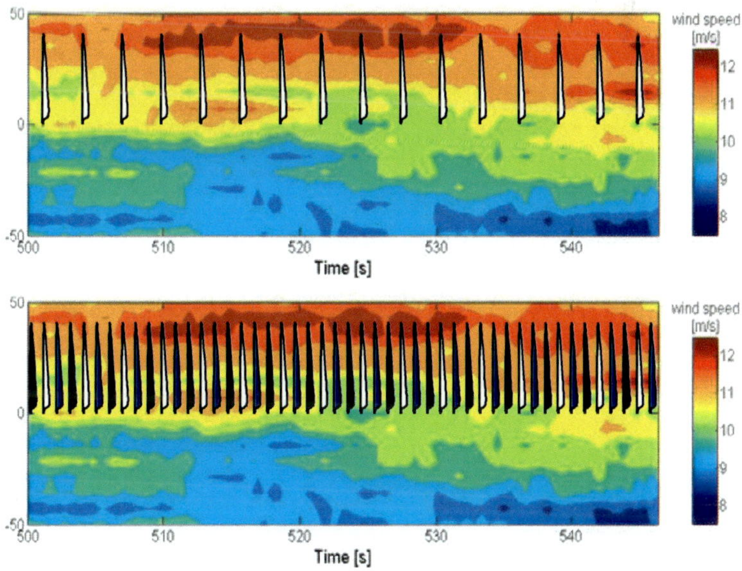

Figure 30. Single blade passing through turbulent eddy (above) sampling 1P frequency
All three blades (black, white, blue) passing alternately through the eddy at 3P frequency

To account for this effect within the turbulence spectrum, the stationary turbulence spectrum, as described earlier, can be transformed into a rotationally sampled spectrum. An example of a stationary and a rotationally sampled velocity spectrum, in the frequency domain, is shown in Fig. 31.

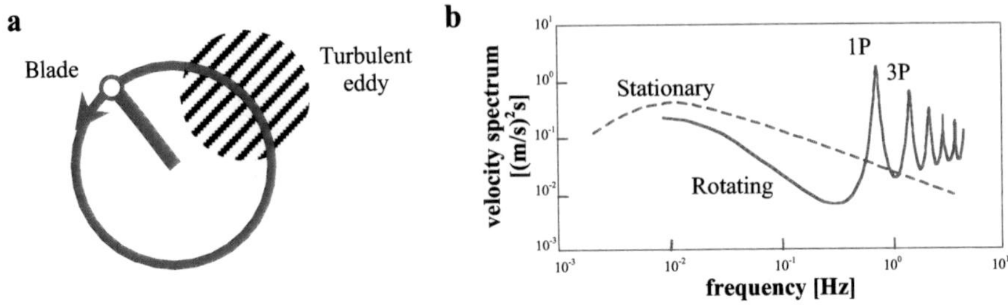

Figure 31. a) Blade passing through a turbulent eddy b) Stationary and rotationally sampled turbulence spectrum

7. DYNAMICS OF OFFSHORE WIND TURBINES

7.1 The basics of dynamics

Modelling of the structural dynamics of the turbine can be illustrated by considering a single degree of freedom mass-spring-damper system as shown in Fig. 32. The single offshore wind turbine can be modelled by a suitable assembly of coupled multi degree-of-freedom mass-spring-damper systems [26].

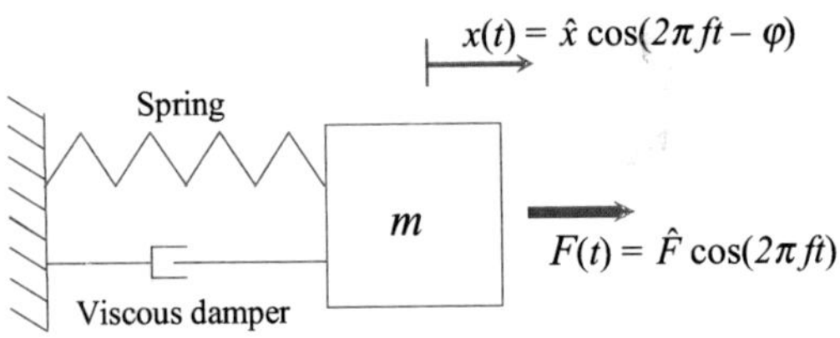

Figure 32. Single degree of freedom mass-spring-damper system

When a harmonic excitation *F(t)* is applied to the mass, the magnitude and phase of the resulting displacement x strongly depend on the frequency of excitation *f*. Three steady state response regions can be distinguished as shown in Fig. 33:

a) *Quasi-static*
b) *Resonance*
c) *Inertia dominated.*

Quasi-static. For frequencies of excitation much less than the natural frequency of the system, the response is quasi-static, as illustrated in Fig. 33a; the displacement of the mass follows the time varying force almost instantaneously, as if it was excited by a static load.

Resonance Fig. 33b shows a typical response for frequencies of excitation within a narrow region around the system's natural frequency. In this region, the spring force and the inertia force almost exactly cancel, producing a response that is a number of times larger than it would be statically. The resulting amplitude is governed by the damping present in the system.

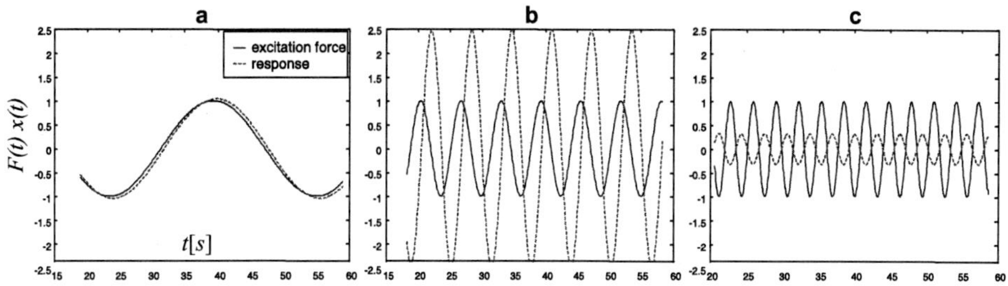

Figure 33. a) Quasi-static b) resonant and c) inertia dominated response Solid line: excitation, dashed line: displacement.

Inertia dominated For frequencies of excitation much greater than the natural frequency, the mass cannot "follow" the excitation any more. Consequently, the dynamic response is small and almost in counter-phase, as illustrated in Fig. 33c. In this case, the inertia of the system dominates the response.

Fig. 33 illustrates the general fact that, in steady state, a sinusoidal input applied to a linear system generates a sinusoidal output at the same frequency, which however generally differ in magnitude and phase.

The magnitude and phase modifying property of linear systems can be summarized by a plot of the Dynamic Amplification Factor (DAF) and the related phase lag. The DAF depicts the ratio between the dynamic response magnitude and the static response magnitude due to the same magnitude of loading. Fig. 34 shows the DAF and phase lag plot of the single degree of freedom system depicted in Fig. 32.

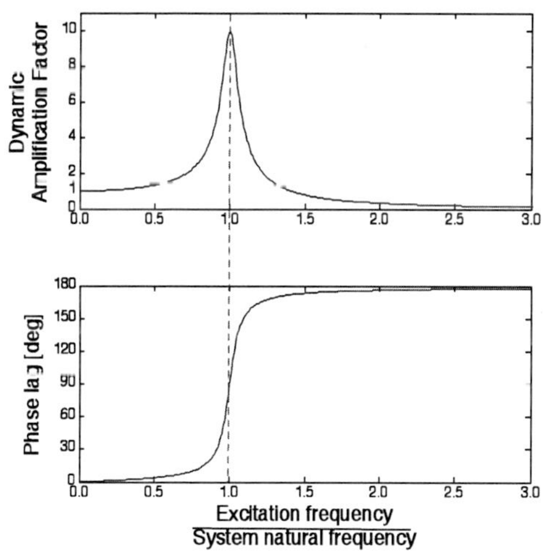

Figure 34. Upper figure: dynamic amplification factor per normalised frequency
lower figure: phase lag versus normalised frequency

The peak in Fig. 34 corresponds to the system's natural frequency. The height of the peak is determined by damping. Therefore, unwanted resonance can be counteracted with adequate damping. In dynamics, the frequency of excitation is at least as important as its magnitude. Resonant behaviour can cause severe load damage and even failure, but it is most feared because of possible *fatigue* damage. For structures where dynamic movement may produce difficulties, detailed knowledge of the expected frequencies of excitation and the natural frequencies of the structure or part of the structure is vital.

The DAF is commonly used in calculations in the preliminary design phase to account for the effect of dynamic response compared to static response (thereby neglecting the phase information). In general, the DAFs are derived from time-domain simulations similar to those in Fig. 34. The important conclusion that can be drawn from this review is that the response of a wind turbine system subjected to time-varying loads needs to be carefully assessed.

7.2 Soft and stiff wind turbine systems

Excitation

To translate the basic model of the previous section to a wind turbine system, the excitation frequencies are examined first. The most visible source of excitation in a wind turbine system is the rotor. As shown previously, the rotor samples the turbulent eddies in the wind field, creating peaks in excitation at frequencies of 1P and 3P for a three bladed rotor.

These two frequencies are plotted in a graph as shown in Fig. 35. The horizontal axis represents the frequency [Hz] and the vertical axis represents an arbitrary response without values. Though higher-order excitations occur, here only 1P and 3P are considered, because these are the primary excitations. To avoid resonance, the structure should be designed so its first natural frequency does not coincide with either 1P or 3P excitation. This leaves three possible intervals:

(i) a very stiff structure, with its first natural frequency greater than 3P is called a stiff-stiff structure,
(ii) if the first natural frequency falls between 1P and 3P, the structure is said to be soft-stiff,
(iii) a very soft structure with its first natural frequency below 1P is called a soft-soft structure.

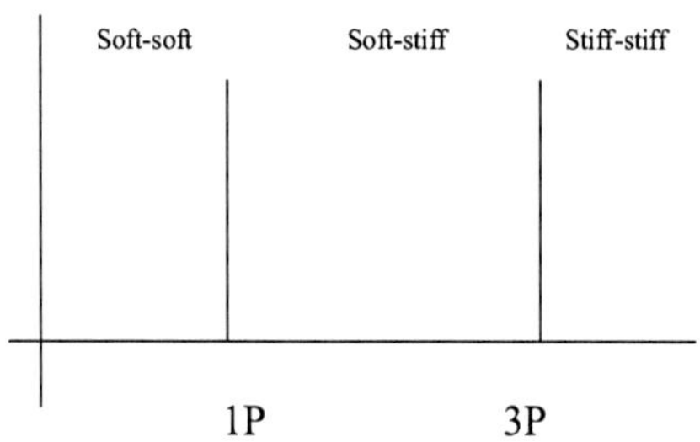

Figure 35. Soft to stiff frequency intervals of a three bladed, constant rotational speed wind turbine

The support structure

A flexible wind turbine can be modelled as a flagpole with top mass mtop, as depicted in Fig. 36. This model resembles the model of the mass-spring-damper system in Fig. 32. The bending flexibility of the tower represents the spring stiffness; the damping is given in the form of a damping coefficient.

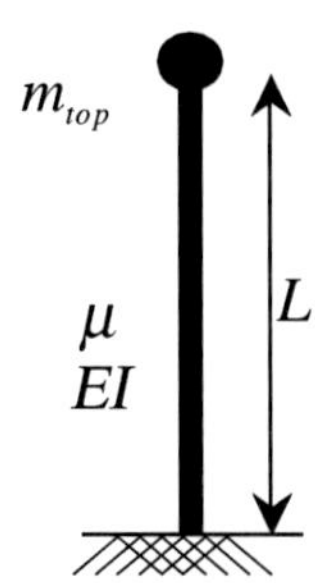

Figure 36. Structural model of a flexible wind turbine system

For this model, consisting of a uniform beam with a top mass and a fixed base, the following approximation for the calculation of the first natural frequency is valid [27]:

$$f_{nat}^2 \cong \frac{3.04}{4\pi^2} \frac{EI}{(m_{top} + 0.227\mu L)L^3} \tag{25}$$

with:

f_{nat}	first natural frequency	[Hz]
m_{top}	top mass	[kg]
μ	tower mass per meter	[kg/m]
L	tower height	[m]
EI	tower bending stiffness	[Nm^2]

As an example, equation (25) is applied to the Opti-OWECS design [11]. This design consists of a 2 bladed constant speed turbine. The rotational frequency 1P is 0.3 Hz, the blade passing frequency 2P is 0.6 Hz. A soft-soft structure would then have (for example) a first natural frequency of 0.25 Hz, soft-stiff 0.5 Hz and stiff-stiff 1 Hz. The wall thickness of the tower is taken to be constant, 75 mm, over the entire height. The system has a top mass of 130 tonne. Equation (25) is applied to determine the diameter D of the tower, corresponding to the specified natural frequencies. The results are listed in Table 4.

Table 4. Required tower diameters by fundamental frequency

Type	f_{nat}	Diameter
Soft-soft	0.25 Hz	2.4 m
Soft-stiff	0.5 Hz	4.2 m
Stiff-stiff	1.0 Hz	7.4 m

As the cost of procurement and handling of large tubular towers is mainly influenced by the diameter and mass, the cost will be least for the "softest" structure. However, a soft structure requires towers supported by cables and controls to prevent inadvertent mechanical response of the tower as forced by bade rotation.

These calculations are for demonstration purposes only, since the support structure of an actual offshore wind turbine will be dimensioned for many more influences. For instance, the soil properties will always be more flexible than the assumed fixed connection of the model in Fig. 36. This flexibility will result in a smaller fundamental natural frequency of the tower, which may need to be compensated by increasing the diameter.

7.3 Design options for support structure dynamics

Variable speed

As described before, variable speed turbines are gaining market share from constant speed turbines. They offer increased energy capture and less dynamic excitation. For example, the Vestas 2 MW turbines have a rotational speed ranging from 10.5 to 24.5 RPM [25]. This means that the interval for a soft-stiff design is correspondingly narrower, as shown in Fig. 37.

Larger turbines

The trend to create larger turbines is still strong. This means that rotor blades become longer and tower top masses (generator) larger. The increase in rotor diameter has a direct effect on the 'soft to stiff' approach. As shown previously, the power performance of a turbine can be represented as a function of tip-speed ratio as shown in Fig. 26. The tip-speed ratio was defined in equation (24) and can be expressed as:

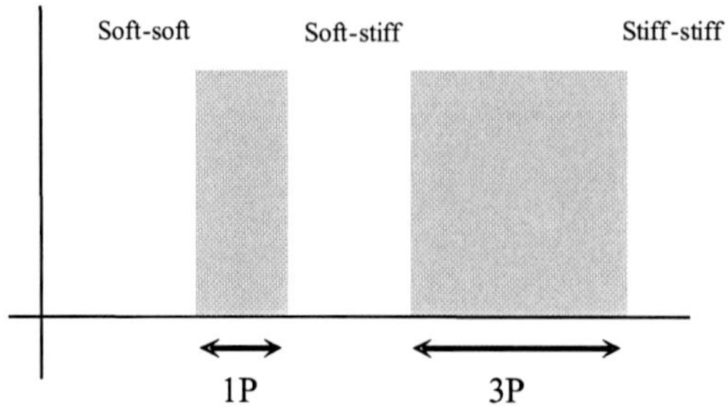

Figure 37. Frequency intervals for a variable speed turbine system

$$\lambda = \frac{V_{tip}}{V_w} = \frac{\Omega R}{V_w} = \frac{f_{1P}\pi D_{rotor}}{V_w} \tag{26}$$

The corresponding 1P rotational frequency is given by:

$$f_{1P} = \frac{\lambda V_w}{\pi D_{rotor}} \tag{27}$$

This means that for a fixed (optimal) tip-speed ratio the rotational frequency will decrease when the diameter increases. The results of equation (27) for a wind speed of $V_w = 11.4$ m/s, $\lambda = 8$ and rotor diameters of 80 m, 100 m and 120 m, respectively, are plotted in Fig. 38.

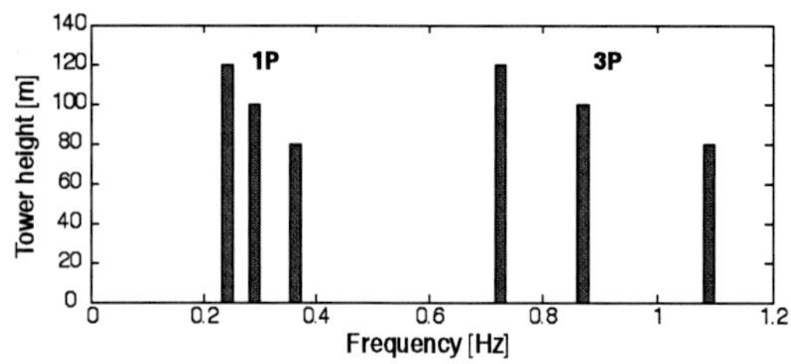

Figure 38. 1P and 3P frequencies for 80, 100 and 120 m diameter rotors operating at constant rotational speed

The increase in rotor diameter also requires a higher hub height. From equation (25) the natural frequency is seen to be inversely proportional to the tower height L squared. This means a large decrease of natural frequency with increasing height for constant tower diameter.

Wave excitation

For offshore wind turbine systems, the waves are an additional source of excitation. Wave frequencies are generally less than the rotational frequency of the rotor. Because waves have various periods, they span a wide range in the frequency band. Fig. 39 shows the range of mean wave frequencies occurring at the NL1 location, the location of the previously mentioned Opti-OWECS turbine near the Dutch coast [11]. The histogram shows the relative occurrence per year of the wave frequencies. From Fig. 39, it is clear that when the offshore wind turbine system is designed with a natural frequency less than the rotational frequency to avoid resonance, it will enter the frequency range where resonance due to wave excitation is important.

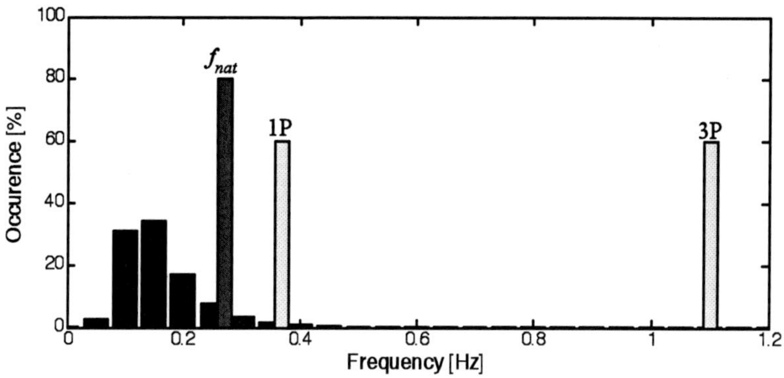

Figure 39. Occurrence of wave frequencies with plotted 1P and 3P frequencies for properties of a 3 bladed turbine after Opti-OWECS off the Dutch coast

7.4 Compensating circumstances

As shown in the previous sections, the design aim would be to create a soft-soft support structure, because it uses less steel and is therefore more const-effective. The trend for the natural frequency of the support structure indeed seems to be moving to this soft range when applying larger rotors and structures and variable speed turbine, with a major risk of resonant behaviour due to wave excitation. Consequently, two important phenomena have to be considered: aerodynamic damping and controllability of variable speed turbines.

Aerodynamic damping

It was shown earlier that when a typical soft-soft support structure is designed to prevent resonant excitation by the 1P frequency of the rotor, it would encounter waves with frequencies near its natural frequency for some 10% of the time (see Fig. 39). Although resonant behaviour will occur, the dynamic response is significantly reduced as the wind loading on the rotor adds damping to the system, which considerably reduces the height of the peak in Fig. 34. The tower top displacement and the total fatigue damage are thus correspondingly reduced. However, this aerodynamic damping is not present when the turbine does not produce energy (i.e. when blades are idling or parked). It has been calculated in [28] that, when compared to a parked turbine, the fatigue life of the Opti-OWECS support structure will be doubled when the turbine is in operation.

Variable speed turbines

Variable speed turbines are equipped with sophisticated controls to keep the system running at optimum speed. Although the variability of the rotation speed narrows the intervals of "safe" natural frequencies for the support structure, the controller can be used to create new intervals. Even though the natural frequency lies in the range of the rotational frequency band, the controller can be programmed to skip rotational frequencies around the natural frequency. This will prevent the rotor from exciting the tower's natural frequency. The tuning of the controller can best be done after installation and measurement of the actual first natural frequency of the support structure, because uncertainties in soil conditions and installation particularities can cause the actual natural frequency to deviate appreciably from the design [29]. This frequency skipping by the control system has been applied successfully at the Utgrunden Wind Farm in Sweden [30].

8. BASIC FATIGUE CONSIDERATIONS

8.1 Introduction

Fatigue is the process of gradual damage done to materials when these are subjected to continually changing stresses. Due to these stress changes, the material slowly deteriorates, initiating cracks which will eventually lead to breaking of the material. Offshore wind turbines are by default subjected to loads varying in time from wind as well as waves. This means that the stress response will also vary continuously, so making offshore wind turbines prone to fatigue. In comparison to another high-fatigue system, the aeroplane, a designer once stated, "during its lifetime, a wind turbine blade receives the equivalent of the loading of a World War II bomber continuously looping 50 m off the ground for 30 years" [31]. Such observations do not only apply to the blades but also to the support structure, only these days it would be an A380 with double wing span 100 m of the ground.

This section gives an overview of the basic principles of fatigue and explains how fatigue is treated in design calculations.

8.2 Fatigue example

When we take an ordinary teaspoon, fatigue can be demonstrated rather easily. By bending the teaspoon back and forth, the spoon will probably break after about 50 repeated bends, as shown in Fig. 40.

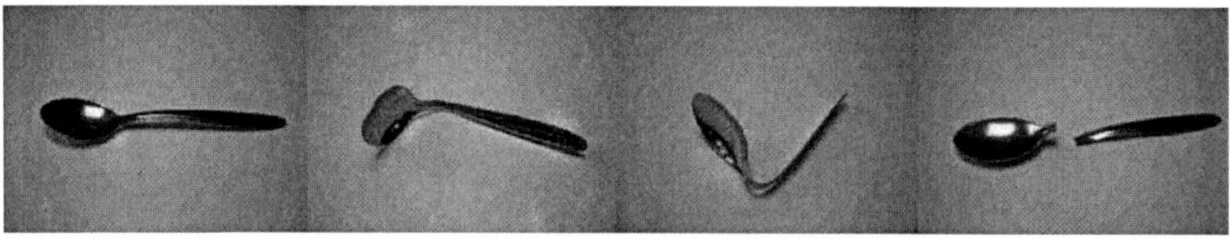

Figure 40. Unbent spoon, bending back and forth until breaking

The breaking is not caused by high tension put on the spoon: it is rather impossible to pull the material to its breaking limit manually. The breaking is caused by fatigue. Every new bending induces small cracks or causes existing cracks to grow. Fig. 41 shows a close-up of the part with small cracks after such bending.

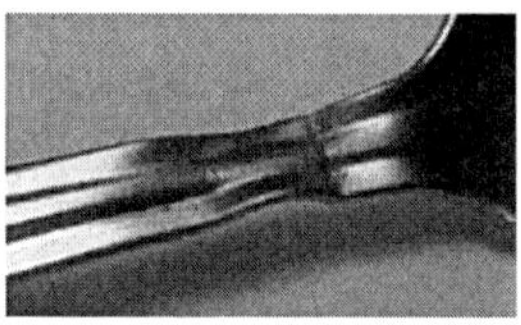

Figure 41. Close-up of bending area showing fatigue cracks appearing

Although this method of demonstration is very clear, fatigue will not only occur when a spoon (or structure) is bent through the plastic limit actually deforming the spoon. Fig. 42 shows the stress-deformation curve of a typical metal. The left hand side shows a linearly increasing deformation with increasing stress following Hooke's law $N_s = E\varepsilon_s$ [32]. When we come to the point marked σ_y for yield stress, the material will start to yield and the deformation increases at more or less the same stress. After releasing the stress, this deformation will remain permanent as in the case of the bent teaspoon. When applying larger stresses than the yield stress, the material generally will be able to endure a larger stress until it breaks.

With the teaspoon, fatigue is induced by continuously loading the spoon beyond the yield stress. With offshore wind turbine support structures, the fatigue process is only acting in the lower end of the linear curve where both deformations and stresses are small. However, when these small stresses are applied a sufficient number of times, fatigue will still occur.

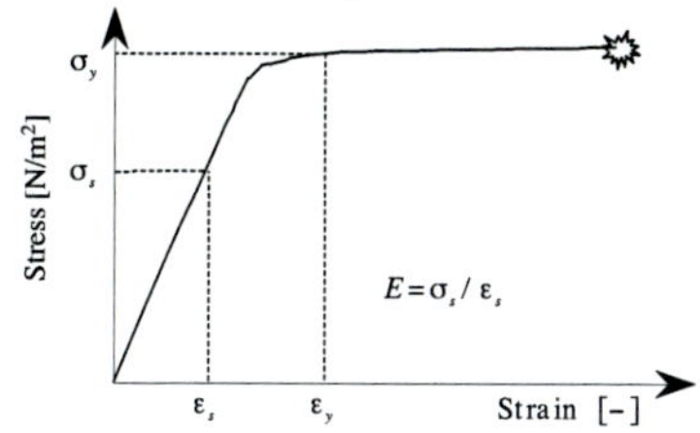

Figure 42. Typical stress-deformation curve for steel

8.3 S-N curve and Miner rule

To be able to consider fatigue in the design process, an empirical design method for the design of steel structures is commonly used. First, for a typical structural detail or a structural connection an *S-N* curve is created. The component is fixed in a testing frame and sinusoidal stress variations are applied to the component. By counting the number *N* of cycles required to fail the component at the applied stress range, a single point on the failure envelope can be determined. For welded specimens, only the stress variation need be considered. The mean stress is unknown and has been shown not to influence the number of cycles to failure to any appreciable degree. Therefore the fatigue loading is expressed by the stress range *S* rather than the stress level.

When the tests are repeated for a large number of identical specimens under different stress ranges *S*, the full failure envelope can be determined. As the empirical data show appreciable scatter, the design *S-N* curve is determined as the mean of the data minus two standard deviations in *N*, as shown in Fig. 43.

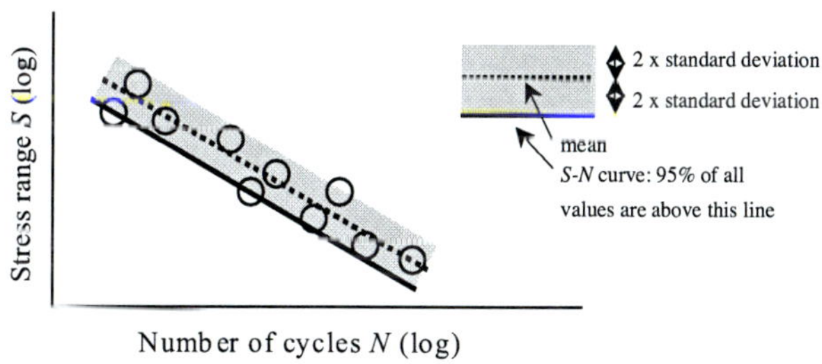

Figure 43. Failure points at stress range S in N number of cycles and the corresponding *S-N* curve

Design standards like those of Germanischer Lloyd and DNV include a number of prescribed *S N* curves for different details. Normal *S-N* curves for steel usually have a branch with slope 3 and for higher cycles a branch with a slope 5 or sometimes 4 (depending on the detail and design standard used), both in the log-log domain. A typical *S-N* curve is shown in Fig. 44. Safety factors can be applied to tailor the generic curve for components in special circumstances such as without corrosion protection or at places where inspection is not possible.

When the S-N curve for the detail under consideration is known, calculations of the stresses that the detail will experience during its lifetime should be performed. When all stress variations are known, they can be binned in number of variations ni per stress range class Si. Taking the associated maximum allowable number of stress variations Ni for each stress range class Si from the S-N curve, the Palmgren-Miner Rule, or just the Miner Rule, can be applied [33]. This rule states that the cumulative fatigue damage Dfat is equal to the sum of ni over Ni for all stress range classes:

$$D_{fat} = \sum_i \frac{n_i}{N_i} \tag{28}$$

The Miner Rule states that the detail will not fail due to fatigue if $D_{fat} < 1.0$.

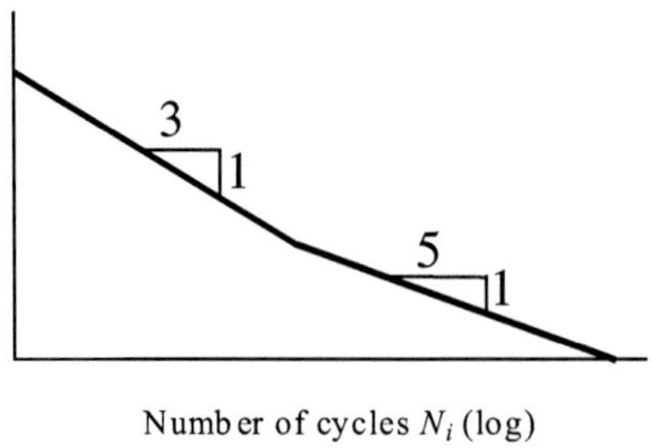

Figure 44. Typical *S-N* curve for a structural detail with slopes 3 and 5 on log-log scales

8.4 Counting methods

The fatigue calculation method for variable stress ranges in the time domain can be summarized by the flowchart in Fig. 45. Calculation of the stresses experienced by the detail being considered under all possible load cases during the lifetime will result in a large number of stress time series. By filtering the number of stress variations for every stress range class, the Miner Sum can be calculated to check whether $D_{fat} < 1.0$.

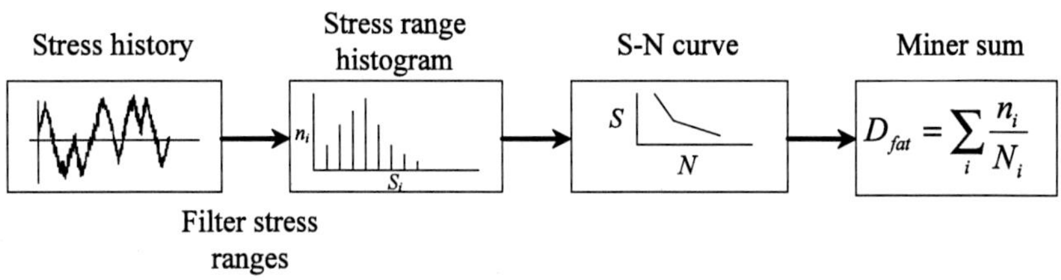

Figure 45. Flowchart of fatigue calculation due to variable stress ranges using *S-N* curve and Miner sum

Not explained in Fig. 45 is how to count the stress range cycles. For this counting, several methods exist. This thesis will not go into detail of the pros and cons of the different methods, but will only discuss methods and their field of application in the time and the frequency domain. The general procedure here is that the method most commonly used for a particular design methodology will be adhered to in this thesis.

Rainflow Counting

The Rainflow Method was invented to consider all peaks without doubling. Rainflow counting [34] [35] received its name because the method resembles rain flowing off a pagoda roof, as shown in Fig. 46. When the stress time series is rotated 90 degrees, the counting algorithm starts:

- at the beginning of the series (point 1)
- at every peak where a previous Rainflow drops away (e.g. points 2, 4, 6,...)
- at every trough (e.g. points 3, 5, 7,...)

Every Rainflow stops:

- at the end of the signal (point 13)
- when it encounters the bottom of a trough (e.g. series 2-3)
- when it encounters a flow from above (e.g. series 3-2')

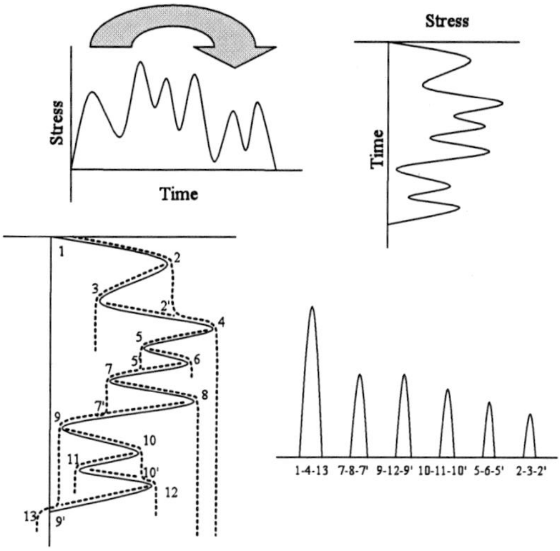

Figure 46. Rainflow counting illustration, with beginnings and endings following the Rainflow Rules

When the method has been executed, the signal is taken apart in a number of half stress range variations: the Rainflow Cycle runs only in one direction each time. The method then combines pairs of half stress cycles: e.g. 1-4 and 4-13 become 1-4-13. The only difference between alternative Rainflow Methods is in the appreciation of the leftovers after combining cycle pairs, but this will not be detailed here. The Rainflow Method has now returned the stress range cycles with number of occurrences which can again be used in combination with the S-N curve to calculate the Miner sum.

9. FOUNDATIONS

9.1 Soil properties

The structure of the seabed is the result of tens of thousands of years of geological and morphological processes. In the North Sea, large parts of the seabed consist of layers of clay and sand. Both soil types are characterised in capacities of loose, medium, dense to very dense for sand and soft, stiff to hard for clay. Fig. 47 shows three offshore locations and the different layers of soil [29].

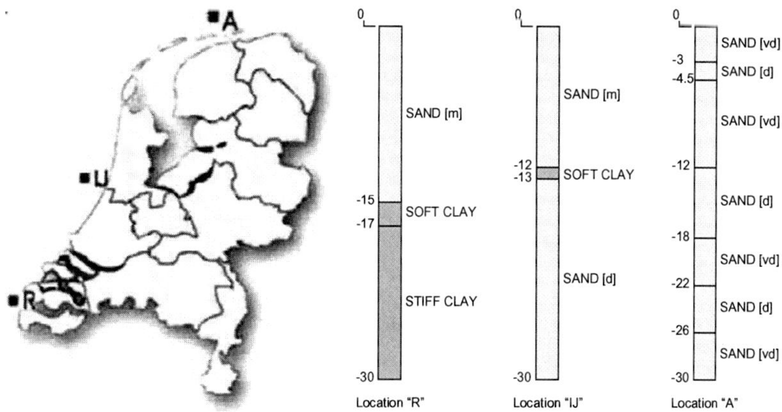

Figure 47. Soil layers for 3 offshore sites with medium [m], dense [d] and very dense [vd] sand layers and soft and stiff clay [29].

The characterisation of loose to dense sand and soft to hard clay only gives a first indication of the ability of the soil to carry load. For design, knowledge that is more detailed is required. This is usually gathered through in-situ sampling and analysis of drilled samples in the laboratory. The first property measured for all types is the density ρ_{soil} [kg/m^3], usually for submerged soil, which is the dry density minus the density of water. A typical value is between 0.4 and 1.0 kg/m^3. For clay, the undrained shear strength su and the strain at 50% of the maximum stress ε_{50} are measured. Table 5 gives an overview of typical values when no reliable soil data is available.

Table 5. Characteristic parameters for clay

Clay type	s_u [kPa]	□$_{50}$ [%]
Soft	0-25	1.5
Firm	25-50	1.5
Stiff	50-100	1.0
very stiff	100-200	0.5
Hard	> 200	0.5

For sand the friction angle ϕ' and the relative density of sand Dr are derived directly from in-situ measurements. The initial modulus of horizontal subgrade reaction, k_s, can then be found with the graph in Fig. 49 [10].

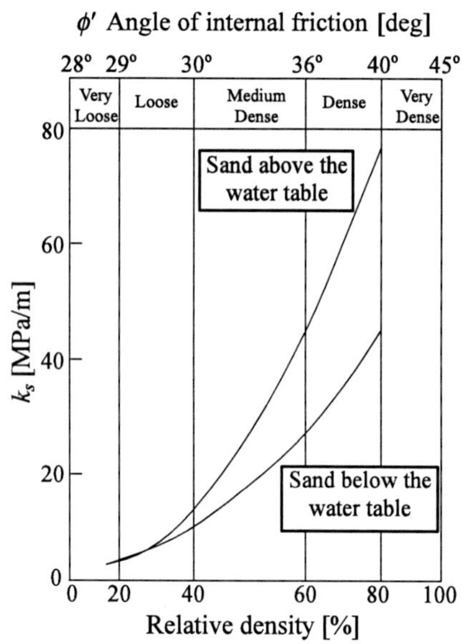

Figure 49. Initial modulus of subgrade reaction k_s as function of friction angle ϕ' [10]

A good standard for soil investigation and characterisation of soil properties at offshore sites is ISO 19901-4 [36], a document in the ISO 19900 series of new offshore structure standards.

9.2 FOUNDATION MODELLING

Transfer of horizontal loads, vertical loads and moments

In the design of offshore support structures, two main directions of load transfer must be analysed. First, the foundation must be able to transfer all vertical loads, the weight of the structure, to the soil.

This is mainly done by *friction:* the soil around the pile takes a small load per area of surface and as long as the load-transfer-area is large enough, the foundation will suffice. Next to the friction on the outside of the pile, the steel rim of the pile and the soil plug inside the pile may also give *vertical bearing capacity*. For monopiles of 4 m diameter and more, the extra pile plug resistance is usually not taken into account due to the large diameter of the pile.

For reducing the effects of *overturning moments*, multi-legged structures mainly rely on characteristics in the vertical direction. Due to the fixation in the frame, the piles of a multi-legged structure deform in an S-shape, mobilising at the same time horizontal soil resistance. The overturning moment is then transferred as axial loads to opposing foundation piles as shown in Fig. 50. For monopiles, all horizontal loads and moments must be transferred directly to horizontal soil reactions, as shown in the right-hand side of Fig. 50. As the pile is not fixed at the top, it is free to rotate and translate. For offshore wind turbines on monopiles, this horizontal load transfer usually dictates the pile length: the pile must be long enough to mobilise enough soil over its length to transfer all loads and prevent "toe kick" (displacement of the tip of the pile).

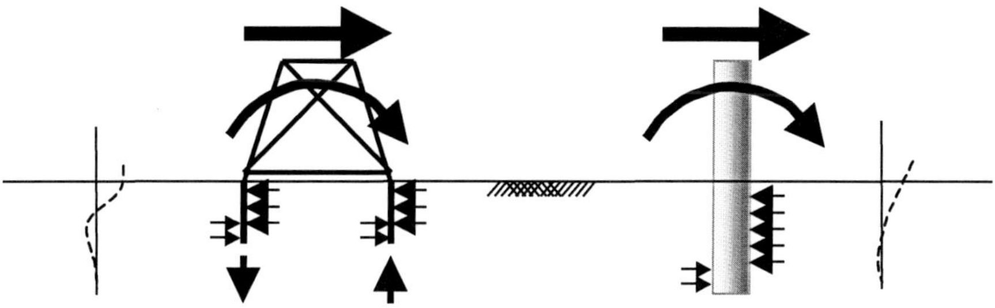

Figure 50. Transfer of horizontal loads and moments in multi- and monopile structures and a schematic representation of the pile deformation

Soil springs

To model the soil reaction loads a set of soil springs is used. Fig. 51 shows the springs for the horizontal and vertical direction as well as for the pile plug [29].

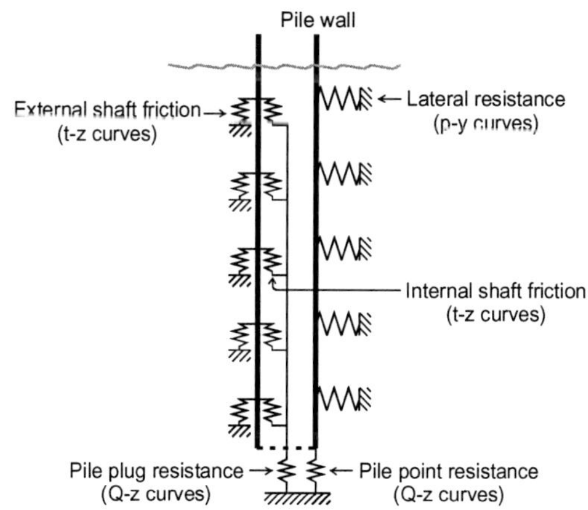

Figure 51. Spring model of pile-soil interaction [29]

All springs in Fig. 51 are non-linear. The properties can be derived from site measurements through calculation methods prescribed in the standards [37] [38]. The typical shape of these curves is shown in Fig. 52. For the first part, left-hand side, the soil reacts linearly and elastically; when the load is released, the soil will return to its original state. Beyond a break point in the curve, deformations will become permanent and the soil starts to lose resistance.

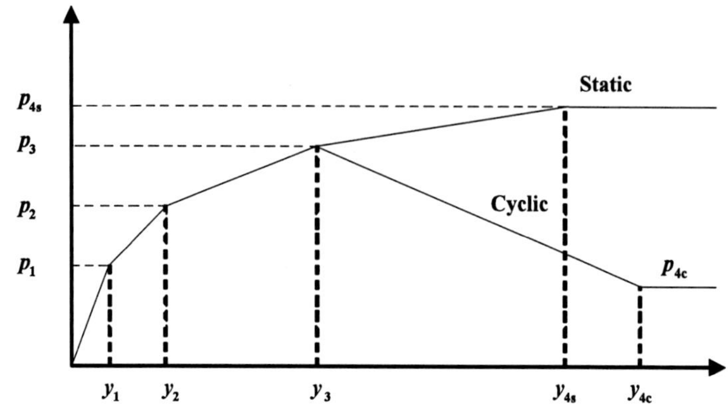

Figure 52. Typical shape of horizontal load *p* and displacement *y* in a *p-y* curve for soft clay under static and cyclic loading according to Matlock [39]

For extreme load cases and foundation design, the full non-linear model must be used. For load calculations of the offshore wind turbine and its support structure, a reduced model can also be used as described next.

Stiffness matrix model for foundation representation

The non-linear spring model can be created in a straightforward way in a finite element program. For fatigue load simulations of the offshore wind turbine however, the full complexity of the non-linear systems is usually not required: most soil reactions remain within the linear elastic range. To reduce calculation time, a stiffness matrix model can be used, which has shown excellent agreement with full non-linear models [29]. The foundation properties are represented by two coupled springs for lateral and rotational reactions as shown in Fig. 53.

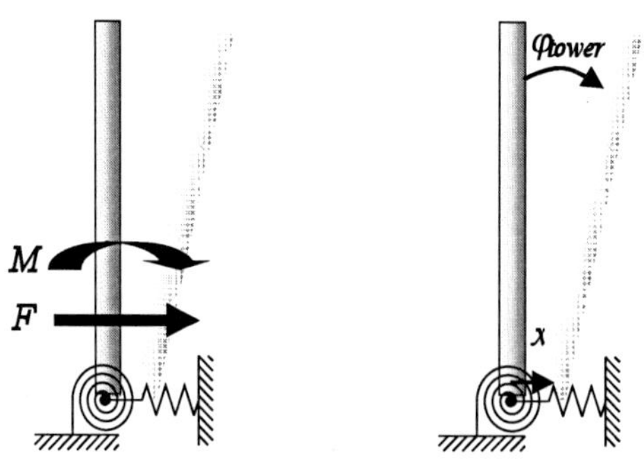

Figure 53. Coupled lateral and rotational springs for foundation representation with loads *F* and *M* and reactions *x* and φ_{towerr}

The spring properties are derived by applying two load sets that are typical for operational conditions of the offshore wind turbine to the non-linear model and using the outcome to derive the spring constants in:

$$\begin{bmatrix} F \\ M \end{bmatrix} = \begin{bmatrix} k_{xx} & k_{\varphi x} \\ k_{x\varphi} & k_{\varphi\varphi} \end{bmatrix} \begin{bmatrix} x \\ \varphi_{tower} \end{bmatrix} \tag{29}$$

Although other models exist, the non-linear p-y model for foundation detailing and the coupled spring model for offshore wind turbine loading proved the most suitable for offshore wind turbine design.

For more details on these and other models and comparison with measurements on offshore wind turbines, the reader is referred to [29] [40].

REFERENCES

[1] Vugts, JH (1995) *Stochastic Processes Lecture notes for X2CT3*, Offshore Engineering, Delft University of Technology

[2] ISO 19901-1 (2005) *Petroleum and natural gas industries – Specific requirements for offshore structures – Part 1: Metocean design and operating considerations* International Organization for Standardization, Geneva, Switzerland

[3] Pierson, WJ, Moskowitz, LA (1964) A proposed spectral form of fully developed wind seas based on the similarity theory of S.A. Kitaigorodshii, *J. Geopys. Res.*, Vol. 69

[4] Hasselmann, et al, (1973) *Measurements of wind wave growth and swell decay during the Joint North Sea Wave Project Deutsche Hydro*. Zeitschr. Riehe, A8

[5] Barltrop, NDP, Adams, AJ (1991) *Dynamics of Fixed Marine Structures*, Butterworth-Heinemann Ltd, Linacre House, Oxford, ISBN 0 7506 1046 8

[6] Wheeler, JD (1970) Method for Calculating Forces Produced by Irregular Waves, *J. of Petroleum Technology*, pp. 359-367

[7] Chakrabarti S.K. (1987) *Hydrodynamics of Offshore Structures* ISBN 0-905451-66-X Computational Mechanics Publications Southampton

[8] Eng-Soon, C et al. (1994) *Laboratory study of plunging wave impacts on vertical cylinders*, Department of Civil Engineering, National University of Singapore

[9] Hagemeijer, PM (1985) *Breaking waves and wave forces in shallow water: a review*, Koninklijke/Shell Exploratie en Produktie Laboratorium

[10] DNV (2004) *Design of offshore wind turbine structures*, Det Norske Veritas, DNV-OS-J101

[11] Ferguson, MC (ed.) et al. (1998) *Opti-OWECS Final Report Vol.4: A typical Design Solution for an Offshore Wind Energy Conversion System*, Institute for Wind Energy, Delft University of Technology

[12] IIMSO for tides, waves etc.

[13] Hoven, I van der (1957) Power spectrum of horizontal wind speed in the frequency range of 0.0007 to 900 cycles per hour, *Journal of Meterology*, 14 pp 160-4

[14] Unknown participant (2004) *Comment during IEA expert meeting on offshore wind energy*, Ringkøbing Denmark

[15] Germanischer Lloyd (2000), *Rules & Guidelines 2000: IV Non-marine Technology – Regulations for the Certification of (Offshore)*, Wind Energy Conversion Systems

[16] Burton, T et al., (2001) *Wind Energy Handbook*, John Wiley & Sons Ltd, England, ISBN 0-471-48997-2

[17] IEC (1999) *Wind turbine generator systems – Part 1: Safety requirements International standard 61400-1*, International Electrotechnical Commission

[18] DS472 (1992) *Code of practice for loads and safety of wind turbine constructions*, Danish Society of Engineers and the Federation of Engineers

[19] www.waveclimate.com

[20] Wieringa, J (1973) Gust factor over open water and built-up country, *Boundary layer Meterology*, 3 pp 424-441

[21] Matthies, HG, et al (1995) *Study of Offshore Wind Energy in the EC*, Joule I (JOUR 0072) Verlag Natürliche Energie, Brekendorf

[22] Manwell, JF, McGowan, JG, Rogers, AL (2002) *Wind energy explained,* John Wiley & Sons Ltd, Baffins Lane, Chichester, West Sussex PO19 1UD Dengland, ISBN 0-470-84612-7

[23] Lysen, EH (1983) *Introduction to Wind Energy*, SWD Publications SWD 82-1, The Netherlands

[24] Betz, A (1920) Das maximum der theoretisch möglichen Auswendung des Windes durch Windmotoren, *Zeitschrift für gesamte Turbinewesen*, vol. 26, p 307

[25] Vestas (2003) V80-2MW *OptiSpeedTM Offshore Wind Turbine,* leaflet on www.vestas.dk

[26] Molenaar, D-P, Dijkstra Sj (1999) Modelling the structural dynamics of flexible wind turbines. In *Proceedings of European Wind Energy Conference and Exhibition*, Acropolis Convention Centre, Nice, France

[27] Vugts, JH (2000) *Considerations on the dynamics of support structures for an OWEC,* Section Offshore Technology, Delft University of Technology

[28] Tempel, J van der (2000) *Lifetime Fatigue of an Offshore wind Turbine Support Structure Section,* Offshore Technology & Section Wind Energy, Delft University of Technology

[29] Zaaijer, MB (2000) *Sensitivity analysis for foundations of offshore wind turbines*, Section Wind Energy, WE 02181, Delft

[30] Kühn, M (2001) *Dynamics and Design Optimisation of Offshore Wind Energy Conversion Systems,* Institute for Wind Energy, Delft University of Technology ISBN 90-76468-07-9

[31] Divone, L (1984) from: Jos Beurskens' personal archives

[32] Verburg, WH (1996) (Over) *Spannend Staal Overspannend,* Staal – Rotterdam, ISBN 90-72830-18-0

[33] Miner, MA (1945) Cumulative damage in fatigue Trans. ASME, *Journal of Applied Mechanics*, Vol. 67, pp. A159-A164, 1945.

[34] Matsuiski, M, Endo, T (1969) *Fatigue of metals subjected to varying stress*, Japan Soc. Mech. Eng.

[35] Dijk, GM van, Jonge, JB de (1975) *Introduction to a fighter aircraft loading standard for fatigue evaluation – FALSTAFF*, National Aerospace Laboratory, NLR, MP75017U, Amsterdam, Holland

[36] ISO 19901-4 (2003) *Petroleum and natural gas industries specific requirements for offshore structures – Part 4: geotechnical and foundation design considerations,* Issued 2003-08-01 by the International Standards Organization

[37] API (2000) *Recommended Practice for Planning, Design and Constrcucting Fixed Offshore Platforms – Working Stress,* Design American Petroleum Institute, 21st edition

[38] ISO 19902 DIS (2004) *Petroleum and natural gas industries specific requirements for fixed offshore structures,* Issued 2004-09-30 by the International Standards Organization

[39] Matlock, H (1970) Correlations for Design of Laterally Loaded Piles in Soft Clay, Paper number OTC 1204, *Proceedings Second Annual Offshore Technology Conference*, Houston, Texas, USA, pp. 577-594

[40] Shaw, VK (2004) Foundation Model and Design for Offshore Wind Turbine Monopiles, *Offshore Engineering*, Delft University of Technology.

Chapter 9:

Offshore Wind Energy in Deep Waters

Andrew R. Henderson

1. INTRODUCTION

With seabed-mounted wind turbines slowly becoming a common feature across the shallower seas of Northern Europe, what are the prospects for wind power in deeper waters, both there and elsewhere in the world? In many regions, potential sites for bottom-mounted windfarms are limited, and the few that exist may suffer from conflicts with other users, such as tourism and fishing. In addition, the costs of bottom-mounted windfarms away from the shallow enclosed seas of eastern Denmark are currently proving to be more costly than previously predicted. This is due to numerous factors, of which many are not specific to offshore wind energy, but the recent history of numerous difficulties with the construction or operation of offshore windfarms has caused the contractors and technology suppliers to add a significant risk premium to the costs to cover the expected problems. However, when the need comes for wind turbines in deeper waters, floating structures will have to be considered and the flexibility in choice of concepts should result in costs that are competitive with bottom-mounted turbines in shallower waters.

This chapter reviews the history of floating offshore wind energy, which to date consists of theoretical studies only. It examines the key challenges and benefits, and attempts to identify the most promising candidate support structures. The general dynamic condition of a floating and operating wind turbine is shown in Fig 1.

2. HISTORY

It is widely acknowledged that wind energy pioneers generated the first electricity over a hundred years ago, however it is less well known that the first proposals and working model for using multiple wind turbines on a floating structure were made not long afterwards in the 1920s [6] [21], Fig 2, though this did not involve the generation of electricity. The vision for modern large-scale offshore wind plants is credited to the late Bill Heronemus [11] [24], an ex-navy engineer and a professor at the University of Massachusetts, who in the early 1970's proposed arrays of wind turbines supported on floating buoys, Fig 3.

Starting in the early 1990s, a number of research projects focused on developing and evaluating various floating windfarm concepts, as briefly summarised below:

- In the United Kingdom, Garrad Hassan and Technomare co-operated in the evaluation of *FLOAT,* a *single turbine spar-buoy concept,* kept in position using eight-point catenary moorings [25]. This was a detailed study, investigating type of wind-turbine (downwind, free-yawing with very-high tip-speed), multiple vs. single turbine structures, sharing of anchoring systems and tower design (lattice type to reduce wind loads and overturning moment). The costs were estimated to be prohibitively expensive, at around twice that of bottom-mounted alternatives. This was primarily due to the conservative design requirements applied for the floater.

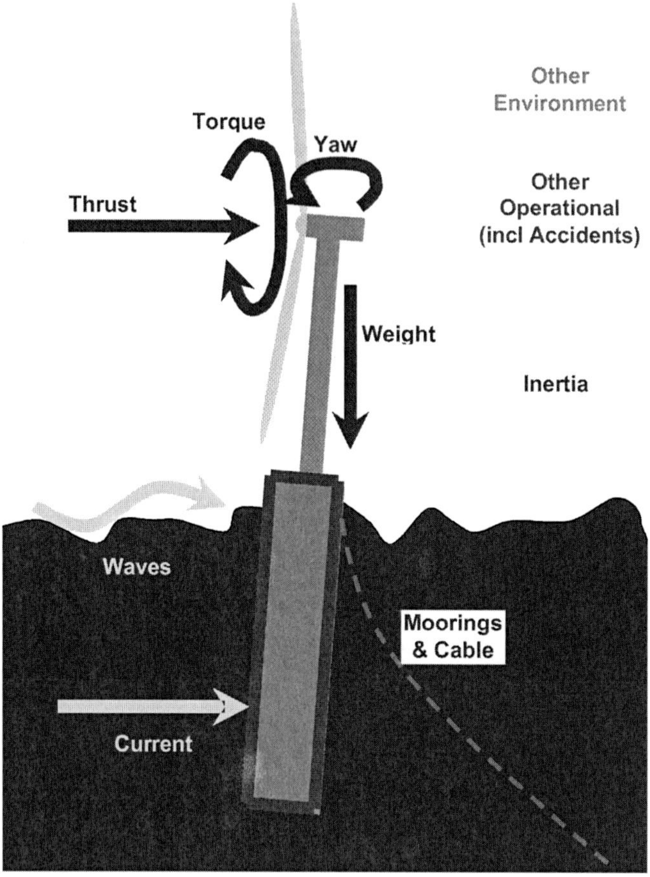

Figure 1. Loads on a Floating Wind Turbine

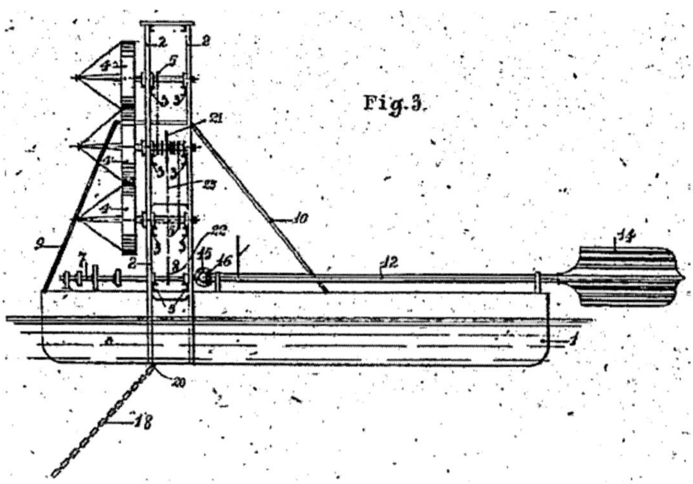

Figure 2. Camille Durey's wind-assisted ship

Figure 3. Heronemus's 'Windships'

- Also in the United Kingdom, a group at University College London investigated the possibilities of locating *several turbines on a single structure, MUFOW*. This had the potential advantages of reduced motion response and shared moorings (hence reduced anchoring costs). The concept was investigated within a PhD programme [7] and an EPSRC research project [9]. The main conclusions were that the MUFOW would be extremely large, in the order of several hundred meters wide and that, since the structures must withstand the extreme wave loads found in regions with an attractive wind resource, it would be very expensive and difficult to design and construct. This was a direct consequence of an increase in turbine size between the development of the original idea and the completion of the research.
- In Italy, a group in Milan investigated placing *individual turbines, each on a toroidal-shaped float* and positioned with tensioned moorings, Eolomar. A complex shape was chosen to minimise wave motion response, but this had the disadvantage of being difficult and expensive to build [3].
- In Japan, the JOIA (Japanese Ocean Industries Association) co-ordinated a group to evaluate the potential for floating wind energy. The first phase was completed in 2001 [15] and a second phase, which included detailed analyses and tank tests was completed in 2003 [26]. Following a reorganisation of Japanese government departments, subsequent research is now being co-ordinated by NMRI (National Maritime Research Institute). The research group's attention is being focused on spars, twin-turbine space-frames and large *multi-turbine floaters* with modular interlocking hexagons [14].
- In Ireland and the United Kingdom, an offshore wind-energy partnership of Ocean Synergy and Sure Engineering, proposed a tensioned *submerged four-floater* concept. This would use gravity anchors and a telescopic tower to maintain stability prior to installation and to avoid the use of expensive installation cranes. Detailed designs and wave-tank tests have been undertaken, with each sub-system constructed in steel or concrete, whichever was more cost-effective [8],
- At Marseille Engineering University (Ecole Supérieure d'Ingénieurs de Marseille) a *tensioned three-leg star TLP* (Tensioned Leg Platform) was investigated, including detailed wave tank tests during 2001-2002 when an attempt was made to model the wind turbine as realistically as possibly [17]

- Massachusetts Institute of Technology (MIT) investigated dynamic response simulations to evaluate concepts for floating offshore wind turbines, for deployment and cost effective electricity generation in water depths between 50 and 200 meters. A tensioned *three-leg star-TLP* (similar to that at Marseille) and a taut-leg buoy will be examined [23]. In collaboration with researchers at the US National Renewable Energy Laboratory, this work integrates existing wind turbine and offshore engineering models to allow fully-coupled analyses for the first time.
- A US DOE funded concept study, to evaluate a *semi-submersible platform* and anchor foundation system capable of supporting a 5 MW turbine, was undertaken by Concept Marine Associates of Long Beach, California. A *positively buoyant concrete platform* would be held in place by moorings, to stay beneath the surface in deep water, so avoiding the most intense wave action. Two primary mooring techniques are included in the study: (i) tension legs connected to a large concrete dead-weight/suction anchor, and (ii) conventional catenary anchors [19].
- In Germany, the engineering consultancy Arcadis, has developed a floating concept suitable for the relatively shallow waters in the Baltic Sea. It is likely that the motivation for a floating support structure was the soft seabed conditions found in many locations in that sea [1] [2]. There has been little publicity, in particular about the technology, but, at such shallow depths, catenary mooring will be difficult and hence some taut or tensioned design is likely, possibly a steel main structure tied down to concrete gravity anchoring;
- In Norway, the supply of electricity from floating wind turbines to offshore oil and gas platforms is being investigated by SWAY [4]. The deep waters require technical solutions very different from those investigated elsewhere become viable, e.g. a deep spar tethered to the seabed either via a single tensioned line or via direct fixing to the seabed with a universal joint [4];
- Finally, also in Norway, the Hydro organisation is proposing a deep spar design, which has the advantages of a simple geometry and reasonably small motion response to the waves. The Hydro concept, HyWind, has a number of interesting aspects but is fundamentally based on proven concepts and technology and hence has greatest prospects of success in the medium term [12] [20].

3. BENEFITS OF DEEPWATER WINDFARMS

Before examining potential floating support structures in detail, consider the main arguments for developing bottom-mounted wind turbines:

i. proven technology for *fixing the structure* to the seabed, using designs and techniques previously developed for the offshore oil and gas industries and for coastal engineering
ii. relatively straightforward offshore *electrical infrastructure*, which can be built using existing types of cables and installation plant,
iii. only small induced motion due to waves, wind and current, the *motion response* being of similar magnitude to onshore turbines,
iv. ability to measure *wind-resource* accurately, using a bottom-mounted hub-height anemometer mast at the project site, though this is currently expensive.

Of course, there are also major challenges for the engineering of bottom-mounted wind farms:

i. even in shallow waters, *costs* increase exponentially with water depth, as more steel and inherently more expensive support structures are needed
ii. Expensive and demanding *installation and access* equipment and procedures needed
iii. removal is difficult and costly, though with such *decommissioning* being far in the future, these costs are significantly discounted

The main advantages for bottom-mounted installations, identified above, apply similarly to floating support structures, namely:

i. Floating support structures would also be based on *proven technology*, there being a wealth of innovative concepts developed and tested by the offshore engineering industry. Cost-effective design and experience applies equally to bottom-mounted as to floating concepts.

ii. The electrical infrastructure has been troublesome for several offshore windfarms, both for the installation as well as the operation. There is no single reason for this. The subsea power cables crossing the Baltic Sea and connecting Scotland and Ireland, have performed well. The use of floating support structures would introduce an additional source of potential failure in the flexible cable used to connect the turbine to the seabed; however, experience of such cabling is being gained within the offshore oil and gas industry and the wave energy industries. Flexible cabling should not be a critical handicap for floating wind-turbines but will require careful attention

iii. Most wind turbine loads are not increased significantly by taking the turbine offshore and locating it on either a fixed or floating support structure. However, this may not be so if the floating support structure experiences excessively motion, particularly in roll or pitch. The design and modelling of new wind turbines relies on previous knowledge and experience; hence, performance data from prototypes will be needed to calibrate and evaluate the computer models.

iv. A floating hub-height meteorological mast is unlikely to be practicable for wind resource measurement, which could assess floating windfarms difficult for third-party financing. Nevertheless, bottom-mounted windfarms have been so financed without previous wind-speed measurements in the immediate area since wind speeds vary less offshore than onshore over short distances. Moreover, there is increased confidence in LIDAR (Light Detection and Ranging; measuring wind speeds by backscattering a laser beam from particles in the air) and SODAR (Sonic Detection and Ranging; measuring wind speeds by backscattering sound waves from particles in the air) systems for reliable source of wind resource measurements without the need for towers.

It is accepted that offshore wind energy is and will remain more expensive than onshore wind energy for the foreseeable future. However, since the capacity of all landscapes to accept windfarms is finite, the real question is how offshore wind energy will compare with other sources of electricity. Several European countries now plan for offshore wind energy being a major source of electricity in the future. Given this situation, it is reasonable to compare bottom-mounted offshore windfarms with floating windfarms:

i. Having a floating support structure allows most of the installation and commissioning to be carried out at the quayside and so such expensive on-site work of bottom-mounted turbines may be reduced.

ii. Access to operational turbines is likely to be similarly challenging as to bottom-mounted turbines, with transferring a person with lightweight equipment to a floating turbine being marginally more difficult. However the floating wind turbine could be returned to port for major repairs

iii. Decommissioning a floating windfarm will be considerably easier than for fixed structures.

Floating offshore wind energy offers a number of other potential benefits, such as:

- The wind resource generally increases with distance from shore, with this benefit increasing for a considerable distance, albeit at a decreasing rate. Nevertheless, to date offshore windfarms have been built relatively close to shore for reasons of economics and access. The economic reason is primarily due to the costs of support structures, which increase significantly with increases in depth and to a lesser extent due to electrical infrastructure costs. A floating wind turbine would not suffer from increased support structure costs, hence the economic benefit of increased wind resources of sites more distant from shore may compensate for the increased electrical connection costs.

- Commissioning of the wind turbine at the quayside, apart from the final cable connection
- Seabed conditions are less important for floating structures than bottom-mounted structures, because smaller loads are transmitted in to the foundations (the loads being carried by the water instead). Anchors are available for all types of seabed, costing a small portion of the total investment. The seabed conditions have a minor impact on the total cost of energy, in contrast with bottom-mounted offshore windfarms.
- Floating turbines must be towed by tugs. The reduced stability of the unmoored structures will require towing floaters in mild sea states.
- Floater design is independent of water depth beyond a minimum value.

Impact on the Wind turbine

Wind loads on a turbine rotor can be extremely large and dominate the design of tower-top components of all forms of offshore wind turbines. However, loads on the tower itself and the support structure, are also influenced by wave loads, which become more important with water depth and severity of sea climate.

Wind turbines mounted on floating support structures may have additional loads due to (i) the wave-induced motion, and (ii) blade gravity forces changing direction relative to the turbine bearings, as the latter oscillate by wave motion. However, analysis has shown that the additional wave-induced loads on a floating wind turbine are not significantly different from the fixed case [10]. Therefore, the wind turbine itself should not be unduly affected, especially since deep-sea waves have long cyclic periods. Nevertheless, the support structure itself may need to be strengthened, since when the turbine is not vertical, the nacelle weight imposes a bending moment on the tower and floater. Of greater concern may be secondary design aspects, e.g. lubrication systems designed for a permanently vertical orientation. In general, careful design should overcome all such difficulties.

With single-turbine support-structures, the largest possible wind turbine should be the most economic. This applies equally for bottom-mounted as for floating support structures, and hence larger wind turbines can be expected in the future, the rate of development being set by the experience and confidence of the designers and the rate of expansion of the offshore wind market.

Minimising wind turbine weight is even more beneficial for floating wind turbines than for offshore bottom-mounted turbines.

4. DESIGN DRIVERS FOR CONCEPT DESIGN

A wind turbine floating support-structure must perform several functions, so the following design-drivers are essential:

i. Maintain the turbine rotor sufficiently high above the sea.
ii. Maintain position within the range required for the power cable.
iii. Counteract the turbine's thrust, torque and yaw loads
iv. Provide a sufficiently stable base for the wind turbine, i.e. to counteract the wave and sea-current loads

The first two requirements are the most straightforward and so the main part of this section will concentrate on the final two requirements. Usually, loads will fluctuate considerably around a non-zero mean and so we consider the mean and dynamic components separately.

For a dynamic system, the response to the mean component is determined only by the stiffness of the system, and the dynamic component is determined by the stiffness, damping and mass together. In theory, it should be possible to tune the properties of the system to respond optimally to the loads, although in practice other design requirements force compromises.

In general, both the mean and dynamic components of turbine-induced loads are significant. In contrast, marine-induced loads are smaller; the dynamic component is dominated by waves and the mean component by sea currents, Fig 1. Soft mooring systems, such as catenaries, provide a reactive force only against these smaller mean loads and allow structures to respond naturally to the larger oscillatory wave loads; this means that a relatively low-load design specification, and hence cost, will be acceptable for such moorings.

Depending on the type of mooring, a floating body moves as a rigid body in up to six degrees of freedom, sway, pitch, roll, surge, heave and yaw, Figure 4. This complicates the analysis, since coupling generally occurs between some of the directions. .

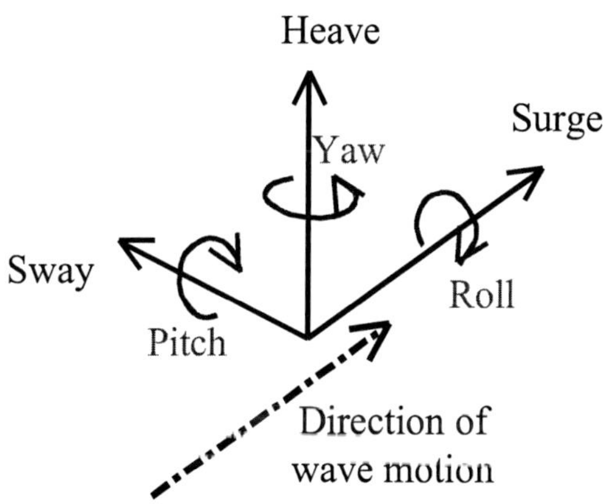

Figure 4. Rigid Motion Degrees of Freedom

The problem requires a solution that satisfies all design drivers and is economically viable. Since offshore wind energy is now only marginally competitive, the economic requirement is challenging and requires compromise in technical design. Returning to the design requirements listed at the start of this sub-section, these can be achieved by:

1. maintaining ***nacelle height***; which is achieved by mainly by buoyancy.
2. ***maintaining station***, as achieved by the moorings.
3. ***resisting the steady component of the turbine thrust***, which can be achieved via any one or combination of the following three ways:
   - *Stiffness from the water plane area*, for example the tri-floater, Figure 9
   - *Stiffness from the moorings*, if taut or tensioned moorings are used, Figure 13
   - *Stiffness due to a large mass deep below the water-line* creating a pendulum effect, for example a spar-buoy, Figure 8.

However each of these methods causes problems for other design objectives, for example:

- *large hydrostatic stiffness* results in short period natural oscillations, causing the structure to resonate with the wave motion. It also requires a large structure at the water plane, thus increasing the total wave loads and large structures tend to be expensive
- *taut moorings* are a relatively innovative and expensive and require elaborate installation procedures

- *increased mass increases costs*, both in terms of the material and increased structural design specification needed for the support of the greater mass

These compromises are illustrated in Figure 5.

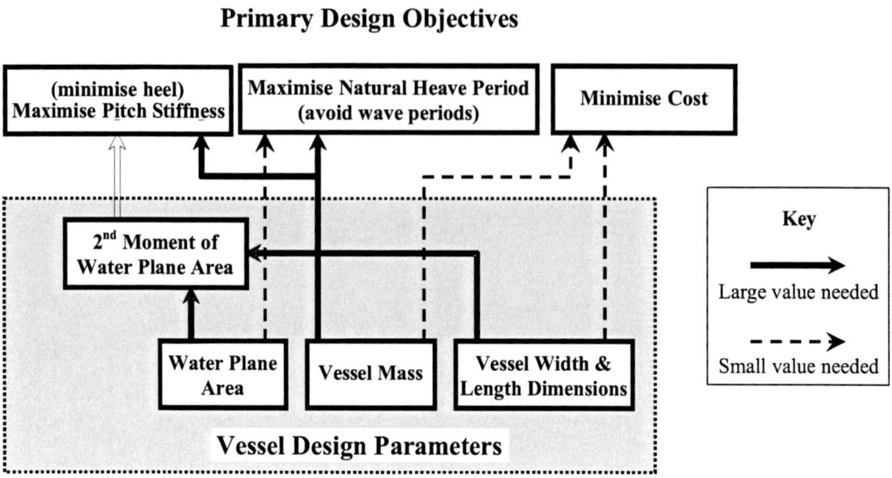

Figure 5. Schematic Design Process for a Spar Buoy

Minimizing the motion due to the wave loads, can be achieved using conventional and less conventional methods. Conventional methods are:

- Reduce the area of the structure projected against the waves near the water-line (where wave loads are largest. Therefore, the main structure either should be above the highest waves or deeply submerged.
- Tune the natural frequencies of the structure away from the wave-dominant frequencies to avoid resonances.
- Increase the vessel inertia.

Less conventional methods include:

- Damping plates, which increase added mass, increase damping, adjust natural frequencies. For example, see Figure 9,
- Vortex-suppression strakes provide a certain amount of damping; strakes are flat plates projecting from to the spar surface, with their main purpose being to minimise vortex shedding under certain conditions when sea currents are strong. (However this is not necessary everywhere.)
- Buoyancy cans at the water-line provide damping and actively tune the natural frequencies; however there is limited experience of this method to date,
- Continuously adjusting the blade pitch can provide damping, with the penalty of increased rotor loads but the possible benefit of more energy production.

The final two design requirements (3 and 4) are complex and require a comprehensive understanding and analysis of the complete system. The challenge is to find an acceptable compromise since the design requirements are contradictorily in places, as illustrated by Fig. 5..

4.1 Wind Turbine Configurations

Early studies into floating offshore wind energy generally concluded that multi-turbine floaters would result in the cheapest energy, primarily because only relatively small wind turbines were available then.

With multi-megawatt wind turbines now in use, the strategy of increasing the production capacity per platform, to reduce the dynamic cable and in-farm electrical infrastructure costs, can be achieved with a single turbine. For example, the early configuration of the MUFOW concept assumed eight 50 m diameter wind turbines, which might typically have a total capacity of around 8 MW on a 400 m long structure. That total capacity was not much more than the 5 MW rated capacity of today's largest individual commercial wind turbines. . The difference is certainly not worth the additional costs of large and complicated structures needed for multiple turbines.

There are a number of other issues that affect multi-turbine floating structures and that have not yet been resolved:

- Structural dynamics of the supporting frame,
- Aerodynamic interaction between closely positioned rotors,
- The physical size of a structure that could support a number of wind turbines, even if they are smaller than the largest currently available turbines,
- Imbalance of thrust loads, during normal operations and in particular if one turbine is not operating,
- Effect on power production if the vessel needs to return to harbour for repairs to a single rotor.

The wind turbine rotor must be able to rotate into the wind, requiring (i) a rotating joint, and (ii) a reaction force from the moorings and the rest of the structure. The rotating bearing is at the nacelle for present wind turbines, however for a floating wind turbine, this could be within the main support structure, at the mooring connection (i.e. a turret mooring system) or possibly at the seabed. The availability of proven technology and the smaller thrust-induced bending moment at the nacelle, means that nacelle rotation should continue to be the common solution, although other solutions continue to be investigated.

Both upwind and downwind rotors are stable when approximately aligned with the wind direction, since they experience a small restoring yaw force that corrects any minor misalignments. However, this effect is stronger for downwind rotors, which also have other advantages, such as allowing greater flexibility in the blades, since the risk of tower strikes is reduced. However, downwind rotors experience a more turbulent wind-field, due mainly to the turbulence shed by the tower, and hence have not been developed commercially at large scale. The tower loads could be reduced with a 'weather-vaning' floater (i.e. the floater swings downwind from its mooring), using an aerodynamically profiled tower instead of a circular tubular or lattice tower.

A weather-vaning floater requires no yaw-mechanism, for example, Figure 12, but the resultant forces from the waves and current are not necessarily in the same direction as the wind, so the turbine may not always be properly aligned with the wind. For multi-turbine floaters, active misalignment of the outermost rotors may be able to control the floater alignment to the wind. Another problem with the weather-vaning concept is the delivery of generated power to the grid; a rotating and watertight electrical connector is required. Such connectors exist in the offshore industry but are complex and expensive equipment.

4.2 Materials and Component Technologies

As onshore, the main construction material for the support structure of bottom-mounted offshore windfarms may be *steel or concrete*. Concrete is lighter, cheaper and more durable than steel, but weaker. It is also possible that a combination of materials would be used, for example, concrete for the base of a spar-buoy to provide additional ballast at a less cost, and steel for the mid-sections of the spar, where loads are largest.

There are two principal types of *moorings*:

- *Catenary moorings*, which rely on the weight of the cable to provide the restoring force. These are the simplest type of moorings and are used by the Spar Buoy, Figure 8; these moorings only provide station-keeping, in that they react against turbine-thrust, wave-drift and ocean current forces. However, response to the dynamic rotor and wave loads must also be provided, for example by inertia in the case of the deep spar buoy or by hydrostatic stiffness in the case of the tri-floater. Three mooring lines are sufficient to maintain the location, however six or more are typically used for oil and gas structures. The positioning of a floating turbine need only to be as accurate as required by the dynamic cable, as opposed to the more stringent requirements of oil or gas-carrying riser pipelines. On the other hand, eight mooring lines do provide redundancy, so station-keeping would be maintained even if one line fails; vessels are said to be slack-moored.
- *Tensioned moorings* rely on tension within the cables to maintain the floaters position. These, provide the most stable platform, since heave, roll and pitch motion is minimal. For example, such moorings are used by the Star-TLP, Figure 14. Usually, a tensioned mooring is exceedingly stiff against vertical loads; while horizontal motion will be limited by the buoyancy forces generated as the vessel is pulled down in to the water. The tendons can be made of a variety of materials, including solid steel tubes, steel wire and synthetic fibre; vessels are said to be taut moored.

In practice, there is a continuum of variation between catenary and tensioned moorings. However, the distinguishing feature of a true catenary mooring, is that the end section should always lie horizontally on the seabed and, in theory, the anchor should only carry horizontal loads, since inclined loads could pull out the anchor.

There are several principal types of *anchoring,* the choice depending mainly on the seabed conditions and the type of mooring. It is particularly important to choose the correct anchor for tensioned moorings, because of the large vertical loads:

- *embedded anchors*: which are the most common type, including traditional ships' anchors. The modern designs have progressed, so the anchors become embedded by drag or suction, and can be designed to carry large vertical component loads,
- *piled anchors*: of drilled or driven piles, of a much smaller diameter than the foundation-piles used for bottom-mounted wind turbines
- *suction anchors*; this is a relatively novel technology and is potentially suitable for sand or clay, but not rock, and has been used with tensioned moorings, e.g. for the Snorre TLP (tensioned leg platform)
- *gravity anchoring*, which basically is a heavy weight, for example a concrete form containing ballast, lying on the seabed

Recently, transmission *cabling* in offshore windfarms has given difficulty, therefore flexible dynamic cables connecting a floater to the seabed will be a cause of further concern . However, the wave energy industry will be using this technology and shore-to-platform power supplies are now utilized by the offshore oil and gas industry. Ideally, subsea switchgear and, possibly, transformers would be used for a deepwater floating windfarm, since exporting the whole windfarms output power over *flexible* high-voltage high-capacity cables is not attractive. Alternatively, it may be more cost-effective to build the offshore sub-station on a fixed-structure, even in deeper waters, since this would involve only a single sea-bed mounted structure for the whole windfarm.

5. CANDIDATE CONCEPTS

As described above, there are two principal forms of moorings: slack or taut, so the candidate floating support structures can be divided into two groups accordingly. In addition, bottom mounted structures can be considered.

5.1 Slack Moored

5.1.1 Disc Buoy

This type of floater is the simplest possible, a round disc floating on the surface of the water, Figure 6. Stability is achieved through the water plane area with catenary moorings, which counter the mean turbine and wave drift loads.

However, calculations show that the disc buoy needs to be large to ensure basic stability, in the order of 40 m diameter [5]. The combination of a large area projected against the waves and a low-frequency natural periodic motion in both heave and roll/pitch, of around 9s and 13s respectively, makes the structure susceptible to wave-induced motion. Hence, this is clearly not a viable design and neither are variants, such as adding a circular skirt to increase the effective ballast.

Derivatives of this concept include barges, ships and catamarans and the conclusions in those cases would be similar. Exception might occur if the barge is required primarily for another purpose, for example desalination or hydrogen generation plant and the wave climate is particularly benign.

Figure 6. Disc Floater

Figure 7. Spar Floater

5.1.2 Spar Buoy

If sites with deeper water are examined, the depth of the disc can be increased and a spar floater, Figure 7, can be utilised. Spar floaters have been used in the offshore engineering industry for several decades, but with dimensions greatly in excess of what would be necessary and economic for a wind turbine. The spar consists of a long vertical submerged cylinder, with ballast at the base for stability. Three mooring lines would be sufficient for station-keeping and for attaching the dynamic electrical export cable.

As the length of the cylinder is increased, the diameter can be reduced, with the roll/pitch stiffness that counteracts the turbine-thrust over-turning moment, remaining constant. Reducing the diameter of the cylinder (i) decreases the area projected against the waves and hence reduces wave loads; and (ii) increases the natural period of oscillation, away from resonance with waves, so reducing the tendency of the spar to respond to the waves.

However, excessive motion-response will remain a difficulty, so requiring one or more of:

- Stiffer mooring lines,
- Heave suppression disc, at the base of the spar, shown in Figure 7, which increases the added mass and the damping (this was proposed for the FLOAT project),
- Vortex-induced vibration (VIV) strakes, which counter such vibrations but also increase the damping,
- Open-bottom damping 'cans', located at the water-plane, which would provide tuneable motion damping; this is a relatively untried solution as yet.
- All these appendages would make handling (supporting and manoeuvring) the structure more difficult.

5.1.3 Deep Spar Buoy

If the water is deep enough, a longer spar can be utilised, Figure 8, which improves performance further. The spar diameter can be reduced, because stability is provided primarily by ballast at the base of the column. Although motion response becomes less of an issue, structural integrity, cost and installation procedures will become more challenging and hence, although deep waters are the most attractive locations for spar structures, challenges remain.

Starting with this simplest of shapes, the uniform cylinder, as installed in the Brent oilfield in 1976, further types of spar are possible, such as:

- narrow-necked spar, with a smaller diameter section at the waterline, experiences reduced wave loads and improved natural frequencies
- the truss spar, which uses less steel, has increased motion-damping and is less sensitive to VIV (Vortex Induced Vibrations),
- the cell spar, which consists of several small-diameter parallel cylinders, bundled together.

Efforts to develop the spar design continue. Because of its relatively simple basic shape, construction costs should have a similar cost lead over jacket and TLP structures, as monopiles do over tripods in the bottom-mounted offshore sector. This advantage will reduce as complexity and appendages are added.

Figure 8. Deep Spar

5.1.4 Tri-floater

With the spar concept, the part of the structure that passes through the water surface consists of a single member. Since, roll and pitch stiffness increases with distance of the water plane area from the axes of roll or pitch, the obvious approach is to separate the single surface-piercing member into a minimum of three members, and spread them out. This is the basis of the tri-floater, Figure 9. The members can be reduced in size, although some of the cost savings would be lost to the material and additional manual work needed for manufacturing and assembling the bracing members.

To reduce the response to vertical wave motion, heave-suppression discs can be placed at the bottom of each floater, as for the spar floater. The natural frequencies are likely to be ~ 15s [5], hence motion response will be an issue.

This concept was concluded to be the most promising floater for 50m water depths in the Dutch sector of the North Sea by the DrijfWind project [5]. It was felt that, within the near-term timescale being considered, the taut moorings were insufficiently mature and hence likely to prove prohibitively expensive. However, this is a conclusion that should change over time.

Figure 9. Tri-Floater [5]

5.1.5 Multi-Floater Variants

There are numerous possible variants of the multi-floater concept, with the DrifWind project [5] identifying the following two as being most promising, but likely to be more expensive than the basic tri-floater:

- the wind turbine could also be positioned on top of one of the three floaters, however this would probably increase steel mass and hence costs
- four floater concept, Figure 10, would have slightly smaller overall dimensions, however the mass of steel would again be greater due to the larger number of cross braces

5.1.5.1 Semi-Submersible Multiple Turbine

The idea of locating the main structure below the surface, hence away from the main wave loads, is frequently used by the offshore industry and called 'semi-submersible' technology. Semi-submersibles

Figure 10. Quadruple Floater

have good wave-response characteristics, hence, from a technical perspective, are eminently suitable for supporting wind turbines. Figure 11 shows a concept supporting five wind turbines investigated at University College London [7]/[9] in the mid-nineties; however, like all multi-turbine concepts, it is unlikely to be the most attractive solution today, now that much larger turbines have been developed.

Figure 11. Semi Submersible Multi-Floater

5.1.6 Dual Row Multiple Turbine

The footprint of multiple-turbine concepts could be reduced by adding a second row of wind turbines above the first. Such arrangements were proposed decades ago for onshore systems, when wind turbine rotors were smaller, for example see Figure 3. Figure 12 shows a concept developed by Lagerwey and Heerema, as reported within the DrijfWind project [5]. This concept is suitable for shallow waters of less than 50 m depth, such as in the Dutch sector of the North Sea, and is moored against a single monopile positioned in the centre of the floaters. Such a floater supporting five 2 MW turbines would require ~ 800 tonnes of fabricated steel (i.e. the unit costs would be significantly more than for a monopile).

Figure 12. Twin-Turbine-Row Multi-Floater

This concept has the advantages of 'weather-vaning', however it has the disadvantage common to all multi-turbine concepts today that single turbines can have large capacity. However, an advanced and complex installation solution would be needed to temporarily remove the pile for the positioning of the floater around it and it would suffer from the general disadvantages of multi-turbine concepts described previously.

5.2 Tensioned Moored

Since their first use on the Hutton TLP (Tensioned Leg Platform) in the North Sea in the mid 1980s, tensioned moorings have been considered for different support structures worldwide, although only about 25 have been built to date. Tensioned moorings can be used, with any of the concepts shown above, to reduce the vertical as well as the roll and pitch (for multiple tendon moorings) motion response. This is achieved in two ways, (i) through the high vertical stiffness, and (ii) by shifting natural periods of motion further away from the periods of wave motion.

The classic rectangular TLP design, such as Hutton, would be too elaborate and expensive for floating offshore wind turbines. An important design objective for offshore hydrocarbon applications is deck area which is irrelevant for wind turbines; however, there are other tensioned-mooring concepts of interest.

An important disadvantage for all tensioned-mooring concepts is a possible lack of stability when floating freely. Stability during towing and installation must be achieved by other means, and the risks of capsizing following moorings failure during operation should be assessed, as evidence of the capsizing of the Typhoon TLP during Hurricane Rita and the resulting total loss of the structure shows. In contrast, failure of catenary moorings would normally result in the structure drifting freely but stably and assuming that a major collision or grounding was avoided, the structure should be recoverably and repairable.

5.2.1 Tensioned Spar

Figure 13 shows a shallow-water application of the tension-moored spar, this variant having been examined within the DrijfWind project [5]. The concept is in theory applicable to any depth of water in the form of a single vertical mooring or several inclined lines. Achieving the design goals in such shallow water concept proves to be difficult, because the diameter at the water plane would need to be

Figure 13. Tensioned Moorings

around 30 m to achieve stability, though this is less than for a catenary moored spar. The lower section could be narrower, perhaps about 15 m diameter, but this would nevertheless be a large structure, with all the associated challenges for handling and manoeuvring.

Similar to a catenary-moored spar, deeper waters would allow a longer spar with improved performance characteristics, including unmoored stability, as has been proposed in Norway [4].

Figure 14. TLP Star

Figure 15: Four-Floater TLP [18]

5.2.2 Mini TLP

A concept that has achieved certain recent success in the offshore industry is the Mini-TLP (also called SeaStar), Figure 14. The success is predominantly due to minimal use of materials and consequential lower costs. However, the design and installation process are more complex and hence inherently more risk, thus the prospects for this concept are likely to be brightest in the longer term.

5.2.3 Four Floater TLP

Finally, a four-floater concept, Figure 15 (such as the Sea Breeze [8]) is a potential solution, but would probably be heavier and would require four tendons to the sea floor, instead of three. This *See Breeze* concept proposed a telescopic tower, given that the complete structure would be stable during tow-out and installation since the nacelle would be retracted. Telescopic towers are in use in other industries, though not with as heavy top-loads wind turbine as nacelles, and have been proposed previously for bottom-mounted offshore wind turbines. However, they require additional steel, which has cost and installed-stability penalties, involve a complicated and risky installation process and have total costs that are likely to be more than conventional crane lifts [27].

5.3 Bottom Mounted in deep water

When examining support structures suitable for deeper waters, bottom-mounted options may still be viable in certain cases, though the prospects are likely to be poor deeper than 50 m. For instance, the Beatrice project has extended the maximum water depth for bottom-mounted wind turbines to around 45 m [16]. Other concepts that have been developed by the oil and gas industry and could be considered, include:

- *Guyed structures*, which allow the weight of the tower to be significantly reduced for a penalty of added design, construction, installation and access complexity. However these have not become widespread even in the offshore petroleum industry,
- *Compliant towers*, which have key natural periods longer than those of the waves and can be of various designs. At its simplest, a compliant tower could be a flexible jacket-type structure designed for deep waters, up to 600m. More complex designs could incorporate a flexible joint at the seabed and additional buoyancy at the water plane, in effect incorporating aspects of simple tensioned-moored floaters,
- Conventional *Jacket structures*, which were used for the Beatrice project, are said to be stiff, in that the natural periods are shorter than those of the waves. The maximum depth is set by the design being able to maintain natural periods below the wave periods without resorting to over-large dimensions or excessive use of steel and is typically around 100m for hydrocarbon applications but will be significantly shallower for offshore windturbines,. Typically they have a tapered profile, with the wider base providing additional foundation stiffness,
- *Tripods*, which effectively are a simplified jacket structure, with the cost savings due to reduced fabrication work likely to be less important than the increased material costs as greater water depths are utilised.

6. CONCLUSION AND FINAL CHALLENGES

The development of offshore wind energy has been slower than the initial optimistic projections suggested. Placing complex rotating plant in remote inaccessible locations is a challenge that most industries make strenuous efforts to avoid. Thus, it could be said that some early setbacks were inevitable; in the offshore oil and gas and space industry, costs have been similarly underestimated and there have been major failures. Wind energy does have the advantage of drawing on the experience of the offshore and coastal engineering industries, but also has the disadvantage that available solutions tend to be excessively expensive for direct application. These issues affect wind turbines on both fixed and floating support structures. The risks are greatest for the first projects, which have today already been constructed for bottom-mounted offshore windfarms, but which are awaited for floating windfarms.

That being said, floating windfarms should not intrinsically be more expensive or risky than bottom-mounted projects, with the exception of the shallowest most-protected sites, such as have been developed in eastern Denmark. There are certainly the additional factors of the motion, but analyses undertaken to date suggest that these will not be critical. A number of challenges remain which will require detailed and costly analysis and testing; however, the high and uncertain but necessary costs for proving the viability of new technologies is common to all and generally public financial support is forth coming when the prospects are sufficiently high, as is the case for deepwater offshore wind.

The concerns of climate change, security of energy supply and reduction of environmental impact that drive the development of offshore wind energy are becoming ever more prominent. Numerous countries, in particular the United States, Norway, Spain and Japan, have aspirations to harvest the offshore wind resources around their coasts, but do not have adequate shallow-water sites suitable for bottom-mounted turbines.

There remain a number of technical and financial challenges; the key requirements are for:

- *Fully integrated computer models*, able to simulate dynamically the wind turbine and the floater. The critical issue is modelling the wind turbine so advanced control strategies can be analysed (see IEA Wind Annex XXIII [13]).
- A comprehensive, and hence costly, program of *wave-tank tests*, examining the towing, installation, operation, survival and repair conditions.
- Application of *offshore engineering technologies*, such as tensioned mooring, in a cost-effective manner appropriate for electricity generation
- *Confidence* that the technology has bright prospects in the near future, so that financial sponsors can invest in the first prototypes and demonstration wind farms.
- The present interest, commitment and rate of development of technology suggest that the first prototypes will be built before the end of this decade. The most promising concepts appear to be the deep spar, for the deepest sites, and small tensioned-moored concepts, for shallower sites. The final choice will depend on the knowledge, experience and confidence of the consortium undertaking the first prototype.

FURTHER READING

Presentations at the two NREL Workshops on Deepwater Wind Energy:

- October 2003, Washington, D.C., www.nrel.gov/wind_meetings/offshore_wind/index.html
- October 2004, Washington, D.C., www.energetics.com/deepwater.html

DrijfWind Research Study Final Report [5], www.offshorewindenergy.org/reports/report_011.pdf

International Energy Agency (IEA) Wind Annex XXIII, Offshore Wind Energy Technology Deployment [13], www.ieawind.org/Annex_XXIII.html

Patents on floating wind turbines on the EPO internet database, http://www.espacenet.com/index.en.htm

REFERENCES

[1] Arcadis Press Release, Weiterer Meilenstein für Offshore Windparks mit “schwimmender Gründung”, www.arcadis.de, 18.11.2004

[2] Arcadis, Anchoring Element for Floating Devices, Patent WO 2007/014670

[3] Bertacchi, P. Di Monaco, A., de Gerloni. M., Ferranti, G., (1994). Eolomar – a moored platform for wind turbines; Wind Engineering Vol. 18, No. 4, p189

[4] Borgen, E., Wind Power Station, Patent WO2004/097217, 11 November 2004

[5] Bulder, van Hees, Henderson, Huijsmans, Pierik, Snijders, Wijnants, Wolf, Studie naar haalbaarheid van en randvoorwaarden voor drijvende offshore windturbines (translation: *Study of the feasibility and boundary conditions of floating offshore wind turbines*), “DrijfWind”, TNO Report 2002-CMC-R043, Netherlands, 2002

[6] Durey, C., Perfectionnements apportés aux appareils moteurs fonctionnant sous l’action du vent (*translation: Improvements to the driving apparatuses through the action of the wind*), Patent FR542172 1922-08-07

[7] Halfpenny, A., Dynamic analysis of both on- and off-shore wind-turbines in the frequency domain; PhD Thesis, University College London, 1998

[8] Hannevig, D., Bone, D., Low cost self-installing offshore wind turbine support structures for deeper water, *Workshop on Deep Water Offshore Wind Energy Systems*, NREL / Department of Energy, Washington, 2003

[9] Henderson, A. R., Analysis Tools for Large Floating Offshore Wind Farms, PhD Thesis, University College London, 2000

[10] Henderson A. R., Leutz R., Fujii T., Potential for Floating Offshore Wind Energy in Japanese Waters, *12th International Offshore and Polar Engineering Conference* (ISOPE), Kitakyushu, Japan, ISBN 1-880653-58-3, May 26–31, 2002

[11] Heronemus, W.E. "The US Energy Crisis: Some Proposed Gentle Solutions", The Congressional Record, Vol. 118, No. 17, Feb. 9, 1972; also presented at the Joint Meeting of American Society of Mechanical Engineers and Institute of Electrical and Electronics Engineers, Springfield MA, Jan. 12, 1972.

[12] Hydro, Concession for floating windmill, Press Release, 13th Sep 2006

[13] International Energy Agency, Implementing Agreement for Co-operation in the Research Development, and Deployment of Wind Energy Systems (IEA Wind), Annex XXIII – Offshore Wind Energy – Technology Deployment, http://www.ieawind.org/Annex_XXIII.html

[14] Kogaki, T., Prospects of Offshore Wind Energy Development in Japan, Modeling and Testing, *Workshop on Deep Water Wind Energy Research & Development Planning*, NREL / Department of Energy, Washington, 2004

[15] JOIA (Japanese Ocean Industries Association), Feasibility Study of Offshore Wind Energy Support Structures, Ocean Development News, Vol. 29, Nr. 6, pp 18-23 (in Japanese), 2001

[16] MacAskill A., "DOWNVInD" Distant Offshore Wind No Visual Intrusion iN Deepwaters, *Workshop on Deep Water Wind Energy Research & Development Planning*, NREL / Department of Energy, Washington, 2004

[17] Molin B., Remy F., Facon G., Etude Expérimentale du comportement Hydro-Aéro-Elastique d'une Eolienne Offshore sur Ancrages Tendus (translation: *Experimental study of the hydro-aeroelastic behaviour of an offshore wind turbine on tensioned moorings*), *Ocean Energy Conference, Brest, France*, 2004

[18] Musial W., Overview: Potential for Offshore Wind Energy in the Northeast, *Offshore Wind Energy Collaborative Workshop*, Washington D.C., February 10-11, 2005

[19] National Wind Technology Center, National Renewable Energy Laboratory, website http://www.nrel.gov/wind/

[20] Nielsen, F. G., Floating Wind Turbine Installations, Patent WO2006/132539, 14th December 2006

[21] Ruer J., Pimenta de Miranda W., Influence of local site conditions on offshore wind turbines design, *Ocean Energy Conference, Brest,* France, 2004

[22] van Santen J. A., de Werk, K., "On the Typical Qualities of SPAR Type Structures for Initial or Permanent Field Development", OTC 2716, Offshore Technology Conference, Houston, Texas, 1976

[23] Sclavounos P. D., Deep Water Floater Concepts for Offshore Wind Turbines. Design, Modeling and Testing, *Workshop on Deep Water Wind Energy Research & Development Planning*, NREL / Department of Energy, Washington, 2004

[24] Stoddard W., 'The Life and Work of Bill Heronemus, wind engineering pioneer', Wind Engineering, vol 26, issue 5, pp 437, 2002

[25] Tong, K., Technical and economical aspects of a floating offshore windfarm, *Offshore Wind Energy in the Mediterranean and other European Seas* OWEMES, Rome, 1994.

[26] Ushiyama, I., Seki, K., Miura, H., A Feasibility Study on Floating Offshore Wind Farms in Japanese Waters, *Offshore Wind Energy in Mediterranean and Other European Seas* OWEMES, Naples, Italy, 10-12 April 2003

[27] Way, J., Bowerman, H., Integrated Installation for Offshore Wind Turbines, ETSU W/61/00617/00/REP URN 03/1649, 2003

Chapter 10:

Access to offshore windfarms

Göran Dalén and Mikael Jakobsson

1. INTRODUCTION

There are several reasons why access to offshore wind turbines becomes a key feature of design and operation as compared with onshore turbines. In approximate order of importance, the main factors are:

1.1 Safety

Both ethical and legislative standards require that safety of human life has priority, which applies to all wind turbines. However, the dangers are compounded at sea for obvious reasons. It is also important to appreciate that the development and deployment of wind energy has been very rapid compared with many other technologies. Initial development of wind power concentrated on land-based turbines and windfarms. Most of such development can be used offshore, but there are also new elements, such as marinisation of the wind turbines and the access facilities. One might think that offshore access techniques for windturbines are well developed, but this is not the case. For instance, offshore oil and gas stations use helicopters as standard for access, but this is not likely to be the norm for windfarms. In fact, we access the relatively small offshore structures on modern windfarms today, rather as we accessed lighthouses in the early 1900's. It is quite new that so many individual and relatively small structures are installed offshore in various environments. Each unit is likely to be accessed by technicians, at a rate of 10 and 20 person-accesses[1] per year. Half of these may occur at periods of strong wind, when sea conditions are severe. For instance, the offshore turbines installed by 2006 required up to about 6000 person-accesses per year and by 2010, this number could rise to exceed 200,000/y. Therefore, the need to develop safer methods to access structures offshore has been put in a new light.

1.2 Economics

Offshore wind energy has very small economic margins compared with offshore oil and gas activities. Therefore, it is difficult to afford the very sophisticated systems developed for the offshore oil and gas sector.

The access system to the structures must be considered, together with the transport from land out to the windfarm. These are closely linked, so it might be difficult to use a particular method of access with existing transport and visa versa. Each project has its own conditions, which may require modifications and different strategies from project to project.

Difficulties of access to repair faults may lead to large financial losses. Perhaps some faults could be rectified by the use of improved conditioning monitoring systems etc, however experience of the first offshore turbines shows poor availability, often caused by trivial faults, but resulting in long periods of shut-down because access has been difficult. The usual difficulties are storm conditions or

[1]Crossing from a boat to an offshore turbine is always likely to be 'person by person', so the unit is 'person-access', rather than, say, 'boat journey'.

large swell; in addition ice has prevented access by sea for windfarms, e.g. in the Baltic Sea The general conclusion is that the expense of stand-by equipment, delayed personnel and loss of generation become key factors affecting the overall economics of a project.

1.3 Other general factors adding to complexity

The access and transport must be suitable for rough environmental conditions, but also simple to use, since 70 – 80 % of the time, the system will be used in "normal conditions". It is tempting to design a system that can handle very rough conditions, but it is important to remember that conditions are usually acceptable, including the times scheduled for routine maintenance.

The task of finding a suitable access system is complex, since it must be suitable for each individual wind turbine and for personnel often having limited experience of working offshore. Future offshore windfarms may consist of up to 200 wind turbines, with each individual turbine requiring maintained at least once per year, preferably during a "calm season", e.g. April to September in north European seas. Moreover, there is a strong case for seeking standard entrances and methods of access throughout the offshore wind industry.

Whatever the difficulties, the overriding constraint is that 'safety always comes first'. A system that is efficient and economical can only be successful if it is also safe for the personnel using it, including vessel crew, maintenance staff, technical specialists and occasional visitors.

1.4 The need to improve availability

The costs and incurred revenue losses for downtime are reasons to improve and develop access technology. The production cost of wind energy offshore also directs the technology towards larger wind turbines, with subsequent increased losses per unit time when shut down. Projects so far, have been installed in near-shore environments, but income generation and the need for more wind projects will require wind power projects in more complex and vigorous environments. Future projects tend to be further out to sea and in deeper waters, all increasing the production cost. Therefore, increased availability and cost effectiveness will require better accessibility.

2. REQUIREMENTS FOR GOOD ACCESS

This section discusses the conditions and requirements for access. These requirements are general and may differ substantially from project to project. The purpose is to establish the different forms, aims and values of a system. The first part considers environmental conditions, and the second, technical functionality. The last part considers cost and its major parameters.

2.1 Environmental conditions

2.1.1 Normal conditions

"Normal conditions" describes periods when the project operates as planned and when sea conditions are acceptable for transport and access. Such periods vary substantially between projects. It is during these times that normal maintenance is performed, including retrofit or additional work. Such periods have most transportation hours and number of person-accesses to the wind turbines. It is during this period, that the access system's value in terms of flexibility and ability to function in a variety of tasks is apparent, e.g. offloading tools, compressors, wind turbine components etc. This period may last for 65 to 80% of the year.

In defining an access system for a project, the need for flexibility of the system during this time is often forgotten. It is during this time that the efficiency of work can be optimised and costs reduced.

2.1.2 Extreme conditions

A good access system is also defined by its function in extreme conditions. The first and foremost condition is for the systems to function in *high amplitude waves*. Therefore systems may generally allow for 1.5 m significant wave height (Hs). The length of the sea waves may also vary dramatically.

In some areas of the Baltic Sea, the waves may be short but difficult *(choppy sea),* for although the wave height is not great, the short wavelength makes conditions more difficult than at longer wavelength

Another condition in the Baltic Sea region is *ice conditions*. A system with ladder access near the splash zone will not be accessible for weeks to months, due to ice build up. In addition, the vessel must be able to go through ice.

In some projects, the impact of *high-speed winds* and *strong currents* in combination is very difficult, causing difficulty in docking to a structure. In these cases, excellent vessel manoeuvrability is needed.

Another general condition affecting an access system is the *tidal condition*. This requires completely different requirements, so, a system for large tidal range (e.g. in a tidal estuary) will differ greatly from a 'normal' 'system.

Some projects also experience *very shallow waters*. This may cause problems in access at low tide and with high waves. Vessels with shallow draft may 'bounce' on the sea floor and severe damage may occur.

An essential general, but difficult, condition is to be able to operate in *poor light and darkness*. This may be necessary in emergencies, but in northern Scandinavia, the winter is very dark and daylight is much shorter than in middle Europe. Here a standard requirement is that access is possible also in both day and night.

2.1.3 Technical requirements

An access system must function in a variety of conditions. There are many aspects of the access procedure that need to be considered, as outlined below.

Ease of docking by the vessel

A system must allow easy docking, as a personal transfer must be smooth and fast. The easiest docking is made by approaching along the main wave direction and into the wind.

Safe position at substructure

When docked, it is of the outmost importance that the vessel rests safely in its position, both for the crew, accessing personnel and transferring equipment. A system with insecure positioning will create stress and thereby contribute to personnel injury.

Personnel transfer

The transfer of personnel needs to be free from stress and not require jumping. Transfer should be smooth and the flow of personnel unrestricted. Climbing must be avoided. The length of time when a person is at risk must be minimised, so lessening the chance of accident, reducing stress and increasing flow. A well-controlled vessel, in a safe position with a short risk period is optimal.

Equipment transfer

An optimal access system allows equipment transfer without the use of cranes or hoisting mechanisms. The use of hand tools and other smaller equipment, which can be transferred easily, will save transfer time and not require lifting equipment for the transfer that could cause additional risk. Time saved by not hoisting improves total logistics and transportation flow in a large windfarm, because several teams may be circulating between turbines.

Reliability

A fundamental factor is the reliability of the access. If use of a poor system increases the possibility of machine failures, the outcome can be severe. The general rule of thumb is "*keep it simple and robust*" and "*don't compromise safety*".

Fast maintenance

When analysing a logistical flow of an offshore windfarm, the need for fast maintenance is clearly important. This includes the support vessel's ability to function for smooth deliver and pick up of equipment and personnel. The vessel has multi-task functions.

Flexibility

An access system must be 'flexible'. For instance, a fundamental benefit is to be able to dock to the structure in several directions. Optimal flexibility is 360 degrees or unlimited docking around the structure, so always allowing the vessel to face into the waves and/or current.

Impact on substructure

Loads and dynamic forces introduced on the substructure by the design of the access system must not harm the functionality or life expectancy of the foundation. Because the access system may affect the project economics, the access design must be decided in the very early stage of a project and performed together with foundation design. Due to the sensitivity of loads affecting the fatigue life of a substructure, an optimal system may not be possible if "added on" to a finished design. Poor design may permanently handicap the access options, including having to use larger and more expensive transport vessels than otherwise needed.

2.1.4 Access System cost

Increased availability of a windfarm results in less costs for standby time and for less loss of generation. Therefore, the acceptable price for access depends not only on the cost of the access facility itself, but also on lost production in downtime and the power price paid. Therefore, a system acceptable in one market may not be acceptable in another.

The following cost components affect the acceptable price of an access system:

- Increased availability, hence gained revenue (days of downtime saved, increased production at power price).
- Vessel size and its fixed (purchase) cost required for access, probably including a backup vessel.
- Operational cost of the vessel.
- Transportation time of technicians; the speed of vessel.
- Saved cost for standby time for technicians and vessel using an improved system
- Flexibility of the vessel system; for example if the vessel has to be hoisted it cannot serve another turbine at the same time.

Local requirements for safety may dictate use of system.

3. COMPONENTS OF AN ACCESS SYSTEM

In order to define and optimize an access system, it is necessary to appreciate its component parts, which should be optimized according to the substructure design. The foundation design will constrain the options for docking. Other factors are:

Transportation

The vessel to be used is normally configured as a compromise between its purpose, the environment in which it is operating, the docking system and the level of safety required. Each design of vessel can then be analysed with regard to the costs and benefits. As offshore windfarm experience has increased, so special-purpose vessels have been designed and constructed, rather than using existing vessels.

Figure 1. Operating vessels in Kalmarsund, Sweden Source: Regional Council of Kalmar

The range of vessels varies greatly by investment and operational cost. In most projects, no single specific vessel is used 365 days per year, since the particular choice of vessel depends on the sea state, the season, and the type of use and load; such changing factors dictate different requirements for transportation on any one day. Therefore, more than one vessel will be used, especially if winter icing is a feature. However, because sea condition and significant wave height tend to determine maximum availability of the windfarms, a special 'project vessel' tends to be purchased or designed for access in the worst sea conditions. Consequently, such a project-vessel may be over designed for most conditions. Therefore, it is necessary to carefully analyze and configure such vessels, depending on the actual need. Other factors may become an issue, such as the noise of the vessel, which may be greater than permitted for the turbines themselves!

Below we describe the general types of vessels available and being considered.

3.1 Vessels and vehicles on the market today

3.1.1 Zodiak inflatable

The Zodiac inflatable rigid-hull support-craft is well known in the offshore oil and gas industries; it is available in various sizes and personnel-carrying capacities.

3.1.2 Single hull

Single hull vessels are by far the most common type of boats used for offshore O & M. This is mainly because many such boats are available on the market and it is easy to modify existing designs. It is also possible that new vessels will be purpose-designed for both installation and O & M service. It could be a cost-effective option to re-deploy the vessel as a maintenance vessel, after completion of the construction phase.

Disadvanatages:
- Large motions in heavy seas (heave, roll and pitch) resulting in limited operations
- In general, provides rather slow access

Advantages:
- Many variations and options
- Inexpensive to purchase
- Low-cost operation
- Can be used with different access systems

3.1.3 Catamarans

Catamarans are suitable for offshore windfarms and are often used. The catamaran, by its design, offers large deck-space and stability. Offshore windfarm developers at several projects, such as the Kentish Flats, Utgrunden I and Arklow Bank, use catamarans for transportation and access.

Disadvantages:
- More expensive to purchase than inflatables and single hulls;
- Needs to be specially designed to absorb loading from access systems

Advantages:
- Stable in vigorous sea conditions;
- Provide more cabin and deck space than basic alternatives
- Can be linked to turbine access systems
- Have high-speed capabilities.

3.1.4 SWATH vessel

One vessel recently developed is the SWATH (Small Water Access Twin Hull). The SWATH vessels are presently available from 19 m service vessels to 60 m support vessels. Speed range is up to 25 knots. Experience to date shows that this design is very stable in strong seas.

Disadvantages:
- Very expensive to purchase
- Needs specialised crews
- Operating costs expensive

Advantages:
- Very steady in harsh sea conditions;
- Can be linked with turbine access systems;

3.1.5 Hovercraft

In principle, hovercraft should be suitable for O & M work in offshore wind parks. No information is available however, at the time of publication, of any systems using or planning to use hovercraft.

Disadvantages:
- Expensive to purchase;
- Needs specialised crews;
- Operating costs high;
- Noisy.

Advantages:

- Very steady in harsh environmental conditions;
- Could be used for personnel to access tower ladders, but only in good conditions.

3.1.6 Semi-submersibles

Semi-subs have a very good track record for offshore drilling operations. However, they require deeper waters than surface craft and so are not usually suitable for bottom-mounted offshore turbines. They are also far too expensive to use, unless very simplified designs are developed.

3.1.7 Helicopter

It is possible to use helicopters for O & M, but only for winching one person at the time onto the nacelle roof. The method is mainly suitable only for emergency operations.

Disadvantages

- Expensive helicopter rental
- High-risk activity
- Access restricted in poor weather conditions
- "Landing deck" required on top of nacelle

Advantages:

- Quick transport of personnel
- Helicopter can be used for moving materials and equipment
- Independent of severe sea states, if wind speed not too great.

Other systems on the market today, but not fully experienced in regular use, are:

(a) *Hoists*; these allow the vessel to be lifted out of the water at the turbine; this requires a suitable hoisting structure on the turbine or its foundation.
(b) *Draw-bridges*; these provide an 'arm' to lock the vessel at a small separation from the turbine, with the draw-bridge functioning also as a walkway to the access platform

3.2 Docking vessels at the turbines

The second component of access is docking, i.e. the method of docking the vessel to the structure. Options include:

(i) pushing the vessel against the structure with propeller thrust,
(ii) tying the vessel to the turbine,
(iii) keeping the vessel at a short distance from the structure.

Most systems of type (i) and (iii) depend on thrust force to make the vessel stable and steady. In using such methods for docking, the mass of the vessel and its power capabilities are important factors for success. Also important is the captain's ability to manoeuvre the vessel to its position and thereafter keep it in position. Here the access transfer system dictates the requirement. The key factor is to keep a vessel as steady as possible during transfer of personnel; the longer this takes, the longer the vessel needs to be steady. If personnel are climbing a ladder, the time for transfer and therefore the risk of an accident, whilst maintaining the craft in position, is much bigger than just crossing at one height. If however personnel move from one horizontal plane to another, the exposure time is least. To reduce the rolling of the vessel, it is essential to head the vessel into the oncoming waves. This is difficult, since very few systems allow for 360° docking. Any deviation increases rolling and subsequently the ability to keep the vessel steady.

3.3 Personnel Transfer

Following docking, there are several methods to transfer personnel from the vessel to and from the structure:

- *Vertical ladder* – most common. Personnel step forward, from the deck of the vessel, on to a ladder fixed to the turbine. They then climb about 5 m to substructure platform at the turbine entrance. On return, personnel climb down and cross to the vessel deck from a backward position.
- *Horizontal direct access* – used with vertical-sided, gravitation-based foundations that stand clear of the water, and by systems such as the SASH system (meaning in Swedish 'safe boarding to offshore structures'). Personnel cross from one horizontal plane to another, with only one stride.
- *Walkway access* – used in systems with inflated or firm walkways, attached either to the vessel or to the substructure. This means that personnel may walk a walkway, creating a "bridge" between vessel and substructure.
- *Hoisted vessel* – enabling personnel to exit the vessel directly on to the wind turbine
- *Hoisted personnel* – personnel are hoisted from the vessel to the turbine by a crane located on the substructure.
- *Helicopter* – helicopter hovers, with personnel on a hoisting harness, descending to or being lifted from the nacelle roof (the turbine rotor is locked stationary).

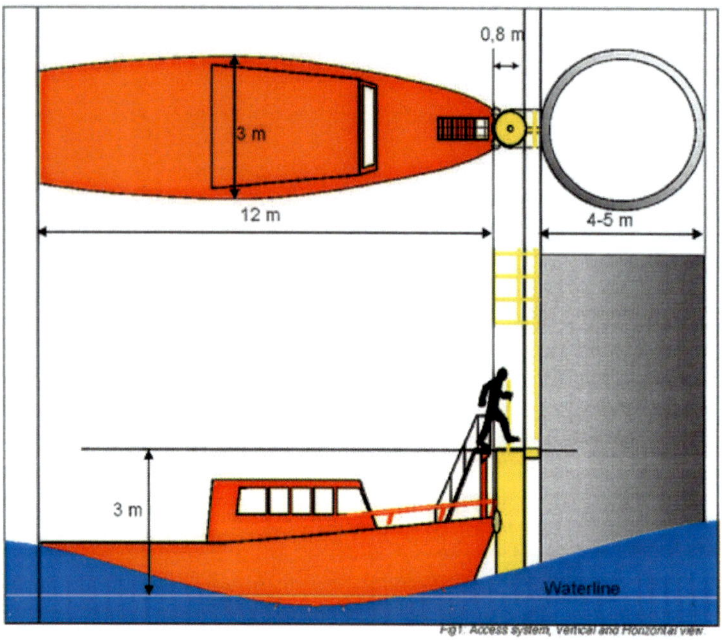

Figure 2. Principle of the SASH system

4. EXPERIENCE AND SYSTEMS ANALYSIS

4.1 Sweden/Denmark/UK

During the 1990's, the first projects were built in very shallow waters (~ 5 m at mean tide) and near shore. Individual projects were built in both Denmark and Sweden. One of the most exposed windfarms during this time was Bockstigen, just off the Swedish island of Gotland. Here several access methods were tried. The project has come to use a hanging ladder to access the machines, due to a combination of the design (a monopile), the fairly severe ice conditions and the bottom geology (rather deep water outside the bank, resulting in rather high seas with choppy waves at the location),

Figure 3. Bockstigen windfarm (G. Britse)

In 2000, the first project with megawatt-scale (MW) turbines was built on the Utgrunden reef in Kalmarsund Sweden. The following year, three additional projects were built with MW machines. Three of the four projects, namely Utgrunden, Blyth Harbour and Yttre Stengrund, had monopile foundations and were in conditions where the shallow sea bottom sloped down towards the deeper sea. The UK Blyth project also introduced the component of tidal waters with a 5 m tidal range. All three projects where built with 'dual J-tubes' (mooring fenders) and a single ladder welded between the vertical tubes (docking has to be in one direction only). See Fig 4.

The Swedish project, Utgrunden I, was built without a protective ice cone around the base of the turbine, whereas at Yttre Stengrund, the 4 m diameter pile has an ice cone. These four projects have operated with the longest experience.

Since these initial projects, other offshore turbines and windfarms have been installed. In Denmark, the two largest projects in the world at construction and each of 160 MW capacity, are Horns Rev and Nysted. These two windfarms have contributed significantly to the experience of accessing multiple offshore turbines in groups. Subsequently, new projects have introduced new challenges, such as extreme wave-heights and tidal range and current. These factors increase complexity for access. The experience gained so far can be classified as:

- *With monopiles,*
  - With and without ice cone
  - With and without tidal waters
  - In water depths down to 20 m

- *With gravitation-based foundations,*
  - With ice cone
  - In water depths down to 10 m

Figure 4. Docking to a doorway ladder in one direction only, however a range of vessel sizes can be used. *Source Mistvind AB*

- *With Jacket foundation*
  - Without ice cone
  - In tidal waters
  - In water depths down to 45 m

- *With wind turbine sizes* mainly in the 1.5 to 3 MW range, but with the Repower 5M (5 MW) machine being the largest type of wind turbine installed offshore to date.

The interest for offshore wind energy in deep water has increased. The Downvind project off Aberdeen, north east Scotland, features two very large turbines (each 5 MW) installed in deep water (45 m), using a jacket foundation (see Fig 5)

Figure 5. The Downvind project, Scotland

The Downvind access system allows safe access without using vertical ladders and is shown in Figure 6. The docking vessel can choose a suitable "access-slot" depending on the tide and the transfer of personnel is done from the vessel to a "staircase" running up around two of the Jacket legs.

Figures 6a and 6b. The Downvind access system on the Jacket Foundation

4.2 Effect of foundation type

There are mainly three classes of foundations from the point of view of access:

1) *gravitational foundations* in shallow waters, with personnel crossing height approximately 3.5 m above water level, on to the concrete platform surrounding the tower,
2) *monopile foundations* having mooring fenders that are becoming standard, with slight modifications to improve ship access between projects.
3) The Jacket foundation described above is a third type but with limited experience so far.

4.2.1 Dissatisfaction

All projects, both with gravity-based foundations and with monopiles, have led to dissatisfaction with the systems used; increased availability has always been sought by modifying the original systems. Although the original designs had the benefit of simplicity, both concepts have shown inherent faults. Consequently, the docking process and the transfer of personnel have become limiting factors. Usually improvements have been sought using larger vessels to achieve better results, with little done to improve the access at the turbine sites themselves

4.2.2 Analysis of the vertical ladder access

In analyzing the monopile access systems with a vertical latter fixed between two mooring fenders, we come to certain conclusions and realise elements of risk. A fundamental requirement of this system is that the ship's bow must be stable and locked in the vertically direction to enable a static point for the crossing of personnel.

The inherent issues are: (i) preventing vertical vessel movement, (ii) obtaining directional positioning, and (iii) access of personnel from a ship's bow to a vertical ladder and vice versa. Consequently, extensive use of special safety equipment and processes are necessary, such as:

- Safety harnesses
- Safety wires and hooks

- Survival suits
- Hoisting arrangements for even the smallest equipment
- Hoisting of injured personnel, even small injuries, on stretchers
- Additional berthing around foundation (3-4 access positions)

The system and its advantages and disadvantages, which can be summarized as follows:

Advantages:
- Simplicity
- No moving parts
- Relatively inexpensive per access point
- System has been in use for ~ 100 years on various structures
- System allows for tidal waters

Disadvantages:
- Limited access direction
- Having extra access points is costly
- Having several access points may handicap substructure design
- Many contact positions for the vessel's bow on vertical fenders increases the risk of vessel vertical movement (slipping)
- Vertical climbing necessary in exposed risk area
- Return to the vessel requires personnel to move backwards without clear vision and footing
- Requires large vessels for use in rough sea

4.2.3 Analysis of the gravitation foundation access

The access design with a gravitation foundation raises different issues from those with a vertical latter. This type of foundation is normally used in waters to 10 meters water depth (see Fig 7).

Gravity foundations have been installed in shallow waters, with a maximum mean-tide depth of about 6 metres. All the sites have the possibility of winter ice conditions. Thus, at water level the foundation has a negative inclination as an icebreaker.

Figure 7. Access to a gravity foundation. *Source: www.middelgrunden.dk*

This negatively-sloping ice cone creates a difficulty for safe access from a vessel, especially in rough seas. It is vital that, the vessel's bow does not slide under the lip of the ice cone, which could result in severe damage. In order to prevent this, and have the access point high enough on the structure, the ship must be relatively large and have a bow–point well above the lip of the cone. If the bow-point is too high, the vessel might slide over the deck of the foundation, risking severe damage and injury to personnel.

However, this system, in comparison with the vertical latter, does allow 360-degree access, which is a fundamental feature for a good access system. Neither does this system require personnel to climb in an exposed area, but to walk onto the structure. This greatly reduces the risk for personnel injury, especially by reducing the time of critical risk to less than a second.

The system and its advantages and disadvantages can be summarized as follows.

Advantages:
- Simplicity
- No moving parts
- The cost of the access system is imbedded in the foundation cost
- Unlimited access directions
- Personnel walk across to the structure, not climb

Disadvantages:
- System does not allow for relatively large tidal range and is sensitive to high/low water levels
- Does not allow for vertical movement of the vessel
- Maximum accessibility requires large ships
- Risk of damage to the ship's bow and ground contact in shallow waters
- Does not allow the use of smaller, and different vessels for access
- A ladder is still required for access from sea level

4.3 A system combining techniques of the two above systems

The SASH system, Fig 8, combines the good points of both the systems described above, i.e.

1) docking to a vertical structure, as with the vertical ladder system with propeller thrust;
2) a horizontal walkway from the vessel to a horizontal plane on the foundation , as with the gravitational foundation

This eliminates climbing a vertical ladder for access.

This system is based on a separate 0.8 diameter vertical 'access pile', mounted 3.5 to 4 m above the main sea level. The pile is secured to the main foundation by two fixings, (i) at the top, 4 m *above* water level, and (ii) at the bottom, 2 m *below* water level. The pile provides a flat upper platform onto which personnel can step horizontally from a walkway on the boat. There is a vertical 'grip stick' in the centre of this platform. The bow on the boat is configured as a 'fendered catcher' on a gantry, so, enabling the boat to dock to the pile and into the wind. The walkway on the boat is adjustable in height, so enabling personnel to access the structure in both directions from one horizontal plane to another. A complete system consists of two access piles, mounted opposite each other on the structure, so providing 360 degrees for docking, on one side or the other.

The system and its advantages and disadvantages can be summarized as follows.

Advantages:
- Simplicity
- No moving parts on the foundation

Figure 8. Prototype SASH at Utgrunden Lighthouse. Source: Regional Council of Kalmar. The person crossing is holding the vertical 'grip stick' on top of the 0.8 m diameter 'access pile'.

- Enables access in rough sea conditions
- 360° access directions
- Personnel walk across to the structure horizontally, with no climbing
- Is combined with a vertical ladder on the 'access pile' for other vessels to use

Disadvantages:
- Influence on substructure loading must be analysed in each case
- Vessel needs special 'fendered catcher' and gantry
- Good manoeuvrability of vessel required when docking to pile
- Only usable if small tidal range; modification needed in large tidal range

4.4 Air transport for general access and emergency rescue

In most projects, helicopter emergency rescue capability is incorporated, with a helicopter hoisting-platform on top of the wind turbine nacelle (not a helipad for landing). Windfarms with complex sea conditions have come to rely on such helicopters for normal access. A good example of this is the windfarm at Horns Rev, were access with a vessel and vertical ladder was more difficult than expected, and the use of helicopters is more frequent than expected. It is highly likely that helicopter support for future offshore windfarm logistics will play a greater role than today.

5. FURTHER TECHNOLOGY

A series of new technologies has been introduced to the market recently using motion compensation systems. The basic principle is that the ship motions are continuously measured and used to actuate hydraulic cylinders of a particular device to counter-act the ship motions. Two possible systems are described below:

OAS

The OAS or Offshore Access System is a flexible gangway installed on the aft deck of a vessel. By actuating the supporting cylinders, the gangway tip is kept stationary in the vertical direction. The tip of the gangway comprises a high-tech gripper system that is extended towards a landing pole connected to the offshore wind turbine. The system reaches behind this pole and is then retracted to activate the gripper. As soon as a connection has been established, the gangway compensation mode is switched of and the gangway hinges passively between vessel and structure. The system is operating in the Southern North Sea on unmanned oil & gas platforms. It has been operational for more than a year and has made offshore access possible up to H_s = 2.5 m. The disadvantage is that it needs a rather large vessel with DP II capabilities, making it rather expensive for the offshore wind market.

Ampelmann

The Ampelmann is a development of the Delft University of Technology. The system uses six hydraulic cylinders to support a transfer deck with a gangway. The core technology was borrowed from flight simulators. By measuring the ship motions and actuating the cylinders, the transfer deck becomes completely stationary in all 6 degrees of freedom. By extending the gangway toward the offshore wind turbine and pressing it against the boat landing or the access platform, personnel can transfer safely from the stationary transfer deck to the offshore wind turbine. The advantage of the Ampelmann system's capability of compensating also the horizontal movements is that the system is more tolerant for station keeping errors: it does not need an expensive dynamic positioning (DP) vessel. The system is designed for working in seas up to H_s = 2.5 m.

6. CONCLUSION

For access to offshore wind turbines, every site is different. Wind direction, wave fetch, tide levels, currents, wave height, distance from harbour, number of turbines, regulations and, possibly ice, are all key parameters. All these factors influence the design of the best access system.

For the transfer of personnel to be effective and reliable offshore, it is important to minimise complexity. In the offshore environment water, salt and ice attack structures and corrode metals and mechanisms. Hydraulic systems, electrical motors and moving parts may handicap the reliability.

Local and national regulations for the safety affect options for climbing ladders in offshore sea conditions. Moreover, the authorities will stimulate the environmental limitations for access. Safety onboard a wind turbine in high wind conditions may limit access, according to the manufacturer's standards.

In addition, optimising the economics requires rapid and safe access. Large windfarms with large cash-flow will be able to develop more advanced systems for increased availability and, thereby, improved economics. For small windfarms, usually located in protected waters, simple and robust systems are likely to be the most economic.

Chapter 11:

Standards and Certification

Axel Andreä, Kimon Argyriadis, Peter Dalhoff, Christian Nath and Silke Schwartz

1. INTRODUCTION

The development of codes and standards for wind energy applications was started at the beginning of the 1980's by official representatives from Canada, Denmark, Germany, the Netherlands, Sweden and the USA [1]. This led to the international standardisation being initiated within the International Electrotechnical Commission (IEC) in 1987. The European work started shortly afterwards by the European Electrotechnical Commission (CENELEC), since electricity generation was seen as the most relevant aspect of wind energy. The responsible committees in each organisation cooperated and were both called 'Technical Committee (TC) 88 – Wind Turbines'. Subsequently, both the IEC and CENELEC have continued to issue Standards, Technical Reports and Technical Specifications within IEC 61400 series. Subjects have included design requirements onshore and offshore, for gearboxes, rotor blade tests, power curve, acoustic-noise, load measurements, power quality and lightning protection. Other topics, e.g. offshore design requirements and a communication standard for control and monitoring, followed.

When the development of standards for wind turbines started, there were projects on the feasibility of offshore wind turbines [2]. It took almost 25 years to develop these initial ideas and actually install 3 MW wind turbines in deep-sea water. The first set of regulations for offshore wind turbines was developed by Germanischer Lloyd and Garrad Hassan within the EU-Project on 'Offshore Wind Energy within the European Union' [3]. These regulations were further developed and issued as GL Regulations in 1995 [4]. These incorporated GL's expertise both in wind energy and in offshore oil and gas. Meanwhile Denmark also issued a certification procedure for offshore wind turbines [5] and Det Norske Veritas (DNV) prepared a guideline for offshore foundations [6].

2. STANDARDS

2.1 Overview and Comparison

Certification of wind turbines started around 1983. It has been applied differently with regard to scope, requirements and depth, firstly in Denmark and then in Germany and the Netherlands. These three countries still lead in the development and application of certification rules, but in recent years a number of other countries and many banks realised the necessity of a thorough evaluation and certification of wind turbines and their installation. Among these countries are China, Greece, India, Spain, Sweden, the UK and the USA.

2.1.1 IEC 61400-3 Design Requirements

IEC 61400-3 [7] is currently a 2005 draft standard and will become international standard by 2008/9. The purpose of this standard is defined as:

> *"The IEC 61400-3 specifies essential design requirements to ensure the engineering integrity of offshore wind turbines. Its purpose is to provide an appropriate level of protection against damage from all hazards during the planned lifetime", [7].*

This standard is one of the IEC 61400 series dealing with design requirements and safety of wind turbines and measurements on wind turbines. IEC 61400-1 is the internationally acknowledged standard for the design requirements of (onshore) wind turbines. IEC 61400-3 should be used in conjunction with the appropriate IEC/ISO standards, and it is consistent with the requirements of IEC 61400-1 [8]. Within IEC 61400-3, emphasis is given to the determination of load assumptions. Details concerning site assessment and load assumptions can also be found. Aspects concerning materials, structures, machinery components and systems (safety system, electrical system) are either only briefly handled or not covered at all. For this, IEC 61400-3 (also IEC 61400-1) states:

> *"When determining the structural integrity of elements of a wind turbine, national or international design codes for the relevant material may be employed. Special care should be taken when partial safety factors from national or international design codes are used together with partial safety factors from this standard. It should be ensured that the resulting safety level is not less than the intended safety level in this standard."*

In conclusion, IEC 61400-3 defines load assumptions and a safety level, but relies on application of national or international design codes for determination of the integrity of the structure, machinery, blades, safety and electrical system.

2.1.2 GL Guideline

Germanischer Lloyd (GL) issued the first regulations for the certification of offshore wind turbines in 1995 [4]. This guideline was completely revised and, from June 2005, was available as a final version [9]. This GL guideline provides state-of-the-art design requirements for the entire wind turbine; it includes certification procedure, loads, materials, structures, machinery, rotor blades, electrical, safety and condition monitoring.

The safety philosophy follows that for onshore wind turbines, i.e. the load safety factors are harmonised with the IEC 61400-1/-3 standards and the safety factors for materials are comparable, but more detailed. IEC specifies safety factors for materials in a general way, without consideration of the particular materials themselves. On the contrary, GL specifies safety factors depending on the material, e.g. the safety factors are more conservative for soil resistance in comparison to those of welded structures, because soil resistance can only be estimated with relatively large uncertainty.

The scope and extent of Type or Project Certification, see Figures 3 and 6, is outlined in the GL guideline. A special feature there is the A and B level for Project Certification. This allows the customer to choose between monitoring all the wind turbines within the windfarm (A level), or monitoring just 25% on a random sample basis (B level), see section 3.1.7.

2.1.3 Danish Recommendations

The mandatory Danish regulations for technical approval of offshore wind turbines [5] were issued in 2001. These regulations are published in the annex of the *"Executive order on the technical certification scheme for the design, manufacture and installation of wind turbines"* [10] and contains instructions and supplementary information about technical requirements for approval of offshore wind turbines. The regulations for offshore wind turbines deal with loads, foundations, occupational safety, lightning, marking, noise emission and environmental impact assessment.

Special offshore requirements for machinery, electrical and safety system are not given in the Danish recommendations, so normal regulations apply.

2.1.4 DNV-OS-J101

Det Norske Veritas (DNV) issued its first standard for the design of offshore wind turbine structures in 2004 [11]. The guideline covers design, construction, installation and inspection; it applies to all structural parts below the nacelle, including foundation and soil. Particular attention is also given to manufacturing and follow-up during the lifetime of the support structure. Design principles and requirements are given for steel and concrete structures, grouted connections, corrosion protection, load effects, transportation and installation. The guideline does not cover the system to be considered within a Type Certification for the offshore wind turbine such as rotor blades, machinery components, electrical equipment, safety system or condition monitoring.

2.1.5 IEC WT01

IEC WT01 [12] comprises rules and procedures for the *"IEC System for Conformity Testing and Certification of Wind Turbines"*. The procedure is not related to offshore wind turbines and it does not specify design requirements. The purpose of IEC WT01 is to define the basic elements, modules and procedures of certification.

Although WT01 was not intended for offshore windfarms, it is applied for Project Certification of offshore windfarms in some cases. However, IEC WT01 is currently modified to include offshore windfarm certification.

Project certification without third party surveillance during manufacturing, transport, installation and commissioning will not reveal problems or failures during these processes. The value of a Project Certificate according to IEC WT01 [12], is the certified assessment of the entire conditions and the design documentation. Since experience with wind turbines shows that manufacturing problems occur occasionally, waiving of third party surveillance leads to greater risk. Repair of failures after installation, not detected during manufacturing, is a very costly and time intensive work, and it leads to a bad reputation for wind energy in general.

Table 1. gives an overview of the standards described and an indication of their content.

Table 1. Comparison of scope of standards/guidelines in the field of offshore wind energy

	Project Certification	Loads	Support Structure	Machinery	Safety, Electrics and CMS
IEC 61400-3		✓	(✓)		(✓)
GL Guideline	✓	✓	✓	✓	✓
(Danish Recommendations		✓	✓		(✓)
DNV-OS-J101	✓	✓	✓		
IEC WT01 (onshore)	✓	(✓)			

✓ = subject is dealt with, (✓) = subject is partly dealt with or reference to other standard)

2.2 Future Developments

The future development of standards and guidelines is directly connected to the technical and economic development of the offshore wind or ocean energy industry. In practice, standards follow design improvements after a delay, although the participants in standardisation (designers, research institutions, and standardisation and certification bodies) cooperate both within research and development projects and within the relevant standardisation committees. In general, the economic situation influences the scope and amount of the standards to be applied and the certification to be performed. This effect influences the technical part of the standards, too, but hopefully not the safety. However, the primary risk of failure is economic, with less risk for life or for pollution, (offshore wind turbines are unmanned structures, containing a relatively small amount of potential pollutants).

Thus, the overall economics influences the amount of work to be performed for certification. For very large projects, a purely probabilistic design analysis may be the best method, by abandoning the classic limit state design with safety factors. However for smaller projects, the "classic" safety-factor design method may be of advantage. Therefore, standards and guidelines covering both methods may be developed and used in the forthcoming years. Moreover, further harmonisation of standards and guidelines will be important.

2.2.1 Wind Turbines

In addition to conventional standardisation, e.g. the development of the gear box standard IEC 61400-4, a number of 'New Work Item Proposals' are being initiated within the IEC. This covers standards for both components and testing.

In the early offshore windfarms, serious malfunctions occurred resulting in considerable retrofit actions, since the turbines were essentially modified onshore turbines. For large windfarms under development and for locations with deeper waters, turbines with larger ratings (>3 MW) are developed and used. The size of a rotor with a diameter of over 100 m raises challenges in all areas influencing turbine safety. Aerodynamics, structural design, transport, installation, operation and maintenance are affected and new engineering solutions applied. The research performed in this sector will lead to improvements also in guidelines and standards. For example within the EU-funded research project *"Integrated Wind Turbine Design (UpWind)"*, all aspects of wind turbine design and analysis were considered. Part of this project is the development of new certification guidelines in conjunction with the work performed within the IEC TC 88 framework.

The development within the IEC is mainly performed by the different maintenance or working groups of TC 88. Current (2005) working groups are listed in Table 2.

Table 2. IEC TC88 Working Groups

Working Groups (WG)	
WG 3	Design requirements for offshore wind turbines
Joint Working Groups	
JWG 1	Wind turbine gearboxes
Project Teams	
PT 61400-14	Declaration of sound power level and tonality values
PT 61400-24	Lightning protection for wind turbines
PT 61400-25	Communication standard for control and monitoring of wind turbine plants
Maintenance Teams	
MT 1	Design requirements for wind turbines
MT 11	Acoustic noise measurement technique
MT 12	Wind turbine power performance testing
MT 2	Safety of small wind turbines
MT 21	Measurement and assessment of power quality characteristics of grid connected wind turbines
MT 22	Revision of IEC WT 01, IEC System for Conformity Testing and Certification of Wind Turbines – Rules and procedures
MT 23	Full-scale structural testing of rotor blades

2.2.2 Offshore Windfarms

With experience gained from the certification process and operation of the first offshore windfarms, continuous development of certification procedures and regulations is expected. This is recognized by all parties involved. Even before IEC 61400-3 becomes a standard, it was clear that further work is needed to cope with the increasing complexity of structures. In the European Union, research for offshore wind turbine design and certification has been supported by several projects within the 6th Framework Programme e.g. *"Integrated Wind Turbine Design (UpWind)"* and *"Distant Offshore Windfarms with No Visual Impact in Deepwater (DOWNVInD)"* projects. Similarly, the US Dept. of Energy and the US National Renewable Energy Laboratory (NREL) perform research mainly focused on deep water systems within their *"Low Wind Speed Technology (LWST)"* offshore wind development project.

A main source of uncertainty in the design of offshore windfarms is the site assessment. Few measurements are available including both wind and wave measurements, as required for the analysis of offshore wind turbines. Often 'hind cast' models are used for the evaluation of site conditions. This situation is reflected in the standards, where the amount and quality of analysis need further refinement. The development of numerical hind cast methods and the outcome of the measurements performed at the German research platform in the North and Baltic Sea (FINO1) and the Danish Horns Rev sites will influence further development of site assessment and consequently the requirements stipulated in the standards. Additionally the measurements and advanced analysis will improve the models for turbulent wind speed and irregular sea states. New models, especially developed for offshore wind conditions, may replace existing models that have been 'borrowed' from onshore experience. Likewise, the shallow-water sea-state description, presently borrowed from offshore oil and gas industry experience, will be refined.

Due to the non-linear and dynamic behaviour of offshore wind turbines, *time domain simulation* is the state of the art for analysing loads on the turbine and the support structure. Interaction of wind and wave loading and structural response require integrated analysis. This method is well established and accepted for stochastic loading and the system response. One weak point is the description of the non-linear kinematics in stochastic wave fields. Further research is needed in this sector, since no widely applied engineering methods exist to consider this effect. Solutions found will directly affect standards and eliminate the deterministic regular waves used so far.

The *influence of the turbine controller* on overall behaviour and loading is considerable. Intelligent systems are developed today leading to active damping of the system and thus minimizing loads. This situation opens a new area of analysis, standards and certification procedures which may include control system software and hardware assessment.

New *foundation concepts* are being proposed to improve economics of larger turbines in very different sites, often in deeper water. Designs considered are steel or concrete tripods, and compliant structures, known from the offshore oil and gas industry. The existing design tools, software and analysis methods may have to be adapted to the needs of a tripod integrated-design. This is considered in several projects, e.g. the DOWNVInD project where *jacket foundation design* has been applied in a full-scale demonstration project with two wind turbines installed off the coast of mainland Scotland at 43 m water depth. Part of this project is the evaluation of certification requirements.

Another alternative considered is the *'suction bucket foundation'*, well known from the oil and gas offshore industry. Its application for wind turbines for transferring bending moments is an extended feature. The demonstration project (Vestas V90, 3MW turbine) in Frederikshavn, Denmark, has been evaluated and will lead to the development of specific rules. A further point of development is the *sea-bed soil behaviour* and its dynamic properties. Soil-pile interaction influences the overall dynamic behaviour of the offshore wind turbine system. The non-linear load carrying description of the soil (p-y curves) was originally developed and standardised for oil and gas structures and the parameters described in the standards were originally limited to pile diameters of 2-3 m. Moreover, little is known about long-term soil behaviour under cyclic load. The development of reliable standards requires further research in this sector.

When analysing offshore wind turbines, scour behaviour is significant. Scour prediction is an inexact science and very often model tests have to be performed. The experience gained from model tests, advanced analysis and experience from existing projects will influence design methods, and the procedures for scour protection stated in the standards. Scour protection methods used nowadays are taken from the oil and gas industry where smaller pile diameters are usually applied. The applicability of analysis methods and effectiveness of protection systems for wind turbine installations with pile diameters up to 8 m has still to be verified.

2.2.3 Deepwater Offshore

The installation of wind turbines in deep water leads to special challenges in design and certification. Not only the use of floating or compliant structures raises new aspects regarding system dynamics, inclination and moorings, but inspection and maintenance, especially of the underwater structure is much more complicated. The International Energy Agency (IEA) has been discussing the need for improvements of offshore analysis codes for all water depths. The committee decided to initiate a task to define coupled turbine/substructure dynamic modelling needs within the IEA Offshore Wind Energy Technology and Deployment, Subtask 2 – *"Technical Research for Deeper Water"*.

In the IEA Annex 23 *'Deepwater Offshore Turbine Modelling'*, participants investigated the capabilities and limitations of present offshore modelling. This includes modelling for both the wind turbine and the oil & gas industries. In addition, there will be assessment of the impact and sensitivity of wave loading on deep-water platforms, extreme breaking waves, ice loading and wave grouping in deep water.

2.2.4 Ocean Energy

There are other renewable energy opportunities offshore. The ocean contains enormous energy resource potential from tidal currents, waves, salinity gradients and ocean thermal energy. Developments in these sectors require standards and guidelines to guarantee reliability and safety. In 2004, at the IEC TC 88 meeting, it was decided to consider ocean energy devices in future tasks and GL set up a *"Guideline for the Certification of Ocean Energy Converters, Part 1: Ocean Current Turbines"* [13]. The European Marine Energy Centre (EMEC) published a standard [14] for testing ocean energy devices in 2004. Supported by the UK Carbon Trust, DNV developed a guideline for the certification of wave energy devices [15].

3. CERTIFICATION OF WIND TURBINES AND WINDFARMS

3.1 General

The European standard EN 45020 [16] defines certification of conformity as the activities of an independent third party to confirm that a product or service complies with a standard or a normative reference. Within the scope of certification of wind turbines, the actions during examination, assessment, measurements and evaluation are covered. Procedures for certification of onshore wind turbines were initiated with the commercial introduction of wind turbines more than 25 years ago and have continuously been updated and developed. In 1989 GL issued a first comprehensive certification procedure for international Type and Project Certification which in its roots still applies today, but was enhanced in order to cover the practical aspects and the experiences and developments in the field.

International standardisation on wind turbine certification procedures started in 1995 within IEC and resulted in the first issue of IEC WT01 [12] published by the Certification Assessment Board (CAB) of the IEC in April 2001. The certification procedure according to IEC WT01 [12] and the procedure according to GL [9, 17] have been introduced internationally and are the most important guidelines for certification of onshore and offshore wind turbines. Therefore, in this chapter, focus is put on these procedures only when describing the certification procedure for onshore and offshore wind turbines.

The certification procedures are subdivided into *Type Certification* (certification of the product) and *Project Certification* (certification of the project or windfarm). According to IEC WT01 [12] and GL [9, 17], Type Certification comprises in general the modules design evaluation, evaluation of manufacturing (quality management and implementation of design requirements in production and erection (IPE)) and prototype testing. Type Certification is more than just certification of a product design based on documentation, since it also includes an inspection of the manufacturing process at the manufacturer's and the sub-suppliers' premises, inspection of critical components and the testing of the prototype. When carrying out Project Certification, conformity is assessed and certified that type-certified offshore wind turbines and particular support structure designs meet requirements governed by site-specific external conditions, local codes and other requirements relevant to the site. Within Project Certification, according to GL [9, 17], the individual offshore wind turbines and windfarms are monitored during manufacturing, transport, installation and commissioning. Periodic monitoring is carried out at regular intervals.

The individual modules of the certification procedure are concluded with *"Statements of Compliance"* (GL) or *"Evaluation Conformity Statements"* (IEC WT01). Certificates are issued upon the successful completion of the relevant Type Certification or Project Certification activities, as described in detail in sections 3.3 and 3.4. Special attention is put on site evaluation for windfarms, considering the configuration and the environmental conditions (wind, waves, earthquake, complex terrain, extreme climate, etc.) at onshore and offshore sites.

3.2 Design Requirements

3.2.1 Safety Philosophy

In the first step of the evaluation of the design of the offshore wind turbine, special emphasis is put on the examination of safety and control. This provides appropriate protection against damage from all hazards during the planned lifetime. For offshore projects, special attention has to be paid to the different environmental conditions and the limited accessibility, although the aim is to maintain a close conformity with the onshore safety philosophy given in existing guidelines and standards [8, 17].

Offshore wind turbines are unmanned structures and the danger of harm to 'third persons' is less likely than onshore. For this reason, it is acceptable to reduce the level of safety. However, reliability of offshore wind turbines is much more important due to the reduced accessibility. In general, for offshore wind turbines only the ultimate and the serviceability limit states are considered. Fault conditions are included in the ultimate limit states. An accidental limit state, as used in oil and gas industry, is not considered since no life or pollution dangers exists. This is a major difference from oil and gas safety philosophy, where the structure has to withstand very rare events which lead to damage but, usually, not to collapse.

In comparison with onshore wind turbines, additional risks have to be taken into consideration for offshore installations, e.g. simultaneous action of wind and wave loading, impact from maintenance vessels. These risks, which are of mayor importance for safety, can be divided into different categories. These can be controlled by the following activities:

1. Loss of structural integrity:
- the loading (simultaneous wind and wave loading, boat impact, ice loading etc) shall be calculated according to accepted standards and codes
- structural components shall be designed according accepted standards and codes
- dual control for the design (independent party or certification body assessment)
- condition monitoring

2. Loss of turbine control
- required redundancy of the turbine control, e.g. control system is supervised by the safety system
- required redundancy of braking systems (independent, multiple braking systems)
- prototype test and commissioning to be witnessed.

3. Personnel safety
- usually operating instructions for offshore personnel and personnel safety is defined within national codes and is not specifically required in certification

4. Nautical and air traffic safety

- indentations of structure, marking, warning systems according to international or national rules, are no required part of certification
- lightning protection as part of the certification.

5. Fire protection

- fire-extinguishing and fire detecting systems were originally not part of certification, but are becoming increasingly important.

To ensure unobstructed operation of the offshore wine turbine the risks have to be observed carefully. The range of remote control possibilities for offshore wind turbines should be enlarged for an adequate possibility of turbine control and due to the fact that maintenance personnel has only limited access to the offshore wind turbine during weather windows and with considerable effort depending on the distance to the coast. In this context, the requirements on the safety system and on remote control of the rotor lock are to be adjusted accordingly.

Onshore, if the emergency system has been activated, fault clearance is only possible with the presence and active involvement of sufficiently qualified personnel at site of the wind turbine. For an offshore wind turbine, fault clearance can be allowed by remote control for a limited number of activations, when there is remote monitoring e.g. by cameras, microphones, by sufficiently qualified personnel. The outside of the turbine can be inspected using a monitoring camera on a neighbouring turbine, a boat or helicopter. The inside of the nacelle should be inspected after the clearance and during the start up, to check that the main components are in place, undamaged and operational. This inspection may also be made using monitoring cameras, microphones or other suitable methods.

For the electrical equipment, there are additional requirements concerning medium voltage (e.g. protective measures, pressure relief, SF6 switchgear and cables), backup power supply, sub stations, transformers and generators.

3.2.2 Loads

Loading of an offshore wind turbine represents a major topic in all related standards and guidelines. Even the installation of structures in shallow water, with its dynamic nature, results in difficulties. The development of standards and guidelines was performed in parallel with first installations and in close cooperation of the participating organisations. The results from EU funded projects such as RECOFF, OWTES and Opti-pile were applied in the development. Many similarities of IEC 61400-3 [7] and the GL guideline [9] can be found.

However, some major differences between standards can be found: While IEC 61400-3 is based on the 3rd Edition of the IEC 61400-1 [8], the GL guideline [9] is harmonised with the GL guideline for onshore wind turbines and the 2nd edition of the IEC 61400-1 in the modified version of EN 61400-1 [42]. This results in different definitions of the external conditions (characteristic vs. mean turbulence intensity) and load case definitions (extrapolation of extreme operating loads or deterministic load cases). A comparison of the governing wind conditions according to GL [9] and IEC 61400-3 [7] can be found in tables 3 and 4.

Table 3. Definition of the type classes according to draft IEC 61400-3 [7]

Wind Turbine Class	I	II	III	S
V_{ref} [m/s]	50	42.5	37.5	Values Specified by the Designer
A I_{ref}		0.16		
B I_{ref}		0.14		
C I_{ref}		0.12		

Table 4. Definition of the type classes according to IEC 61400-3 [8] and GL [9]

Wind Turbine Class		I	II	III	S
V_{ref}	/m/s	50	42.5	37.5	Site Specific
V_{ave}	/m/s	10	8.5	7.5	
A	I_{15}		0.18		
	a		2		
B	I_{15}		0.16		
	a		3		
C	I_{15}		0.145		
	a		3		

GL and IEC include a new turbulence class "C", dedicated to offshore conditions. The differences in the wind speed turbulence-intensity definition are not as extreme as the values in the tables impose. Figure 1 shows the variation of the turbulence intensities over mean wind speeds.

The description of the marine conditions is based on the standard approach used for the offshore oil and gas industry. Special problems arise when analysis is performed in the time domain for shallow waters. In these cases, it is essential to consider the influence of the shallow water and bottom inclination on the wave field and its distribution. The short-term wave distribution is distorted and the maxima distribution is not Rayleigh distributed, as it is in deep water conditions. GL [9] includes the option to use modified wave spectra according to Texel, Marsen, Arsloe (TMA) [44] to consider the different frequency contents of the waves in stochastic wave field simulations. The different distribution of the maximum wave elevation can be considered by applying shallow water distributions as, for instance, that proposed by Battjes [18].

One major problem in establishing the design environmental conditions for offshore wind turbines is the combination of the external conditions to derive design loads. Contrary to other offshore structures where wave loads usually dominate loading, offshore wind turbine support-structures may equally be loaded by *wind, wave and sea ice forces*, with currents of minor importance in shallow waters. Analysis has shown that, in most cases, wind and wave loading are the two main sources of loading. In some areas like the northern Baltic Sea, sea-ice loading, combined with extreme wind may be the design driving actions. It is generally accepted, that sea-ice does not occur in combination with waves, at least in the near shore conditions where offshore wind turbines are erected.

Several methods on the combination of external conditions exist, showing a different amount of conservatism and data to be considered. In general the approach of "the less the information on real conditions, the greater the conservatism in combination" has to be followed in the design of offshore wind turbines. A classification for the extreme event analysis may be seen in Table 5.

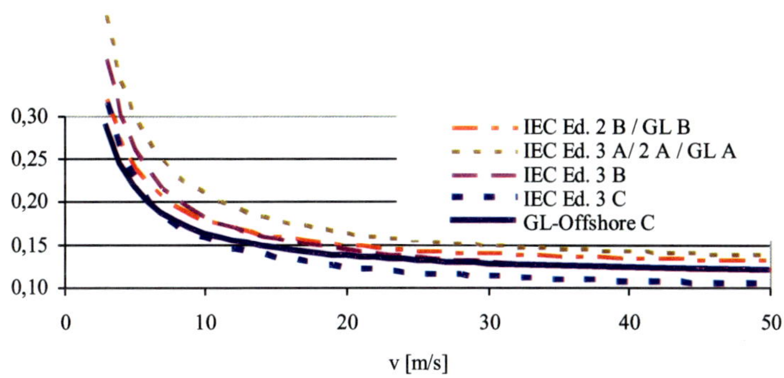

Figure 1. Turbulence intensity according vs. wind speed according to different standards/ guidelines.

Table 5. Comparison of possible combinations of extreme wind and wave events

Method	Example	Complexity	Loads	Probability
Add extremes	50-year gust + 50-year design wave	Very simple	Very conservative	Very unlikely
Extremes occur during same storm with 50-year recurrence period	50-year gust + reduced design wave reduced gust + 50-year design wave	Simple	Conservative*	Very low, but may happen
Extremes during the same storm have different individual long term probability of occurrence	50-year gust + 10-year design wave 10-year gust + 50-year design wave	Simple	Conservatism or not depends on the site**	Site specific
Combination of probability density functions for individual storms	Mean wind speed and sea state with a combined probability of 50-years are derived, theoretical assumption for individual extremes during the storms corresponds to: n1-year gust + n2-year design wave, ... ni-year gust + nj-year design wave	Difficult, individual probability distributions needed, many realisations	Accurate***	Realistic***
Combination of probabilities of individual storm events	Extreme values are combined according to their existing joint probability of occurrence The extremes are analysed as a function of the number of storms	Very high	Best fit	Realistic

* Usually only the two extreme cases of the joint probability during the storm are considered, which is correct for linear loading. This may lead to some errors if intermediate combinations of higher order loads dominate. In practice this case occurs seldom and the error is small.

** The combination 50-year + 10-year may have to be changed at some sites to avoid over- or non-conservative results. The method is used for the North Sea.

*** Provided the short term analysis takes care of shallow water and other site specific events

In GL [9] and in IEC [7], the assumption is made that extreme mean wind and extreme sea state occur at the same 50-year storm. The short-term fluctuations are not correlated. Of course, if better knowledge from measurements exists for the site, these real conditions should be applied.

For *Type Certification* without knowledge of any specific site, GL [9] gives some simple methods to estimate the wind-speed to wave-height relation, by using the JONSWAP wind-generated wave spectrum in conjunction with the TMA filter. This approach is sufficient for Type Certification, where only the structure itself is considered, but is seldom applicable for site-specific analysis.

For the relatively shallow sites of offshore wind turbines, *non-linear wave kinematics* has to be considered. Unfortunately, there is no engineering method widely applied to consider non-linear kinematics in stochastic wave fields. To overcome this drawback, GL [9] recommends that both stochastic wind and wave field simulations are performed, or that deterministic analysis is used. The load cases using a stochastic model of the environment only consider linear wave kinematics, but result in a proper representation of the systems dynamics. In contrast, the load cases with deterministic gusts and periodic waves consider non-linearity correctly, but lack dynamic response considerations.

For the inclusion of breaking waves, the wave and sea bottom conditions leading to wave breaking need to be defined, indicating the type of wave breaking mechanism. GL [9] gives a method, developed by J. Wienke [19] to calculate the impact load from breaking waves.

3.2.3 Support Structure

3.2.3.1 General

During the lifetime of an offshore wind turbine, the structural integrity depends on several factors. The *serviceability and functional* considerations may influence the structural design, the quality of materials and construction, and the necessity to *withstand the loading* described above (ultimate limit states). '*Serviceability limit-states*' include the maintenance of a safety gap (distance of rotor blade to tower, or of rotor blade to guy wire), limited vibration amplitudes and accelerations.

The support structure for offshore wind turbines comprises *the foundation, the substructure and the tower*; see Figure 2. There are different design options for foundation, e.g. monopile, gravity or a variety of jacket platforms (braced structures).

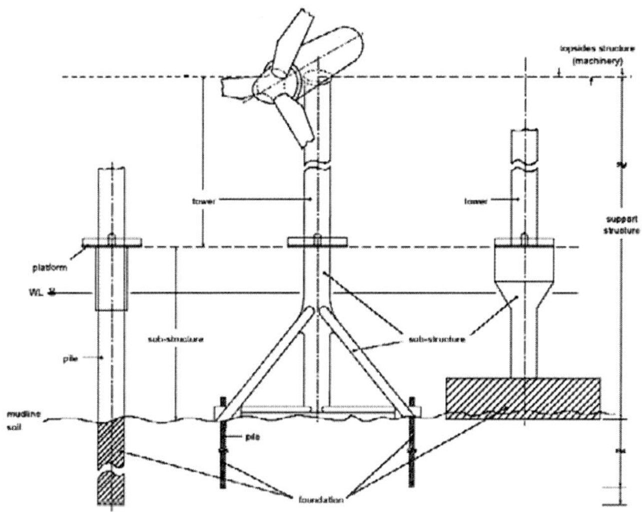

Figure 2. Definition of offshore wind turbine structures

3.2.3.2 Aspects of Design Requirements

The evaluation of the support structure is performed based on *site-specific loads*. Therefore, the dynamic behaviour of the complete structure has to be taken into account, including the mass distribution of the support structure, the nacelle and the rotor, and the soil stiffness. The state-of-the-art is to determine the section forces of the support structure by time domain computations, considering simultaneous load effects and structure elasticity. If forces for fixed offshore wind turbines can only be determined with the aid of a quasi-static analysis, a *dynamic amplification factor*, which represents the dynamic response of the offshore wind turbine on varying wind and wave loading, has to be taken into account.

Rotor induced vibrations, where the exciting frequency of the rotating rotor is close to the natural frequency of the complete system (proportional difference less than 5%) or to the wind-induced transverse vibrations, should be avoided adequately, e.g. by damping devices, by oscillation monitoring and shut-down.

Once the loads have been derived, the analysis of *ultimate and fatigue strengths* should be carried out for all load-carrying structures and connections, with consideration of the quality grades of materials used, the welding specifications and the manufacturing tolerances. Structural elements subjected to predominantly compressive or shear stresses, which may also result from bending or torsion, are to be investigated for overall and/or local buckling failure. Buckling calculations may result in special manufacturing instructions regarding tolerable imperfections.

The material safety factors for steel structures are identical to those used onshore, see Table 6.

Table 6. Partial material safety factors (GL [9] and IEC [7])

	Fail-safe	non fail-safe
Accessible, regular maintenance and surveillance	1.0	1.15
Not accessible, no maintenance and no surveillance	1.15	1.25

Besides the investigation of *ultimate strength, fatigue strength and buckling*, the investigation of the *loss of static equilibrium,* (e.g. overturning of the foundation) and acceptable deformations has to be carried out.

An example of ultimate limit states for concrete support structures is the crack-formation limit state. Crack widths shall be kept within acceptable limits. The stress or strain limit state is for instance defined by the limitation of tensile and/or compressive stresses in concrete structures, especially pre-tensioned members, or the precondition that steel structures shall not undergo plastification under characteristic actions.

3.2.3.3 Connections

Connections within the support structure may be designed as *bolted, grouted or welded connections*; special care shall be taken about stress concentrations, imperfections in production and fatigue. Grouted pile connections are well known from the offshore oil and gas industry. However, in offshore wind energy grouted pile to structure connections experience bending moments and shall be assessed by suitable analysis, e.g. FEM, or by tests. Due to limited experience, the safety factors for grouted connections are considerably more stringent than for most materials. Fatigue calculations show large variations in resulting damage due to the sensitivity of non-linear effects. The suitability of the grout mix shall be proven prior to the installation. The compressive strength shall be confirmed by laboratory tests of grout samples which were mixed and cured under field conditions. Parametric formulae applicable for the design of axially loaded grouted connections are provided by the American Petroleum Institute (API) [21] and the UK Department of Energy.

Procedures for *bolted connections and tubular joints* are given in [9] and [21]. For bolted flange connections, special care shall be taken to guarantee evenness of contact surfaces to avoid overstressing of bolts. For tubular joint connections, attention shall be paid to ensure quality of welding which is of major importance for this type of connection. GL guidelines [9] give detailed guidance on execution and testing of tubular joint welds. If possible, connections in the splash zone should be avoided. Cast components may be an alternative to welded joints.

3.2.3.4 Soil

The soil parameters may vary considerably within the windfarm. Therefore, the determination of the soil properties has to be carried out with special attention for an adequate number of locations within the windfarm at an early state of the design. The selection of the foundation type, the assembly procedure and the soil preparation (scour protection) depends on the soil capacity and soil properties at each wind turbine location. Consequently, different foundation designs may be needed within one windfarm.

Guidance for deriving the horizontal and vertical soil parameters are given, for example in the *"Standard for Geotechnical Site and Route Surveys"* of the Bundesamt für Seeschifffahrt und Hydrographie (BSH) [20]. Recommendations for the type and extent of site investigations are stated in this standard, too. Assessment criteria for foundations can be found in [9]. For horizontally loaded piles, the *"vertical tangent"* criteria has been introduced, which means that the pile should remain vertical and fixed under extreme loads. This requirement replaces the *"zero toe kick"* criteria of former standards, as used in oil and gas industry [21].

Special care has to be taken on the short and long-term interaction between the *support structure and the soil*. The interaction of structure and soil may behave non-linearly resulting in a possible reduction of the foundation stiffness as a function of load level and type. The dynamics of the installed offshore wind turbine influenced by water depth variations and different soil stiffness in windfarms have to be verified and checked against the dynamics included in the load assumptions. If necessary, the load assumptions have to be revised and the design of the support structure reviewed. Further, the following risks should be investigated: settlements and seabed movements, liquefaction of the soil, the possibility of soil instability for normal or soft consolidated clays and loose sand.

3.2.3.5 Scour

Scour is erosion of the seabed by currents or waves, which is particularly likely to occur if structural elements disturb the currents. Scour is likely to reduce vertical and lateral support for foundations, e.g. causing undesirable settlement of gravity foundations. The possibility of scour or undermining around the foundation shall always be investigated and if necessary avoided. This can be accomplished by a scour protection feature which should function during the whole lifetime of the offshore wind turbine. Inspection is needed at regular intervals, since analysis alone is not reliable enough.

The *scour protection* has to be developed and tested by model tests. As an alternative, the foundation has to be treated as partly unsupported. If reliable data are unavailable for the specific site conditions, the scour depth around pile foundations may be estimated as 2.5 times the pile diameter for design purposes. Less conservative estimates may be made, but have to be verified by regular monitoring and inspection. In any case, the scour protection has to be inspected and scour erosion found has to be refilled. The period between inspections may be adjusted with the experience gained at the particular site.

3.2.3.6 Corrosion protection

Offshore wind turbines are exposed to aggressive environmental conditions and marine climates. Corrosion protection depends on the selection of suitable materials, appropriate coatings, protective films and regular inspection. The selection of mechanical and electrical components must include their liability to corrosion, and the effect of inadvertent corrosion on their function, e.g. jamming of rusted joints or failure of sensors.

Corrosion may be allowed only for components of minor significance, e.g. for structures having a short service life or where regular inspections and repairs are possible.

Corrosion damage may be prevented by the following corrosion protection measures:

- *designing* the systems and components through the application of suitable structural measures according to EN ISO 12944 part 3 [22]
- *influencing the characteristics* of the 'reaction partners' and/or modification of the 'reaction conditions'
- *separating metals* from surrounding electrolyte by protective layers ("passive corrosion protection")
- *electrochemical action*, e.g. cathodic protection ("active corrosion protection"); which is always required for submerged structural members of offshore wind turbines.

3.2.3.7 Material selection

Selection of materials is a vital part of design. The materials of structural members shall be suitable for:

- the environmental conditions
- the loading
- long-term durability.

For steel, detailed requirements are presented in [9] and [17]. Beside requirements on the strength of materials to withstand extreme and fatigue loads, the selection criteria are e.g. thickness of products (relating to welding criteria), design temperature and importance of the structural members. Further limitations regarding chemical composition and impact energy should be mentioned.

For concrete structures, major importance lies on protection of the reinforcement steel against corrosion. This will be ensured by adequate concrete composition and concrete cover and limiting the maximum crack width.

3.2.4 Machinery

3.2.4.1 General

For the machinery part of the offshore wind turbine strength calculations for all components and connections have to be carried out with respect to ultimate limit states and serviceability limit states. The strength calculations will be based on the design loads considering applicable analysis methods for the respective components. Specifications of all hydraulic fluids and lubricants, cooling systems, heating mechanisms and material quality have to be examined, too. Additionally special requirements for offshore application have to be met. The *"marinisation"* of the machinery components includes among others air flow, air filtration, heating/cooling, salt filters, corrosion protection and protection classes and encapsulation of the nacelle due to the offshore environment.

Because of the likelihood of poor accessibility for repair, the calculation procedure for pitch and yaw drives has been enhanced and the material safety factors against tooth root fracture and pitting are increased compared to onshore sites. Gearbox requirements include: (i) a running-in period under partial load have to be observed, and (ii) design against unacceptable loads caused by pendulousness of the rotor during standstill of the offshore wind turbine. The latter requires suitable methods. e.g. parking brake on the low-speed shaft or a rotor lock which can be activated by remote control.

3.2.4.2 Offshore environmental conditions

The components inside the offshore wind turbine that are necessary for the operation require specific atmospheric conditions. These conditions should be defined, together with the way these requirements are met inside the nacelle, hub or other spaces.

Relative humidity inside the offshore turbine of more than 70% should be avoided to keep the corrosion rate at a normal level. This can be achieved by keeping the inside temperature 5 K higher than the outside temperature. The compliance of the atmospheric requirements should be monitored by the control system of the offshore wind turbine.

For all *components unprotected against corrosion* and for all operating materials (e.g. lubricants, oils) which are in direct or indirect contact with the offshore atmosphere it has to be ensured that these are adequate for the use in an offshore atmosphere and that the function will not be disturbed.

For components protected against corrosion, corrosion classes according to EN ISO 12944 [22] are introduced. Following that standard outside components, fittings, sensors etc. shall be protected against corrosion according to class C5-M, inside surfaces, directly exposed to outside air shall be protected against corrosion according to class C4 and inside surfaces, not directly exposed to outside air, shall be protected against corrosion according to class C3.

3.2.5 Differences Between Onshore and Offshore Applications

Compared with onshore wind turbines, offshore turbines suffer additional load impacts from waves and currents which coincide with the loads from the wind. The nature of each type of load is physically different, therefore simultaneous analysis of wind and wave loads within the load simulations leads to problems on how to combine these loads as acting on the wind turbine [23].

For load analysis, the *loads resulting from wind, waves, currents and ice drift* are to be considered. If no data are available for the sea climate parameters at the specific site, assumptions may be derived from the respective standards and guidelines listed in section 2.1. These documents have been issued especially for offshore wind turbines, whereas the IEC standards, the European standards and the Eurocodes define general requirements for wind turbines and structures. The standards and guidelines in section 2.1 define requirements for the design and components of offshore wind turbines. The loads with respect to the normal and extreme load cases are derived by combining the normal and extreme wind events with reduced wave loading, or by combining the extreme waves with reduced wind loading.

Environmental conditions offshore require some modifications in the design of wind turbines. It is widely accepted, that offshore wind turbines are not to be designed to the same safety class as onshore turbines, because the *public safety offshore is regarded differently*. Less rigorous safety factors may be applied. As the surrounding area of offshore wind turbines is not inhabited, noise emission and the impact on the landscape might also be less important. More effort has to be put into the reliability of the wind turbines, remote controls and enhanced intervals of inspection. If an onshore turbine cannot be put into operation by the remote control, a service team may inspect the defect turbine within hours. A similar defect offshore may result in the situation that the turbine needs to be put out of operation for a long time if weather conditions do not permit access. Remote controls for offshore turbines therefore should be more comprehensive than for onshore turbines. For instance, more attention is needed for scheduled inspection of support structures, requiring underwater survey of scour protection, marine growth, and corrosion protection.

One of the most important design changes for offshore turbines is *improved corrosion protection*, including the inner part of the nacelle, cranes, boat landing, splash zone and facilities for access by helicopters. In order to reduce cabling costs, it may be necessary to install transformers in each of the wind turbines, preferably even within the nacelle. Special air conditioning inside the nacelle is another feature to avoid problems arising from the offshore environment.

3.3 Type Certification

3.3.1 GL Guideline

A very important part of the Type Certification is the evaluation of the design documentation, a thorough design review with respect to the requirements defined in the relevant codes and standards. The testing of the prototype wind turbine represents the practical aspects of the Type Certification, whereas the manufacturing evaluation confirms the manufacturer's quality management system to be in accordance with EN / ISO 9001:2000 [24] and the ability of the manufacturer to manufacture the components and assemble the turbine according to the design requirements. An overview of the procedures for Type Certification according to GL [9, 17], applicable for onshore and offshore wind turbine certification, can be seen in Figure 3. Once all modules have been concluded by Statements of Compliance and the Final Evaluation was carried out successfully, the Type Certificate will be issued by the certification body.

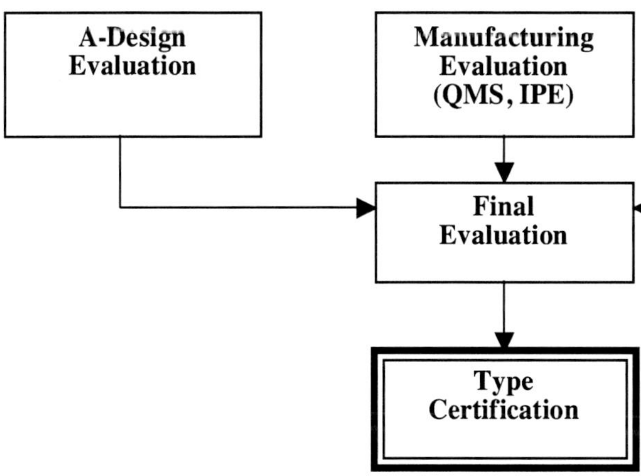

Figure 3. Modules of Type Certification

3.3.1.1 Design Evaluation

The evaluation of the design is performed along the steps ("elements") of evaluation, as listed in Table 7. It is generally carried out in two sequential steps. The first part covers all aspects of the safety and control, the load assumptions and the load calculations. The loads for offshore wind turbines should be analysed applying aero elastic codes, stochastic wind fields and modal or finite element analysis techniques [25]. Additionally, the loads have to contain both the aero elastic and the fluid-structure interaction of the submerged part. The latter may have a considerable influence on the structural response for certain types of support structures [26] and in some cases directly influence the loading of the machinery.

Table 7. Elements of Design Evaluation

•	Load Assumptions
•	Safety System and Manuals
•	Rotor Blades
•	Machinery Components
•	Nacelle Housing and Spinner
•	Tower and Foundation / Support Structure
•	Electrical Equipment and Lightning Protection
•	Condition Monitoring System

During the second part of the Design Evaluation, all components (rotor blades, machinery, tower electrical equipment and optional foundation/support structure) should be examined on the basis of the previously approved loads and the relevant standards and regulations. At the end of the Design Evaluation, *manuals and procedures* for transport, erection, start-up, commissioning, operation and maintenance are checked for suitability, completeness and compliance with the assumptions in the design documentation. Static rotor blade testing [27] forms an integral part of the Design Evaluation of the blade. The prototype test of the main gearbox on the test bench should be completed successfully as a requirement within the evaluation of the machinery components. Lightning protection [28] will be assessed under the scope of the electrical equipment. A flow chart of the Design Evaluation is shown in Figure 4.

The Design Evaluation is subdivided into C-, B- and A-Design Evaluation.

> *The C-Design Evaluation* (prototype assessment) is issued for the *prototype of a wind turbine*. As a rule, power and load measurements should be performed on the prototype, and the results should be compared with the calculated values. Modifications of the control system are permissible, provided that the resulting loads do not change considerably. The C-Design Evaluation requires usually a complete plausibility check of the loads, the rotor blades, the machinery components and of the tower and foundation / support structure. Before going offshore, prototypes of offshore wind turbines are usually tested onshore. National or local regulations may require that the tower and foundation and/or other components be subject to a complete analysis. The final step of a 'positive' evaluation is to issue a Statement of Compliance for the C-Design Evaluation. which is valid for test operation for a maximum of 2 years or 4000 full load hours. At the end of this period, the B-Design Evaluation should then be obtained as soon as possible.

> To attain the *A- or B-Design Evaluations*, a complete examination of the design analyses including load calculations for the respective offshore environment and support structure. All required material and component tests as indicated above are required, together with witnessing of the commissioning of one of the first wind turbines. Following completion, the certification body will issue a Statement of Compliance for A- or B-Design Evaluation.

The B-Design Evaluation may be issued for *items that are still outstanding*, providing these are not directly safety-relevant. The B-Design Evaluation has a validity period of one year. During the validity period, all installed wind turbines of this type shall be reported to the certification body and requirements as per the outstanding items shall be resolved during that period.

The A-Design Evaluation is issued if *no outstanding items are present and the gearbox tests on the test bench have been completed*. The A-Design Evaluation is valid indefinitely, as long as the design is not modified with respect to the requirements.

The examination of a support structure (tower, substructure and foundation) is optional within the scope of the A- or B-Design Evaluations for offshore wind turbines. The dynamic influence of the support structure is to be considered in the load assumptions. After successful evaluation the Statement of Compliance for the Design Evaluation and the respective Certification Reports on load assumptions, safety system, rotor blades, machinery components, tower, substructure and foundation, electrical installations, commissioning, hub and nacelle cover will be issued.

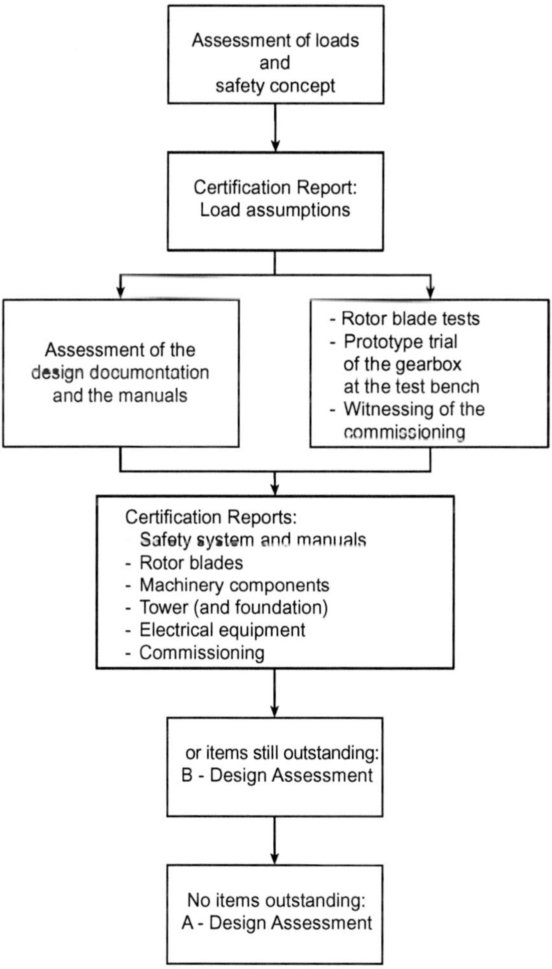

Figure 4. Procedure for the A- and B-Design Evaluation

3.3.1.2 Manufacturing Evaluation

The evaluation of the manufacturer's quality management covers the whole range of activities necessary to confirm the quality of the product. The certification of the manufacturer's quality management system (QMS) according to DIN EN ISO 9001 [24] (including design) covers a large portion of these requirements.

However, the link between quality management and product quality needs to be specially addressed. It shall be ensured that the requirements stipulated in the technical documentation with respect to the components are observed and implemented in production and erection (IPE). The respective IPE evaluation requires a verification to be carried out by inspection, to check that at least one wind turbine is manufactured following the design requirements under certification (topics to be assessed among others: work shop drawings and instructions, purchase specifications, installation instructions, fabrication methods, qualification of personnel, material certificates, random checks on fabrication, etc.).

For all relevant components, a single fabrication inspection is required. For multiple component suppliers, the inspection may be limited to one selected supplier if the component specification is identical for all the other sub-suppliers. If the inspection reveals fabrication problems or deviations from the certification documentation, more than one inspection will be necessary. Therefore, the requirements for conformity assessment of designs, workshops and special fabrication techniques remain a necessary part of the (type) certification procedure.

3.3.1.3 Prototype Testing

In order to validate the design calculations, to optimise control, performance and noise behaviour and to verify the performance of the safety and control systems, prototype testing becomes an integral part of the design and certification process. The wind turbines, at which measurements or gearbox trials are carried out, should conform as much as possible to the design for which the Design Evaluation is issued. Compliance should be confirmed by declaration of the manufacturer. The elements of Prototype Testing listed in Table 8 should be verified by measurements based on the relevant standards. For the incorporation of measurements in the certification process, either, (i) the measurements shall be performed by independent bodies accredited according to ISO / IEC 17025 [29],or, (ii) the certification body or an accredited laboratory shall witness and approve the measurement set-up, the calibration and carry out plausibility checks.

The *prototype tests* performed at a prototype turbine onshore are generally also sufficient for the offshore turbine for prototype testing. If a prototype turbine onshore is used for the prototype tests, it is recommended that a site close to the seacoast be chosen, so the marine atmospheric and wind conditions are comparable.

Load measurements are a central part within the evaluation of the prototype testing. The measurements are compared with the calculated loads of the Design Evaluation. In this context, mainly transient load situations (such as start and stop procedures, emergency shutdown) and the operational loads for the investigation of the fatigue loading are validated. As the wind conditions at the prototype

Table 8. Elements of Prototype Testing and the Respective Standards

Elements of Prototype Testing	Codes or Standards to be applied
Power Curve	IEC 61400-12 [30]
Noise Emissions	IEC 61400-11 [31]
Actions, loads and stresses, dynamic behaviour	IEC TS 61400-13 [32]
Electrical Characteristics	IEC 61400-21 [33]
Blade Tests	IEC TS 61400-23 [27]
Prototype trial of the gearbox at the wind turbine	GL [9, 17]
Commissioning, Safety and Function Test	GL [9, 17]

site, onshore or offshore, generally deviate from the generic conditions assumed in the Design Evaluation, new load simulations using the site-specific wind characteristics may have to be performed for a proper comparison.

3.3.2 IEC WT01

IEC developed a *certification system for wind turbines*, described in WT01 [12]. One objective of the IEC system and the system described in section 3.3.1 is to facilitate international trade of wind turbines. Figure 5 gives an overview of the modules for Type Certification according to this system. Once all modules have been concluded by an evaluation report, the documentation of the findings will be resumed within the final evaluation. The final evaluation also covers the assessment of the manuals and the preparations to issue the Type Certificate.

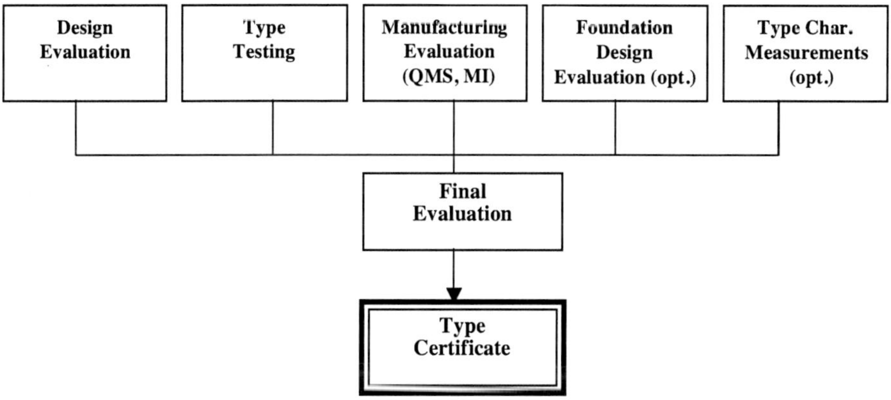

Figure 5. Modules for Type Certification according to IEC WT01 [12]

3.3.2.1 Design Evaluation

The module Design Evaluation is subdivided into the elements, listed below:

- Evaluation of Control and Protection System
- Evaluation of Loads and Load Cases
- Evaluation of Structural Mechanical and Electrical Components
- Evaluation of Component Tests
- Evaluation of Foundation Design Requirements
- Evaluation of Design Control
- Evaluation of Manufacturing Plan
- Evaluation of Installation Plan
- Evaluation of Maintenance Plan
- Evaluation of Personnel Safety

The assessment of the documentation of the elements mentioned above must be carried out by an accredited certification body. The control and protection system being the supervisory part of the wind turbine is to be assessed with respect to its fail safe design, the system logic and hardware implementation. Focus is to be put on the analysis of the braking system. The control and protection system evaluation forms the platform for the load case description and consequently the load calculation itself.

After assessment of the loads, the structural and mechanical components shall be verified against their strength, stability and fatigue resistance. The examination of the electrical components is also covered by this element. As no IEC standards are available for the design and evaluation of the machinery components and the support structure GL [9, 17], or other accepted guidelines may be applied to fill these gaps.

If certain material data are not available or if functionality of components is questioned, the certification body may require *material or component testing* for design values.

The documentation of the foundation/support structure design requirements should contain a summary of information to enable a design of the foundation/support structure covering the requirements stipulated in the Design Evaluation. Among others, the design loads (ultimate and fatigue), geometry of tower-foundation/support structure connection, required soil properties and tolerances must be documented and are to be assessed by the certification body.

Evaluation of the *manufacturing plan* forms the link to the practical aspects in manufacturing. Critical processes shall be identified. The manufacturing shall be specified by indicating the human resource requirements, skills, tooling, required fixtures and quality check points, all accompanied by good quality record keeping.

The *installation plan* must enable, for a typical site, the required design assumptions for installation to be met. The installation plan shall focus on the same items as the manufacturing plan, but, additionally, include personnel safety, environmental protection, commissioning procedures and technical specifications for construction works.

The *integrity of the turbine* over time depends on the maintenance plan. Here, the scheduled maintenance actions, including inspection intervals and routine actions, are the central items. The manuals for maintenance and the operating instructions are vital.

Personnel safety forms the closing part of the Design Evaluation. Compliance with these requirements shall be assessed by the certification body. Among the topics addressed are the climbing facilities, access ways and passages, platforms and floors, fixing points, electrical earthing systems and emergency stop buttons.

The *successful evaluation* of all elements will be concluded with the Design Evaluation Conformity Statement of the certification body.

3.3.2.2 Type Testing

The purpose of Type Testing is to verify power performance and aspects that are vital for safety and need additional experimental verification, and aspects that cannot be reliably evaluated by analysis. Type Testing comprises the elements indicated Table 9

Table 9. Elements o Type Testing and the Respective Standards

Topics for Type Testing	Codes or Standards to be applied
Power Performance Measurements	IEC 61400-12 [30]
Safety and Function Test / Commissioning	-
Load Measurements	IEC TS 61400-13 [32]
Blade Test (static and fatigue)	IEC TS 61400-23 [27]
Other Tests	-

The demonstration of the control and protection system functions should be evaluated and the certification body shall verify if the turbine under test displays the behaviour predicted in the design. *Power performance measurements* are carried out to verify that the measured power curve and predicted annual energy production for the wind turbine type comply with the design values. The measurement procedures shall be in accordance with IEC 61400-12 [30]. *Load measurements* are carried out in order to enable a verification of the turbine loading under specific conditions. The certification body shall assess if the loads measured on the turbine (which shall be structurally and dynamically similar to the turbine submitted for Design Evaluation) are similar. As normative reference, IEC 61400-13 [32] should be applied for load measurements.

The purpose of the blade test is to verify the structural design of the blade and to assess the suitability of the manufacturing process. Full scale testing shall be carried out with respect to fatigue and static loading. Guidance for the test is given in IEC TS 61400-23 [27].

If necessary, the certification body may require other tests. Such test may be e.g. electromagnetic compatibility.

3.3.2.3 Manufacturing Evaluation

The Manufacturing Evaluation consists of the two elements, *"Quality System Evaluation"* and *"Manufacturing Inspection"*. The purpose of the manufacturing evaluation is to assess if a specific wind turbine type is manufactured in conformity with the design documentation. The requirement for the quality system evaluation is satisfied if the quality system is certified to comply with EN ISO 9001:2000 [24]. If not, the certification body shall evaluate the system of the applicant.

For the *manufacturing inspection*, the certification body shall verify that at least one wind turbine is manufactured according to the design under certification. The inspection must include the following items:

- verification that the design specifications are properly implemented in the workshop drawings, instructions, purchase specifications etc,
- verification of fabrication methods, procedures qualification of personnel,
- review of material certificates,
- other items as specified.

3.3.2.4 Foundation Design Evaluation

A *foundation or support structure design* is optional in the Type Certification procedure, according to IEC [12]. If inclusion is desired, the certification body shall assess whether the foundation and support structure design is in accordance with the foundation specifications detailed in the design documentation and with the agreed standards and codes. If relevant, the documentation should include the concrete lay out and construction sequence plans for verification.

3.3.2.5 Type Characteristic Measurements

Optional Type Characteristic Measurements may complete the evaluation process for the Type Certification. The optional elements in that respect are the power quality measurements and the acoustic noise measurements. Again, measurements require an accredited test laboratory and they shall be in compliance with the respective standards, such as IEC 61400-21 [33] for power quality and IEC 61400-11 [31] for acoustic noise. The test reports shall be evaluated by the certification body for conformity with the applied standards, measurement conditions, instrumentation calibration and analysis.

3.3.3 Comparison of the Type Certification Procedures of IEC and GL

Further to GL certification [9, 17], additional requirements specified in the following sections are needed for the Type Certification of IEC WT01 [12].

3.3.3.1 Design Evaluation

Design Evaluation according to IEC WT01 [12] requires additionally:

- *Evaluation of manufacturing plan*
- *Evaluation of installation plan*
- *Evaluation of personnel safety.*

The content of a manufacturing plan is partly contained in the drawings, which are part of the Design Evaluation according to GL [9, 17]. However, additional documentation is required describing the manufacturing process and the quality management procedures. The installation plan is part of the erection manual required for Type Certification, according to GL [9, 17]. Surveillance of the erection and installation will be treated within the Project Certification, as the installation plan is part of the transport and erection surveillance module. Personnel safety is an important part of the certification procedure and should be assessed [34]on the basis of EN 50308 [35]. A comparison of the activities in Design Evaluation of IEC WT01 [12] and GL [9, 17] is given in Table 10.

Table 10. Comparison of the elements of Design Evaluation of IEC WT01 [12] and GL [9, 17]

Elements of Design Evaluation	IEC WT01 [12]	GL [9, 17]
Evaluation of Control and Protection System	X	X
Evaluation of Loads and Load Cases	X	X
Evaluation of Structural Mech. and Elect. Components	X	X
Evaluation of Static Blade Test	Within PT	X
Evaluation of Component Tests	X	X
Evaluation of Foundation Design (optional)	X	X
Evaluation of Design Control	X	Within ME
Evaluation of Manufacturing Plan	X	Partly *)
Evaluation of Installation Plan	X	Within PC
Evaluation of Maintenance Plan	X	X
Evaluation of Personnel Safety	X	Optional
Witnessing of Commissioning	No	X

*) defined by specifications, drawings, schematics, etc

(ME-Manufacturing Evaluation, PC-Project Certification, PT-Prototype Testing)

3.3.3.2 Prototype Testing

Prototype Testing differs between GL and IED within the following elements:

- *Evaluation of dynamic blade testing*
- *Optional in IEC-WT01: Noise measurements*
- *Optional in IEC-WT01: Power quality measurements.*

The evaluation of the dynamic blade test is based on a full-scale fatigue test. In GL [9, 17] this requirement is reduced to fatigue tests of specimen structural parts. However, a full-scale static blade test is required in both. The GL blade test [9, 17] is part of the design evaluation of the rotor blade. The noise and power quality measurements are mandatory within the GL Prototype Testing [9, 17], however IEC WT01 [12] treats these optional. Table 11 compares the requirements.

Table 11. Comparison of the elements of Prototype Testing of IEC WT01 [12] and GL [9, 17]

Elements of Prototype Testing	IEC WT01 [12]	GL [9, 17]
Safety and Function Test / Commissioning	yes	Yes
Power Performance Measurements	Yes	yes
Power Quality / Electrical Characteristics	Optional	Yes
Noise Emissions	Optional	Yes
Load Measurements and Dynamic Behaviour	Yes	yes
Evaluation of Dynamic Blade Test (fatigue)	Yes	No
Evaluation of Static Blade Test	yes	within DE*

(*DE- Design Evaluation)

3.3.3.3 Manufacturing Evaluation

The evaluation of the Quality Management System as part of the Manufacturing Evaluation is based on IEC ISO 9001:2000 [24] and is mandatory within both GL and IEC certification procedures, Table 12. The manufacturing inspection requires the inspection of at least one wind turbine manufactured according to the design under certification and for the verification of fabrication methods, review of material certificates and design specifications. These actions are mandatory according to GL [9, 17] within the scope of the IPE Evaluation (Implementation of Design Requirements in Production and Erection) within the Type Certification process. The IPE Evaluation covers also the requirements for manufacturing evaluation as per IEC WT01 [12].

Table 12. Comparison of the elements of Manufacturing Evaluation IEC WT01 [12] and of GL [9, 17]

Elements of Manufacturing Evaluation	IEC WT01 [12]	GL d [9, 17]
Quality System Evaluation (Design Control)	Yes	yes
Manufacturing inspection / IPE	Yes	Yes

3.4 Project or Windfarm Certification

3.4.1 GL Guideline [9, 17]

Project Certification covers the aspects of *Site Assessment, Monitoring of Fabrication, Transport and Erection, Witnessing of Commissioning and Periodic Monitoring*, see Figure 6. Upon successful assessment of these modules, the Project Certificate will be issued. Project Certification is carried out for wind turbines with Type Certification, whether erected onshore or offshore. The purpose of Project Certification is to evaluate whether type-certified wind turbines, and particular support structure designs, are fit for the external conditions, the applicable construction, the electrical codes and the other requirements relevant for the specific site. Evaluation is carried out even if the wind and sea conditions, other environmental conditions, electrical network conditions and soil properties at the site are more benign than those defined in the design documentation. Basically, Project Certification is intended for projects covering more than one single wind turbine, such as windfarms onshore and offshore. This certification may include all necessary installations, such as measuring masts, power cabling, power transmission, transformer stations, platforms etc.

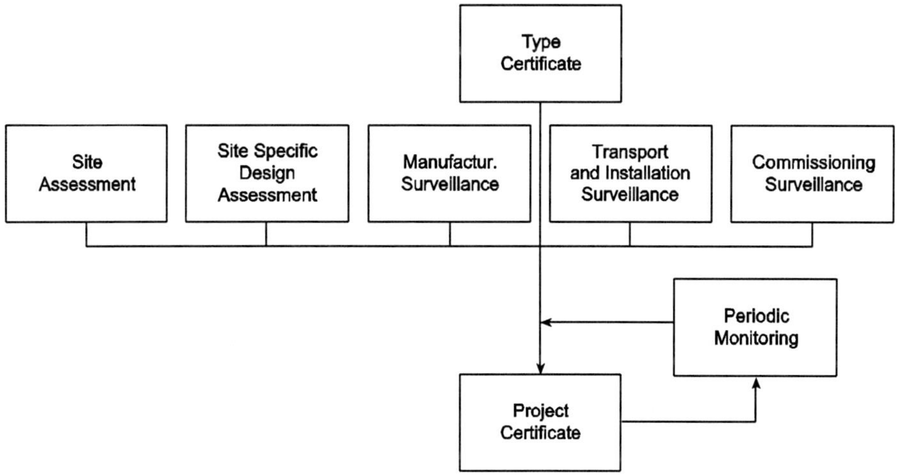

Figure 6. Modules of Project Certification, GL [9, 17]

3.4.1.1 Site Assessment

Site Assessment can be carried out based on GL [9, 17]. For the selected sites the following conditions should be evaluated by measurements and correlated with long term records of meteorological stations or derived from theoretical or other data (e.g. in case of earthquake, from standards):

- *Wind conditions*
- *Sea conditions*
- *Soil conditions*
- *Other environmental conditions (currents, temperature, marine growth, ...)*
- *Electric conditions*

Soil properties at the intended site should be assessed in accordance with the local conditions (subsoil, building codes) and by a geotechnical expert.

Within the Site Assessment, the local site conditions should be checked against those used within the generic Design Evaluation as part of the Type Certification.

The effects of the local external conditions and the local influence of other wind turbines (*wake induced turbulences*) and structures at the site should be investigated. It shall be shown that the external conditions are not critical in relation to those assumed in the design of the wind turbine. If there are conditions at the site that lead to larger loads or to more unfavourable situations than those assumed in the design, a design evaluation shall be performed applying the site-specific conditions. In case the conditions at the site lead to larger loads, additional analysis or modification of the turbine design shall be carried out to demonstrate the integrity of the design for the site in question. Special offshore wind turbine variants, "hot climate" or "cold climate" versions, or e.g. equipment may also be assessed. The generic Design Evaluation will then be enhanced to a Site Specific Design Assessment.

3.4.1.2 Site Specific Design Assessment

Based on the external conditions at the site, the Site Specific Design Assessment should be carried out by the following assessment steps:

- *Site specific load assumptions*
- *Comparison of site specific loads with those from Type Certification*
- *Site specific support structure (e.g. tower, sub structure and foundation)*
- *Modifications of the machinery part and rotor blades in relation to Type Certification, if modifications exist*
- *Stress reserve calculations for the machinery part and rotor blades, if load comparison indicates larger loads than those for the type certified machinery components*
- *Maintenance plan (e.g. weather window, access).*

3.4.1.3 Manufacturing Surveillance

Before surveillance of manufacturing begins, certain quality management (QM) requirements shall be met by the manufacturers. As a rule, the QM system should be certified according to ISO 9001 [24], otherwise the QM measures shall be assessed by the certification body. The extent of the surveillance of manufacturing, and the number of samples to be surveyed, depend on the standard of the quality assessments. These must be agreed with the certification body. Prior to the manufacturing surveillance, the necessary evaluations and approvals of workshops, materials and procedures shall be carried out. In general, the following actions and approvals will be carried out by the certification body:

- *Valid QM-certificate according to ISO 9001 [24]*
- *Quality management: QM-manual, QM-procedures- and working instructions*
- *Welding shop approval*

- *Qualification verifications of the welding personnel (e.g. welder exam according to EN 287 [36] or ISO 9606 [37])*
- *Qualification verification of the welding supervision personnel (EWE)*
- *Qualification verification and equipment of the personnel for non-destructive testing*
- *Welding procedure tests (WPAR)*
- *Approval of welding fillers*
- *Welding procedure specifications (WPS)*
- *Design drawings.*

Design drawings and WPS are examined for conformity with the design documentations (drawings and preliminary welding instructions), which were submitted for design evaluation.

Surveillance can be subdivided into the activities performed at the manufacturing workshops and those carried out at the site of erection. The extent of the surveillance at the manufacturers' workshops depends largely on the relative importance, the material used and the fabrication process of the component produced. Production is in general monitored on a random basis. Important processes and final tests should be always witnessed.

In general, the following actions and approvals are to be carried out:

- *Review of the material certificates according to ISO 10474 3.1C GL [38] or EN10204 3.2 [39] of the steel manufacturer*
- *Checks of the records of non-destructive tests*
- *Inspection and testing of materials and components*
- *Inspection of the manufacturing processes and the results of the non-destructive tests*
- *Inspection of other quality documentation*
- *Scrutiny of QM records such as test certificates, tracers, reports*
- *Surveillance of production, including storage conditions and handling, by random sampling*
- *Inspection of the corrosion protection*
- *Inspection of the electric power system*
- *Witnessing of the final tests*
- *Final inspection of finished components.*

3.4.1.4 Transport and Installation Surveillance

At the site of erection, the important steps during support structure and wind turbine installation must be monitored. Prior to this monitoring, the *transportation of the components* from the manufacturer's works to the relevant site must be surveyed. Before work begins, an installation manual should be prepared, which contains all actions under consideration of the special circumstances of the site. This will be checked for suitability for the assessed design and for the transport and erection conditions (climate, job scheduling, etc.) prevailing. Furthermore, a *site plan* showing the locations of the offshore wind turbines should be prepared, together with plans of the electrical installation showing how the plant will be connected to the utility grid. The extent of the monitoring depends on the quality management standards of the companies involved in transport and erection. As a rule, the following activities are to be carried out:

- *Identification and allocation of all components of the offshore wind turbine in question*
- *Inspection of the components for damage during transport*
- *Inspection of the job schedules (e.g. for welding, installation, grouting, bolting)*
- *Inspection of prefabricated subassemblies, and of components to be installed, for adequate quality of manufacture, insofar as this has not been done at the manufacturers' works*
- *Surveillance of important steps during the erection on a random basis (machinery, rotor blades and support structure inclusive lifting operations, pile driving, grouting)*

- *Inspection of grouted and bolted connections, surveillance of non-destructive tests (e.g. welded joints)*
- *Inspection of the corrosion protection system*
- *Inspection of the scour protection system*
- *Inspection of the electrical installation (run of cables, equipment earths and earthing system)*
- *Inspection of sea fastening and marine operations.*

3.4.1.5 Witnessing the commissioning

Commission witnessing forms an integral part of the certification process during the transition from construction to operations. This should be performed for all the turbines in an offshore windfarm to confirm that each is ready to operate and fulfils all necessary standards and requirements. During commissioning, which is performed according to the previously approved procedures, all components related to operation and safety should be inspected and/or tested. Before commissioning, the start-up manual, together with all the tests planned, should be submitted to the certification body. The commissioning is performed under surveillance of the certification body. The witnessing by the inspectors covers a minimum of 25% of offshore wind turbines during the actual commissioning, see 3.4.1.7. The remaining turbines must be inspected after commissioning and the relevant records scrutinized.

Crossing to an offshore turbine from a support craft

In the course of commissioning, selected tests from the commissioning manual should be carried out, focusing on the safety tests and the turbine behaviour, as assessed for compliance with the design documents. Compliance with any restrictions stated in the Evaluation Reports is assessed as far as possible. This includes the following tests and activities:

- *Functioning of the emergency stop buttons*
- *Triggering of the braking systems by every condition possible during operation*
- *Functioning of the yaw system*
- *Behaviour at loss of load*
- *Behaviour at overspeed*
- *Functioning of automatic operation*
- *Visual inspection of the entire offshore installation*
- *Checking the logic of the control system's indicators.*

In addition to the tests, the following items should be examined or viewed during commissioning:

- General appearance
- Corrosion protection
- Damages
- Conformity of the main components with the certified design and their traceability and numeration within the records.

3.4.1.6 Periodic Monitoring

To maintain the validity of the Project Certificate, maintenance of the offshore wind turbine must be carried out in accordance with the approved maintenance manual, and the condition of the installation monitored periodically by the certification body. Maintenance should be carried out and documented by authorized persons. The time interval for maintenance is dependent on several parameters, e.g. the relative importance of a component, the condition of the overall system and the maintenance procedures or other specific requirements. Periodic Monitoring intervals are to be defined in the inspection plan and agreed with the certification body. These intervals may be varied depending on the condition of the wind turbine.

As a rule, the Periodic Monitoring interval is two years. This interval may be varied depending on the condition of the installation. Any damage or major repairs must be reported to the certification body. To maintain validity of the certificate, any alterations have to be approved. The extent to which this work is supervised shall be agreed. The maintenance records will be perused by the certification body.

Inspecting blade fixings

Inspection reports should include checks regarding scour protection, seabed level, underwater structure and splash zone. The structure within the splash zone should be inspected visually for corrosion, marine growth and damage, e.g. from collision. Where damage is found that could extend further down, diver inspections may be required. Plate thickness measurements may be required where there is evidence of excessive corrosion. Surfaces of concrete must be inspected for cracks, abrasion, spalling (cracking) and any signs of corrosion of the steel reinforcement and embedded components. This is particularly important in the splash zone, in ice conditions, and where repairs have been carried out previously. Cleaning of the surface may be necessary. The type, location and extent of corrosion control (i.e. coatings, cathodic protection system, etc.) and its effectiveness, and repairs or renewals should be evaluated.

The Periodic Monitoring comprises the following assemblies and the respective scope of inspection, Table 13:

Table 13. Scope of inspection for Periodic Monitoring

Assembly	Inspection for / possible defects
Rotor blade	Surface damage, cracks, structural discontinuities. (Close-up inspection: visual and structural examination using suitable methods (e.g. tapping, ultrasonic testing)). Pre-tensioning of bolts. Damage to the lightning protection system.
Drive train	Leakages, unusual noises, condition of the corrosion protection, greasing, pretensioning of bolts, Condition of the gearing (oil analysis, if relevant), Damage to the lightning protection system.

continued over

Table 13. continued

Assembly	Inspection for / possible defects
Nacelle and structural components	Corrosion, cracks, unusual noises, greasing, pre-tensioning of bolts, lightning protection
Climate control, dehumidification and air filters	Function, contamination, dirt
Hydraulic system, pneumatic system	Damage, leakages, corrosion, function
Support structure (tower, sub-structure and foundation)	Corrosion, corrosion protection (e.g. cathodic protection), damages and deformation, cracks, abrasion, spalling, pre-tensioning of bolts, marine growth
Mono pile or intersection pieces	Sealing (visual inspection)
Safety devices, outside lightings, sensors and braking systems	Functional checks, compliance with the limiting values, damage, wear
Control system and electrics including transformer station and switchgear, Condition Monitoring System	Terminals, fastenings, functional checks, corrosion, dirt
Helicopter hoist, boat landing, fenders	Fastenings, function, corrosion, cracks, dirt, damages and deformation
Emergency shelter	Visual inspection of outside lighting and emergency shelter equipment and rescue at sea equipment and possible backup power supply (if applicable).
Perusal of documentation	Completeness, observance of the conditions, construction according to certified documents, test documents, maintenance carried out at regular intervals. If applicable: execution of modifications / repairs according to approval.

Periodic Monitoring should be carried out by wind turbine experts approved by the certification body. The experts must have the necessary technical knowledge for the evaluation of the complete wind turbine. The relevant training and a continuous exchange of experience should be proven. An accreditation according to EN 45004 [40] or EN 45011 (ISO/IEC Guide 65) [41] or equivalent is required, or the aptitude of the experts checked by a competent examination board. The experts shall be independent and shall have access to the relevant technical documentation of the offshore wind turbine.

3.4.1.7 A and B Levels of Project Certification

The Project Certificate will be issued after successful accomplishment of the steps described. Concerning the surveillance of manufacturing, transport, installation, commissioning and periodic monitoring, there are two different levels of Project Certificate, the A-Project and the B-Project Certificates.

For the *A-Project Certificate*, surveillance is required for all of the offshore wind turbines. Monitoring should cover the support structure and essential parts of machinery, blades and electrical system.

For the *B-Project Certificate*, surveillance is to be undertaken for only 25 % of the offshore wind turbines on a random sample basis. Monitoring should cover the support structure and essential parts of machinery, blades and electrical system. If the surveillance should reveal major failures, deviations from the certified design or deviations in the quality management, the number of turbines to be monitored must be doubled.

3.4.2 IEC WT01

Project Certification according to IEC WT01 [12] comprises the modules Site Assessment, Foundation Design Evaluation, Installation Evaluation and Operation and Maintenance Surveillance, see Figure 7. A Project Certificate will be issued upon successful evaluation of the site assessment and the foundation/support structure design evaluation.

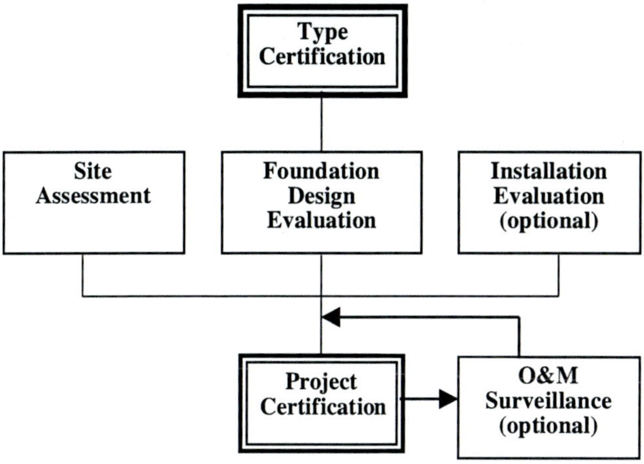

Figure 7. Modules for Project Certification according IEC WT01 [12]

3.4.2.1 Site Assessment

The purpose of the Site Assessment is to verify that environmental, electrical and soil properties at the site accord to the parameter values defined in the design documentation. The same requirements as in section 3.4.1.1 apply. The certification body requires that qualified personnel (meteorologists, engineers or geologists) carry out the evaluation and the reporting of the external conditions at the site. A Site Assessment Conformity Statement concludes the satisfactory evaluation of the site parameters with respect to the certification.

3.4.2.2 Foundation Design Evaluation

If the Foundation Design Evaluation was not carried out within Type Certification, the foundation/support structure design shall be included in the Project Certification. The design of the foundation/support structure must evaluated with respect to its suitability at the site in question and with respect to the standards and normative documents agreed between certification body and the applicant. If a conformity statement has already been issued for the foundation/support structure design within the Type Certification, the evaluation may be limited to local sea-bottom/soil/ground conditions and the applicable local codes and standards.

3.4.2.3 Installation Evaluation (optional)

The purpose of the Installation Evaluation is to verify that one or more offshore wind turbines have been installed and commissioned in conformity with specific standards and with the design documentation. The Installation Evaluation is subdivided into the elements "Installation Quality System" and "Surveillance/Audits". The certification body evaluates that the installation quality system is in agreement with the installation plan (see section 3.3.2.1) and other installation or construction plans for civil and electrical works. Here the effectiveness of the installation quality system is evaluated by systematic surveillance, if the party responsible for installation is not operating a quality system that meets the requirements specified in ISO 9001:2000 [24]. Part of the "Surveillance/Audits" element is the commissioning of at least one turbine at the site.

3.4.2.4 Operation and Maintenance Surveillance (optional)

The purpose of the optional Operation and Maintenance Surveillance is to establish that a specific wind turbine, or group of wind turbines, at a specific site are operated and maintained in conformity with the relevant manuals included in the design documentation. Surveillance must be carried out at regular intervals on the basis of an agreement between certification body and applicant. Items to be surveyed include:

- *maintenance has been carried out by authorised and qualified personnel in accordance with the maintenance manual and at the intervals specified*
- *control settings have been checked with respect to the limiting values specified in the design documentation.*

3.4.3 Comparison of the Project Certification Procedures According to IEC WT01 and GL

With respect to Project Certification according to IEC WT01 [12], the following differences compared to a Project Certification according to GL [9, 17] can be dscerned. If the (optional) foundation design evaluation of IEC WT01 [12] is handled within the Type Certification, only a Site Evaluation needs to be carried out to receive a Project Certificate. The modules Installation Evaluation and the O&M Surveillance are optional, see Figure 7.

It is obvious, that the certification according to GL [9, 17] requires a number of additional aspects to be handled when carrying out Project Certification. An integral part is the maintenance of the Project Certificate, which requires periodic monitoring as a mandatory module. The time interval of the monitoring is established on the basis of an agreement with the applicant, see section 3.4.1.6. This monitoring is called *Operation and Maintenance (O&M) Surveillance* in the IEC WT01 [12], but covers the same actions as the *Periodic Monitoring* described in section 3.4.1.6. However, IEC WT01 [12] treats this as an optional module.

It is similar for the Installation Evaluation, which is optional in IEC WT01 [12] and mandatory according to GL [9, 17] as "Surveillance During Transport and Erection". A general comparison of the differences is given in Table 14.

Table 14. Mandatory and optional modules within Project Certification according to IEC WT01 [12] and GL [9, 17]

Elements of Project Certification	IEC WT01 [12]	GL [9, 17]
Site Specific Design Evaluation / Site Evaluation	yes	yes
Foundation / Support Structure Design Evaluation (mandatory)	yes	yes
Surveillance during Production	No	yes
Surveillance during Transport and Erection Monitoring / Installation Evaluation	Optional	yes
Surveillance during Commissioning	No	yes
Periodic Monitoring / O&M Surveillance	Optional	yes

Benefits from Project Certification are the following:

- *Project Certification is requested by lenders and authorities and performed by certification bodies to minimise risks.*
- *Both Design Evaluation and Monitoring activities are necessary to identify insufficiencies in design, production and installation.*
- *Reliable turbines are necessary to ensure project returns, requiring attention has to be paid to machinery quality and adequate testing.*

3.5 VALIDITY AND RE-CERTIFICATION

The Type Certificate has validity for two years (GL). However, the Type Certificate will lapse before this period if the certificate for the quality system is no longer valid. During the validity period, all installed wind turbines of the type shall be reported annually to the certification body.

Upon expiry of the validity period of the Type Certificate, re-certification should be performed on request. After completion of the process, the certification body issues a Type Certificate with a reference to the re-certification and a validity period of two years. For the re-certification, the following documents and listings are evaluated:

- *valid drawings*
- *all modifications to the design of components considered in the Design Evaluation and, if applicable, documents for evaluation of the modifications*
- *alterations to the QM system since the last audit*
- *all wind turbines of the type installed (at least statement of the type with precise designation of the variant, serial number, hub height, location)*
- *all faults and failures of the wind turbines installed.*

If modifications were made to the structure, these are to be examined and a revision of the Statement of Compliance for the A-Design Evaluation is issued upon successful assessment.

The Project Certificate according to GL [9, 17] is valid indefinitely and only lapses if the Periodic Monitoring is not carried out. Major modifications, conversions or repairs not approved by the certification body will also lead to loss of the certificate's validity.

The validity period of certificates according to IEC WT01 [12] should be established individually but must not exceed 5 years for both the Type Certificate and the Project Certificate. During the validity of the Type Certificate, an annual report should be submitted to the certification body. The report must include information on deviant operating experience known to the certificate holder and minor modifications. With respect to the Project Certificate, modifications to the site or the wind turbine shall be reported to the certification body for renewal of validity or extension of the certificate.

4. RISK EVALUATION

4.1 General

A significant number of offshore windfarms are currently being planned in Europe, USA and Canada. For example, for German areas in the North Sea and the Baltic Sea, it is proposed to install wind turbines of each up to 5 MW rated power, with up to 25,000 MW of total power. It is necessary to study the effects of those windfarms with regard to the safety of shipping in order to estimate the related risks to people, ship traffic and the environment. A guideline for risk analysis for offshore windfarms has been issued by GL in 2002 [46].

Within the approval process, detailed risk analyses have to be submitted, which enable the responsible authorities to carry out an evaluation on the basis of comprehensible criteria. Measures for risk minimisation, such as the *provision of salvage tugs or ship traffic control*, can be evaluated for their efficiency to achieve the highest safety standard for people and the environment. Additionally, risk analyses provide the necessary safety information for insurers and operators of windfarms.

During the approval process, the responsible authority stipulates acceptance values for collision frequency and risk. On the basis of the calculated *"collision frequency"* and *"quantity of harmful substances"*, an evaluation is possible. By comparing the acceptance values with the calculated values, a decision regarding the approval can be made. The effects of risk-reducing measures can be demonstrated and thus have a direct influence on the acceptance of the planned windfarm.

The methods of risk analysis can also be applied to wind turbines onshore. This is of particular interest for exposed locations with larger than usual potential risk to people and the environment. Examples are wind turbines next to busy roads or general traffic routes, and wind turbines close to chemical plants, power plants, etc. Here accidents of wind turbines and the related hazard due to falling parts have to be taken into account to evaluate and reduce potential risks to people and the environment to an acceptable value.

In section 4.2 a brief description of risk analysis and minimisation methods is given. A more comprehensive overview of methods to calculate and minimise ship collision risks is given in [54] summarizing the results of the EU-funded research project SAFESHIP.

4.2 Risk Analysis

4.2.2 Risk Assessment Methods

The risk associated with hazards originating from a technical system is defined as *the product of the frequency of an undesirable event multiplied by the expected consequences of such an event*. In order to assess the risk, i.e. to answer the question *"is this technique safe?"*, two different aspects shall be considered resulting in two additional questions. The first question *"what can happen?"* can be answered by risk analysis, that is a rational, structured and documented analysis. Answering the second question *"what is allowed to happen?"*, leads to the risk evaluation that is more subjective (see Figure 8). In the following focus is given to the risk analysis part.

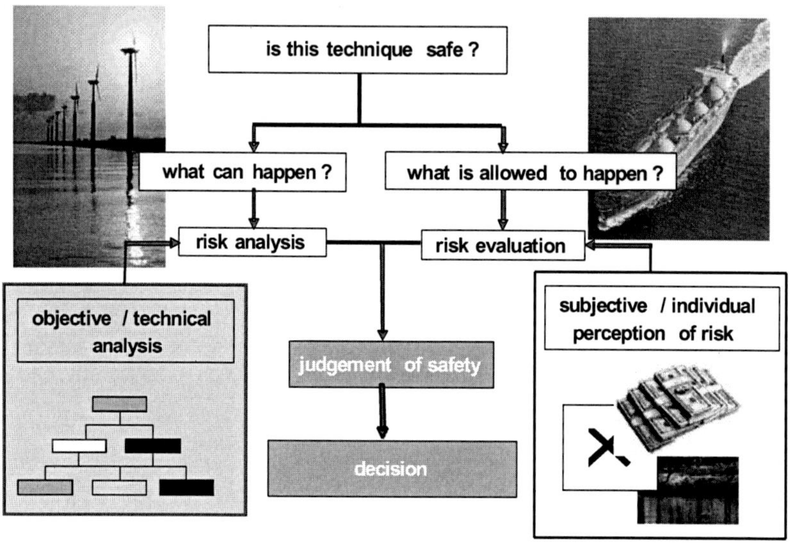

Figure 8. Assessment of risks

Risk analysis combines the frequency of an undesirable event with the consequences of such an event. The frequency of an undesirable event can be determined by statistical data of similar systems acting under similar conditions and assuming that such systems have the same failure behaviour. Statistical methods using historical data can be applied only if the failure behaviour of a similar system is known. That requires that a certain number of system failures have to be registered in a certain period to ensure a solid statistical basis for the statistical results. If system failure is very seldom, the required data can be derived from known statistical data of system components by subdividing the system into subsystems and components. From that logical structure, the failure behaviour of the whole system can be deduced.

The nature of consequences of an undesirable event can vary considerably. The possible consequences may be subdivided into 5 groups:

- *consequences for health and life of human beings*
- *environmental consequences*
- *economical consequences*
- *social consequences*
- *ethical, political and legal consequences*

It is a huge effort to determine all consequences completely and therefore it is necessary to define a frame that includes only consequences which, in connection with the related frequencies, deliver a significant contribution to the overall risk.

In [45], a study is made of available knowledge on maritime risk analysis which could be applied for the purpose of 'ship with windfarm' collisions. Methodologies such as Formal Safety Assessment (FSA) and Safety Case (SC) are described. Of particular relevance is the guideline for risk analysis for offshore windfarms [46].

4.2.2.1 Collision frequency

The first studies to calculate frequencies of ship collisions with fixed offshore installations were made in the seventies in Japan with a method based on historical data of accidents in Japanese waters.

Considering the collision of ships with fixed offshore installations like windfarms, two different scenarios should be distinguished:

- *collision of powered ships*
- *collision of disabled ships (called "drifters")*

This distinction is necessary because both scenarios have different causes, progresses and outcomes. Because of the different nature, different methods are necessary to calculate the frequencies of those events.

Existing models which cover both scenarios are COLLRISK [47], COLWT, [48] and SAMSON, [49], see [45] for a full description. All models require the use of ship traffic data, for which many data sources are available [45]. The approaches to model collision frequencies for powered ships and disabled ships are described below.

4.2.2.1.1 Powered ships

The probability of a collision depends on:

- *Probability of a ship being on a collision course (based on ship routeing pattern and the dimension and orientation of the structure being assessed)*
- *Probability of the vessel not recovering from collision course, i.e. that no action is taken to correct the erroneous course, the so-called "causation probability".*

The probability that ships are on collision course is given by the area underneath the appropriate distribution function of the ship traffic, which is bounded by the projection of the respective wind turbine (or transformer station) and half a ship's width on both sides of the projection. The projection is parallel to the ideal shipping lane, see Figure 9.

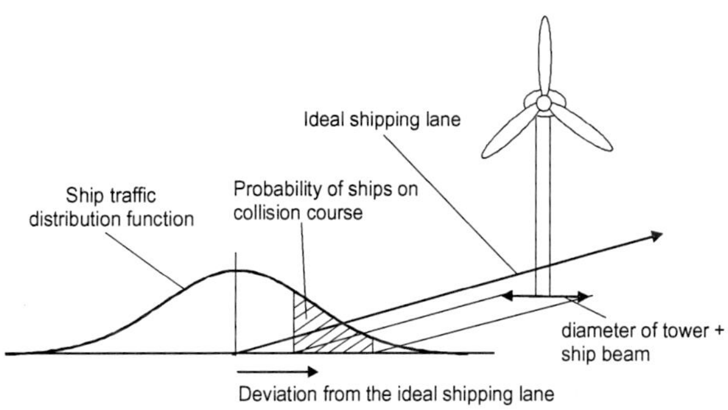

Figure 9. Parameters for the calculation of the probability of ships on collision course, [54]

The probability that no action is taken to correct the course can be modelled by using a calibration factor, based either on historical accident data with other offshore installations or by an analytical approach (Fault Tree Analysis, Baysian Networks). The use of statistics from historical accident data is of limited value, because collision accidents with windfarms have not yet occurred and comparable collision accidents (e.g. collisions with offshore oil and gas platforms) have been very rare. Moreover, such accidents differ from a potential accident concerning a windfarm. Further, new navigational equipment like the Automatic Identification System (AIS) or Vessel Traffic Systems (VTS) were not available in the past. An additional disadvantage of using historical accident data is that they can only be determined for large geographical areas. Thus, these causation probabilities are average values for large areas neglecting the individual characteristics of the windfarm and its neighbourhood.

Application of analytical models to determine the causation probabilities can solve these problems but the number of influences and especially the complex human behaviour in operating a ship makes establishment of such a model a difficult task. An example of the Baysian Network under development is shown in Figure 10.

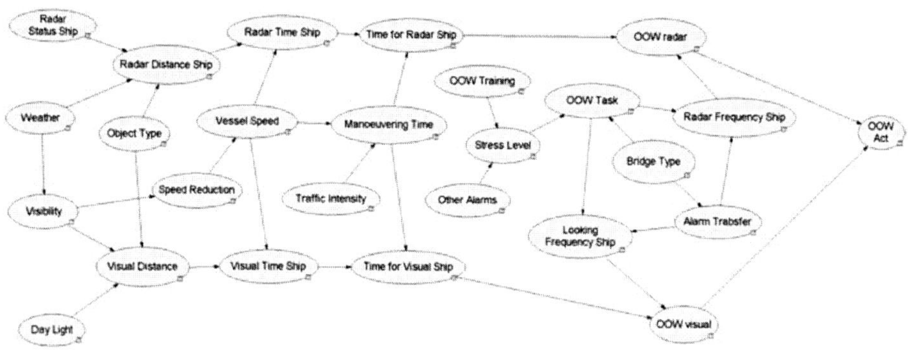

Figure 10. Example of a Baysian Network to establish the causation probability, [54]

4.2.2.1.2 Drifting ships

If the propulsion or the steering system of a ship fails, the ship is called "disabled" or a "drifter". That means the ship operator has no influence on the motion of the ship. Therefore, there is a certain probability that disabled ships drift into the area of windfarms and collide. Influencing factors are:

- *failure rate of propulsion systems*
- *time to repair before collision (many failures are identified and corrected)*
- *emergency anchoring*
- *recovery by salvage tugs*
- *drift motion of the ship, depending on the existing wave, current, and wind conditions and vessel size and type.*

Considering the collision scenario of disabled ships with windfarms, the drift duration is important (i.e. the time from the failure of the propulsion system or the steering system up to the time of collision of the disabled ship with the windfarm). The probability that the failed system can be repaired by the crew depends on the time available. Support from salvage tugs depends also on the available time. The drift time can be calculated by the drift velocity which is also the main influence for the success of emergency anchoring manoeuvres. The sensitivity of risk reducing measures (repair, emergency anchoring, salvage tugs) is to be considered. Preliminary calculations were performed for the Baltic Sea and the North Sea. These calculations showed that the influence of salvage tugs on the collision frequency mainly depends on the location and the capacity of the salvage tugs.

4.2.2.2 Collision consequences

4.2.2.2.1 Overview of collision consequences

The consequences of a collision of a ship with an offshore windfarm are illustrated in Figure 11:

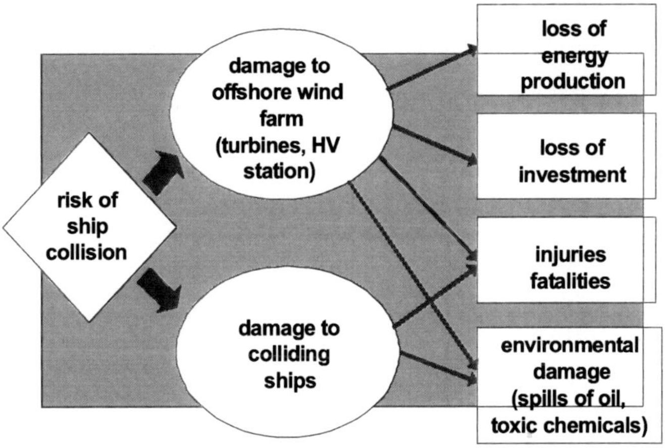

Figure 11. Consequences of a ship collision with an offshore windfarms, [54]

Potentially severe consequences are:

- *damage to the HV station. This could result in a long idle period for the whole windfarm, with extensive loss of potential energy production*
- *collision during construction or maintenance activities which could result in injuries or even casualties of employees (occupational safety issue)*
- *collision resulting in spills of oil or other ecotoxic chemicals.*

4.2.2.2.2 Computation of damage

The computation of the consequences of a possible collision of ships with offshore windfarms can be performed by finite element procedures, both regarding possible damage of the ship structures or of the windfarms. The scope of analysis is the overall dynamic behaviour in a collision scenario and not the detailed estimation of the local deformation in the impact zone.

To model the damage of colliding ships, the following analysis was performed in [45, 54]. Two ships of different size have been modelled and set to drift against a stiff pillar resembling a wind turbine. The resulting damage was calculated, and it was found that the extent of the damage did not cause rupture of the ships' side-casings. From these results, it has been estimated that the damage to double-sided ships following a drifting collision with a wind turbine is so small that it will not cause the vessel to sink. This may not hold for ships with single hulls. It is noted, however, that the damage estimates provided here are quite conservative, since the wind turbine is assumed stiff and so does not absorb any collision energy.

Additionally modelling of the damage of wind turbines was performed. A model of the whole structure with 3D analysis for a 2 MW wind turbine including soil and ship is shown in Figure 12.

The model of the wind turbine comprises the transition piece consisting of volume elements, the tower and the nacelle with rotor blades modelled with shell elements. The soil properties are considered as non-elastic beam elements using P-Y curves. The monopile is modelled with shell elements. Soil conditions were chosen from a windfarm in the Dutch North Sea. Scour is ignored because of a scour protection system.

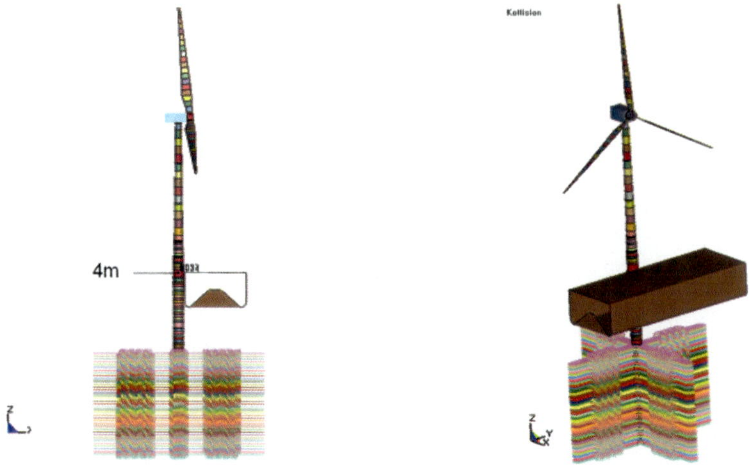

Figure 12. Global model for side impact with wind turbine, ship and soil springs

Five different ship types were modelled by means of rigid shell elements. The mass, different drifting velocities and moments of inertia were considered. One impact scenario for each load case was simulated and the time dependent moments and forces were traced. To evaluate the results, the following conditions were reviewed: plastic deformation, behaviour of the nacelle and rotor, yield or shell buckling, and failure of foundation soil.

The simulation results for all ship types showed the following tendencies:

- *The chronology of events regarding failure is: yaw bearing, main shaft, top of the blade, mono pile below mud line and blade root.*
- *The calculations indicate that the yaw bearing and buckling of the upper tower section are highly sensitive.*
- *In case of a mono pile failure, it is likely to fail below mud line.*
- *At a drifting velocity of 4 knot, the section loads in nearly all considered parts exceed the maximum design loads of the wind turbine.*
- *With the considered drifting velocities of 1, 2 and 4 knot, the turbine is likely to be pushed away and not to fall towards the ship.*

4.2.3 RISK REDUCING METHODS

4.2.3.1 Project content

Here "risk reduction" is defined as passive and active measures incorporated in the set-up, design and operation of the windfarm which reduces ship collision probability and ship collision damage.

- *"Reduction of collision probability"* includes techniques, such as marking lights, painting, buoys, installation of mooring buoys to eliminate the risk of anchored ships drifting to the farm, training of crews, anchor procedures etc. In particular, attention is paid to the more advanced Automatic Identification System (AIS) and advanced radar technology based solutions for hazard identification, communication, emergency management and related decision support. Since collisions with maintenance vessels have a high probability, attention is also paid to use of service vessels with reduced collision probability or extent of damage, such as the use of (light-weight) SWATHs (Small Waterplane-Area Twin Hull) for supplying and repair work.

- *"Damage reduction possibilities"* include classic fender techniques and more advanced techniques specifically developed for offshore windfarm applications. Damage reduction solutions for offshore windfarms will be different from classical applications: for individual (mono-pile) wind turbines one has to balance the wish to absorb the impact of small vessels without damage and the requirement of quick collapse of the monopile when the colliding ship is an oil/chemical tanker to avoid puncturing of the tanker. Attention is therefore also paid to the design of wind turbines and HV stations for reduced ship and windfarm damages. Ideas are the use of underwater HV stations and deeper trenching for cables to prevent damages from fishing /anchoring vessels and risk of exposure of the cable.

Figure 13 gives an impression of costs associated with oil spill volume; data have been obtained from International Tankers Owners Pollution Federation (ITOPF).

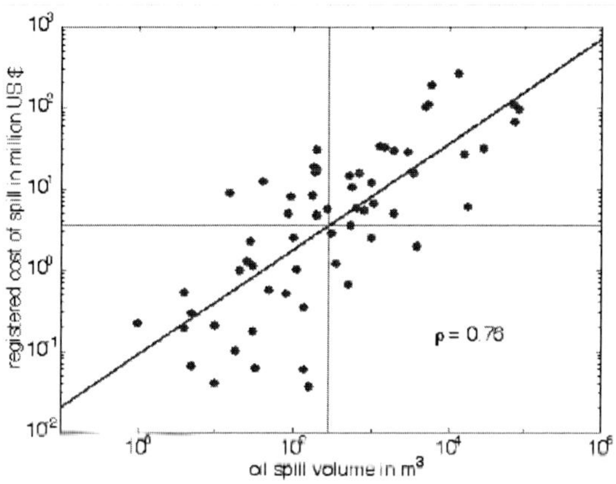

Figure 13. Costs of oil spills, used in the feasibility cost model

Regarding measures to reduce the collision probability, the introduction of the Automatic Identification Systems (AIS) for ships has been one of the most promising technologies.

4.2.3.2 Reduced collision probability by Automatic Identification Systems AIS

In addition to the "traditional" mandatory lighting/colouring requirements, cost-effective collision avoidance is in particular related to the implementation of AIS.

The AIS is a *transponder system* which has been developed to improve navigational ship-to-ship safety. The system can send and receive ship information such as ship identification, ship characteristics, position, heading, speed, rate of turn and other safety-related information from other ships and fixed AIS stations. The system communicates digital in VHF frequency, which means that the effective range is mainly dependent on antenna height. The system complies with international standards, to ensure that the vessel can communicate with other vessels or with shore-based AIS stations all over the world. The overall purpose of AIS is to identify vessels and their sailing characteristics, assist in target tracking, provide information to assist in collision avoidance and finally to reduce verbal mandatory reporting from ships to shore stations. Some of the advantages of AIS compared to radar systems are that AIS provides fast, accurate and detailed information about other nearby vessels and that AIS is not influenced by wave and rain clutter as is the case for radar systems. Moreover, AIS supplements the radar information and consequently, due to the mandatory implementation of AIS, the maritime community has undoubtedly observed an increase in navigational safety.

The increased navigational safety is evaluated in [51] by using a risk model which evaluates the navigating officer's ability to react in time when the vessel is on collision course with another vessel.

Use of AIS reduces the probability that the navigating officer's does not react in time by a factor of 2. The overall risk reducing effect is found to be 55 %.

IMO regulations [61] require all ships above 300 tons (except fishing vessels) to be equipped with AIS transponders. With these data, maritime authorities can greatly improve the vessel traffic management, with obvious advantages also for the safety of offshore windfarms. Ships on potential collision courses can be warned by coast guards and measures can be taken in time. AIS and its use by coast guards for vessel traffic management will reduce (with an estimated factor of ~1,25) the risks of ship collisions with offshore windfarms. However, further advantages of AIS can be obtained if also the windfarm operator makes optimal use of AIS. Several improvements are possible, both in a passive and in an active way, see "improvements" indicated in Figure 14.

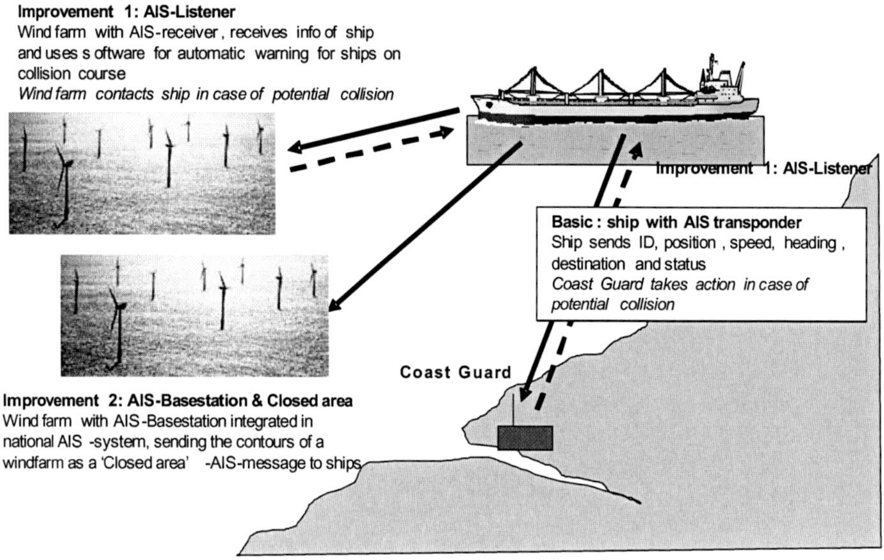

Figure 14. Use of AIS (Automatic Identification System) to reduce risks of collision of ships with offshore windfarms [54]

Today, the general opinion is that only maritime authorities, such as coast guards, should actively use the AIS data. In addition, a solution has to be found for the fishing vessels which are now exempted from the IMO requirement [61]. Fishing vessels are in particular regarded as a potential danger because they will be attracted to *the increased fishing stock within windfarms*.

Another improvement of safety of offshore windfarms concerns extending the GSM-network by using special directional GSM-masts at the coast-line or a special GSM-basestation at the offshore windfarm, so standard *cell-phones* can be used within the area of an offshore windfarm. This improves the safety of offshore windfarms and windfarm operators by allowing easy, fast and early warning for potential collisions.

4.3 Due Diligence

Due diligence covers project control to demonstrate the viability of the project to investors, financiers or other interested parties. The elements include *technical control, financial viability, quality of project partnership/contracts and the interfaces between them.* For most large projects (not only in wind energy), it is common practice that the financiers ask for due diligence investigations. This activity is usually divided between the legal advisor, the financial advisor and the technical advisor, with the aim to study and prove that the legal, technical and financial risks are within a range that makes the project bankable. The challenges and opportunities facing wind energy projects are overwhelming and these are caused by several external conditions and factors. Figure 15 lists some of these.

Figure 15. Wind Energy Project – Conditions, External Factors and Market Challenges

Investors, financiers and other stakeholders look for rationality in long-term investments, which is a relative term, especially if there is much pressure on the viability of wind energy projects. Consequently, the advisors try to guide these investors by thorough investigations of all aspects and by encouraging certification of all activities related to the establishment and operation of a windfarm. This enables financially healthy and technically feasible projects to be realised, which in turn positively influences investor confidence.

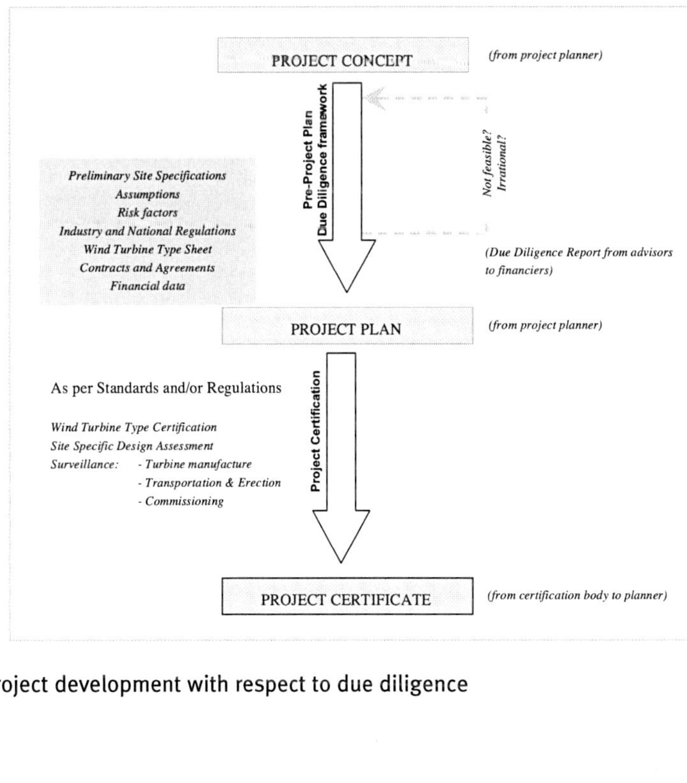

Figure 16. Project development with respect to due diligence

Prior to a comprehensive Project Certification process, which bases itself on a finalised project plan and detailed technical specifications, see Figure 16, it is advisable to carry out an analysis of several important 'hard' and 'soft' factors based on the initial project concept. This test of financial and technical viability includes:

- *Critical assessment of assumptions in the project concept*
- *Assessment of energy yield prediction*
- *Compatibility checks of wind turbine technology, wind park management systems, power control and transmission equipment, emergency systems in relation to existing and forthcoming regulations*
- *Feasibility tests considering location, political, legal and logistic factors, etc.*
- *Assessment of risk factors related to operator track-record, wind turbine track-record (turbine reliability and availability), equipment suppliers, supply chain bottlenecks, power transmission and grid issues, weather forecasts*
- *Verification of power purchase agreements and assessment of eligibility for green trading certification*
- *Verification of contracts with optional juristic implications on breach of trust*
- *Scenario analysis considering probable risks with implications on initial capital costs and cash flow projections*
- *Optional financial and technical rating of equipment suppliers, windfarm operators and finance partners.*

The due diligence assessment is based on the project concept submitted by the windfarm planner. Estimates and projections from similar projects/databases, knowledge of the wind energy industry structure and related risks, are used to evaluate critically the different scenarios. Valid windfarm concepts are then eligible for financial commitments of investors. This due diligence, tailor-made especially for windfarm projects, concludes with a report containing:

- *A feasibility study of the windfarm concept*
- *A list of assumptions and risks*
- *A list of possible obstacles to realisation/financial health of the windfarm project*
- *A capital requirement and cash flow projection for probable scenarios.*

Financial closing marks the end of the due diligence stage, paving the way for the next stage – Project Certification. Project Certification of windfarms is based on the finalised project plan and is a comprehensive technical analysis involving state of the art measurements, computations and simulations and monitoring of all important steps in production, erection and commissioning. Refer to section 3.4 for detailed descriptions of Project Certification.

4.4 Condition Monitoring

Because of the size and capacity of the turbines, the loading of the individual components of the turbines has increased. With regard to the investment and operational costs, the economical operation of the turbines is necessary, which means that the downtimes should be minimized. Operators and manufacturers of wind turbines try to obtain availabilities of more than 97%, which are contractually established and guaranteed.

For offshore wind turbines, the same reliabilities are required as for onshore wind turbines, although the difficult access in the offshore area will probably imply significant downtimes in the case of faults or failures.

a) *Continuous condition monitoring of the wind turbines is indispensable in order to comply with the demands of*
b) *minimizing downtimes,*
c) *predicting future damage and limiting it locally to prevent secondary damage at other components, and (iii) complying with the demand of scheduling the downtimes of the turbines which result from maintenance requirements.*

One way of doing this is applying a Condition Monitoring System (CMS).

With a CMS, it is possible to detect relevant changes in the condition of the monitored components of the turbine. These changes represent deviations from the normal operational behaviour and may result in the premature breakdown of the components. The CMS measures vibrations and impact sound at the components of the wind turbines, for example, at the components of the drive train and the tower, and collects operational parameters, as power output, speed, oil and bearing temperatures (see Figure 17).

The recorded data are then compared with the established threshold values for the corresponding component. If the CMS detects that the threshold value was exceeded, an alarm message will automatically be sent to the corresponding control centre. The control centre will carry out an evaluation of the measured values, in order to take the necessary measures.

A guideline for the certification [55] describes the requirements for the *certification of Condition Monitoring Systems* for wind turbines and control centres. It establishes a basis for the development and installation of CM systems in wind turbines and establishes rules for the application of the measured values, for example, for the evaluation, interpretation and storage, and procedures to be carried out when the determined threshold values are exceeded.

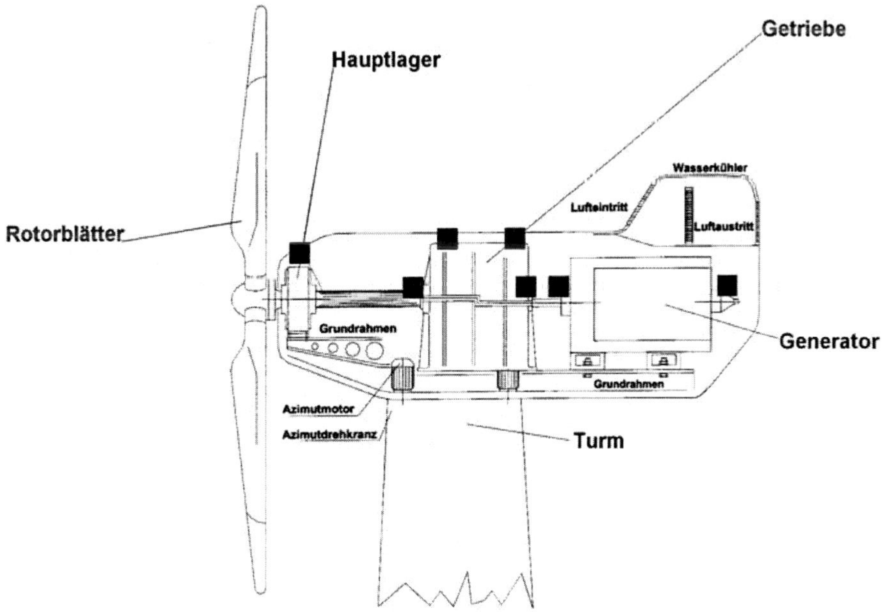

Figure 17. Typical positions for vibration measurements

4.5 Other Requirements

4.5.1 Other Components of Offshore Windfarms

In previous sections the wind turbine and its support structure has been looked at concerning site specific conditions. Other important structures/components of a windfarm are:

- *Meteorological Mast*
- *High voltage transformer station*
- *Cable/ grid connection*
- *Hoisting devices, cranes, extensions.*

For this purpose, the previous mentioned procedure of Project Certification can be applied correspondingly. The certification can be carried out based on the relevant standards and guidelines, e.g. [56 – 60].

4.5.2 Analysis and Assessment Beyond Project Certification

It is possible that technical observations other than Project Certification are required:

- *Evaluation of the safety concept of a windfarm (e.g. safety of navigation, labelling, lighting, emergency plans)*
- *Risk analysis concerning collision of ships with wind turbines with respect to environmental risk (see section 4.2)*
- *Assessment of a decommissioning concept*
- *Personnel safety (e.g. for helicopter landing decks and helicopter hoists/platforms, manned electric power substations, boat landing, ladders, etc.)*
- *Examination of arrangements for avoidance of environmentally hazardous substances respectively their entry into the sea environment (e.g. ship friendly design of the sub structure)*

For insurance companies, it is common practice to request a warranty survey for transportation and erection of the offshore windfarms and their components.

5. SUMMARY AND CONCLUSIONS

The chapter on standards and certification covers three main topics, namely description of the relevant standards, evaluation of the certification procedures and risk assessment with respect to offshore windfarms.

The standards and regulations described are those related to offshore wind turbine structures and windfarms. Most of these cover only a part of the complete offshore windfarm. However, comprehensively, the standards contain relevant information for the complete system, from the loads on the blades and the support structure to the electricity fed into the grid.

The section on *certification* starts with a complete description of the design requirements, especially emphasising offshore applications. This part is concluded with a comparison of onshore and offshore requirements.

Type Certification is described by applying the procedures formulated by IEC and GL.

Comparison of the two sets of standards are made, stating the main differences.

The internationally valid certification procedures for Project Certification are also issued by IEC and GL. Although the differences in type certification are only marginal, the comparison of the two procedures shows considerable differences in Project Certification. Some modules are optional for IEC, but required by GL. The latter is especially due for all kinds of surveillance.

The section on *Risk Evaluation* covers all kinds of risks related to planning, development, approval and operation of offshore windfarms.

Within the risk evaluation, *Due Diligence* can be considered the governing evaluation, relying on all the detail criteria for technical, financial and legal risks.

A detailed description concerns the risk analysis of *ship collision* with windfarms, both for the assessment and the risk reducing methods.

Within the scope of risk reduction, *condition monitoring* is described, including the vibration sensing and the control centre actions.

A section on activities beyond Project Certification, considers aspects other than the wind turbines themselves, including *safety of personnel and the environment*, and *decommissioning*.

Considering the vast *experience in the offshore oil and gas industry*, it is clear that Certification is also essential for offshore windfarms. The harsh environment, the limited accessibility and the enormous financial risks require special measures, not only for the design and control of offshore windfarms, but also for certification procedures far exceeding the requirements onshore. This is not only necessary for the support structure, where a number of newly developed design features are being applied, but also for the wind turbine and the rest of the windfarm to take into account the modifications of existing designs, the mass production effects in offshore wind energy and the high level of required safety and availability.

In many places, offshore windfarms are being planned close to shipping lanes, near to *environmentally sensitive areas* and without the experience of years of operation.

To design, build and operate windfarms in a safe and reliable way, a *proper risk evaluation* including Type and Project Certification forms an indispensable basis to achieve the former.

6. REFERENCES

[1] Beurskens, J.: *"Recommended Practices for Wind Turbine Testing – 6. Structural Safety"*, Review of (draft) standards/codes of practice, Preparatory Information, International Energy Agency, 1st Edition 1988

[2] Östergaard, C: *Randbedingungen zur Seeaufstellung großer Windkraftanlagen*, Technischer Bericht Nr. STB-925, Germanischer Lloyd, Hamburg, 1982

[3] A. D. Garrad, H. G. Matthies, et. al.: *Study of Offshore Wind Energy in the EC*, Verlag Natürliche Energie, Brekendorf, Germany, 1995.

[4] Germanischer Lloyd: *Regulation for the Certification of Offshore Wind Energy Conversion Systems*, Hamburg, 1995.

[5] The Danish Energy Agency: *"Recommendation for technical approval of offshore wind turbines"*, December 2001

[6] Det Norske Veritas: OS-J101: *Design of Offshore Wind Turbine Structures*, Oslo, 2004

[7] IEC 61400-3: *Wind Turbine Generator Systems – Part 3: Design Requirements for Offshore Wind Turbines*, Committee Draft for Voting (CDV), 2005

[8] IEC 61400-1: *Wind Turbine Generator Systems – Part 1: Safety Requirements*, Ed.2, 1999-02

[9] Germanischer Lloyd: *Guidelines for the Certification of Offshore Wind Turbines*, Edition 2005

[10] Executive Order of the Danish Ministry of Economic and Business Affairs no. 1268 dated 10.12.2004: "Bekendtgørelse om teknisk godkendelsesordning for konstuktion, fremstilling og opstilling af vindmøller" ("Executive order on the technical certification scheme for the design, manufacture and installation of wind turbines")

[11] Det Norske Veritas (DNV) offshore standard DNV-OS-J101, *"Design of Offshore Wind Turbine Structures"*, draft January 2004

[12] IEC WT 01: *IEC System for Conformity Testing and Certification of Wind Turbines, Rules and Procedures*, 2001-04

[13] Germanischer Lloyd: *Guideline for the Certification of Ocean Energy Converters, Part 1: Ocean Current Turbines*, Draft Edition 2005

[14] EMEC: *Performance Assessment for wave energy conversion systems in open sea test facilities*, European Marine Energy Centre Limited, 2004

[15] Det Norske Veritas (DNV): *Guideline on Assessment and Application of Standards and Recommended Practices for Wave Energy Conversion Devices*, Draft November 2004

[16] EN 45020, 1998 *Standardisation and related activities – general vocabulary*

[17] Germanischer Lloyd: *Guidelines for the Certification of Wind Turbines*, Edition 2003 with Supplement 2004

[18] J. A. Battjes, H. W. Groenendijk, "Wave height distribution on shallow foreshores", *Journal of Coastal Engineering*, Vol. 40 (2000), pages 161-182

[19] J. Wienke, *"Druckschlagbelastung auf schlanke zylindrische Bauwerke durch brechende Wellen – theoretische und großmaßstäbliche Laboruntersuchengen"*, Thesis, Technical University Braunschweig

[20] Standard for Geotechnical Site and Route Surveys, *Minimum Requirements for the Foundation of Offshore Wind Turbines*, by Bundesamt für Seeschiffahrt und Hydrographie (BSH), Hamburg and Rostock 2003)

[21] API RP 2A LRFD: *Planning, Designing and Constructing Fixed Offshore Platforms – Load and Resistance Factor Design*, Washington D.C., 1994

[22] DIN EN ISO 12944-3, Publication date: 1998-07 *Paints and varnishes – Corrosion protection of steel structures by protective paint systems – Part 3: Design considerations* (ISO 12944-3:1998); German version EN ISO 12944-3:1998

[23] Matthies, H.-G., Meyer, M., Nath, C.: *Offshore Windkraftanlagen*; Kombination der Lasten von Wind und Wellen, DEWEK 2000

[24] DIN EN ISO 9001, 2000-12 *Quality management systems – Requirements* (ISO 9001:2000); Trilingual version EN ISO 9001:2000

[25] D. Quarton, F. Rasmussen, K. Argyriadis, C. Nath: Wind Turbine Design Calculations – the State of the Art, *Proceedings, European Wind Energy Conference*, Göteborg, 1996

[26] Kühn, M.; etc. al: *Structural and Economic Optimisation of Bottom-Mounted Offshore Wind Energy Converters*, EU-JOULE III, JOR3-CT95-0087, Final Report, August 1998

[27] IEC TS 61400-23: *Wind Turbine Generator Systems – Part 23: Full Scale Structural Testing of Rotor Blades*, 2001-04

[28] IEC TR 61400-24: *Wind Turbine Generator Systems – Part 24: Lightning Protection*

[29] DIN EN ISO/IEC 17025, Publication date: 2005-08 *General requirements for the competence of testing and calibration laboratories* (ISO/IEC 17025:2005)

[30] IEC 61400-12: *Wind Turbine Generator Systems – Part 12: Wind Turbine Power Performance Testing*, 1998-02

[31] IEC 61400-11: *Wind Turbine Generator Systems – Part 11: Acoustic Noise Measurement Techniques*, 1998-09

[32] IEC TS 61400-13: *Wind Turbine Generator Systems – Part 13: Measurements of Mechanical Loads*, 2001-06

[33] IEC 61400-21: *Wind Turbine Generator Systems – Part -21: Power Quality Requirements for Grid Connected Wind Turbines*, 2003-02

[34] Dutch Pre-standard NVN 11400-1: *Wind Turbines – Part 0: Criteria for type certification – Technical criteria*, 1st edition, April 1999

[35] DIN EN 50308, 2005-03 *Wind turbines – Protective measures – Requirements for design, operation and maintenance*; German version EN 50308:2004

[36] DIN EN 287-1/A2:2005-09 *Qualification test of welders – Fusion welding – Part 1: Steels*

[37] ISO 9606 , Publication date:1994-08 *Approval testing of welders – Fusion welding*

[38] ISO 10474, Publication date: 1991-12 *Steel and steel products; inspection documents*

[39] DIN EN 10204, Publication date: 2005-01 *Metallic products – Types of inspection documents*

[40] EN 45004

[41] DIN EN 45011, Publication date: 1998-03 *General requirements for bodies operating product certification systems* (ISO/IEC Guide 65:1996)

[42] EN 61400-1: *Wind Turbine Generator Systems – Part 1: Safety Requirements* IEC 61400-1:1999, modified), February 2004

[43] Bows, Günther, Rosenthal, Vincent: "Similarity of the wind wave spectrum in finite depth water – 1. Spectral form", *Journal of Geophysical Research* 90 (1985) CI, 975-986

[44] E. Bouws, H. Günther, W. Rosenthal, C.L. Vincent, "Similarity of the Wind Wave Spectrum in Finite Depth Water" *Journal of Geophysical Research*, vol.90, 1985.

[45] State of the Art of Risk Models, Deliverable 5 form the SAFESHIP project, GL-AG, GLW, MARIN, 03-10-2003

[46] GL (2002): *"Richtlinie zur Erstellung von technischen Risikoanalysen für Offshore-Windparks"*, Germanischer Lloyd, Hamburg

[47] Anatec (2003): *"Description of Shipping Risk Models"*, Anatec UK Limited, Ref. T9999-Internal

[48] Otto, S., Nusser, S., Braasch, W. (2002): *"Collision risk of Ships with offshore windfarms and the danger of pollution of coastal regions"*, (German), Germanischer Lloyd Offshore and Industrial Services GmbH, Report No. GL-O 01-234, Hamburg

[49] Tak, C. van der (2003): *"Collision Risk Method"*, MARIN, Report No. 18056.620/1

[50] E.S. Ravn, P.Friis-Hansen, *Predicting Collision damage and Resulting Consequences*, paper ICCGS Conference, Japan, October 2004

[51] M. Lützen, P. Friis Hansen; *Risk reducing effect of automatic identification system (AIS) implementation.* Paper presented WMTC, San Francisco, October 18-20, 2003

[52] Berthold R. Metzger; *Wind Energy – Give Your Business a Cutting Edge*, PWC, Berlin (Germany)

[53] *"Anforderungen an Condition Monitoring Systeme für Windenergieanlagen"* (Requirements for Condition Monitoring Systems for Wind Turbines) by Allianz Zentrum für Technik GmbH [3].

[54] H. den Boon, H. Just, P. Friis Hansen, E. Sonne Ravn, K. Frouws, S. Otto, P. Dalhoff, J. Stein, C. van der Tak, J. van Rooij; *Reduction of Ship Collision Risks for Offshore Windfarms – SAFESHIP*, EWEC London, 2004

[55] Germanischer Lloyd: *Richtlinie zur Zertifizierung von Condition Monitoring Systemen in Windenergieanlagen*, Edition 2003

[56] Germanischer Lloyd Rules and Guidelines, *IV – Industrial Services, Part 6 – Offshore Installations*

{57] Germanischer Lloyd Rules and Guidelines, *IV – Industrial Services, Part 6 – Offshore Installations, Chapter 6 – Guidelines for the Construction and Classification/Certification of Floating Production, Storage and Off-Loading Units*, Edition 2000

[58] Germanischer Lloyd Rules and Guidelines, *IV – Industrial Services, Part 6 – Offshore Installations, Chapter 7 – Guideline for the Construction of Fixed Offshore Installations in Ice Infested Waters*

[59] Germanischer Lloyd Rules and Guidelines, *IV – Industrial Services, Part 8 – Pipelines, Chapter 1 -Rules for subsea pipelines and risers*, Edition 2004

[60] Germanischer Lloyd Rules and Guidelines, *I – Ship Technology, Part 5 – Underwater Technology*, Edition 1998

[61] International Maritime Organization (IMO): Publications Catalog, www.imo.org

Chapter 12:

Offshore Wind Turbine Foundations in Deep Waters

F.G. Cesari, T. Balestra and F. Taraborrelli.

1. INTRODUCTION

The search for better wind conditions for increased wind energy generation has led to sites placed in the sea. When the sea depth reaches 30 to 50 metres, the sites are said to be in "deep waters". Wind turbine foundations for these relatively large depths have to be carefully studied and designed. The foundation types for offshore wind turbines (OWT) are discussed in this chapter to demonstrate the economic and technical feasibility and to allow their installation for windfarms. The distinction between shallow and deep waters is clarified, each referring to a specific form of substructure. The methodology for this definition will emerge from the analysis of economic acceptability of the different foundation types.

The support device for an offshore wind turbine is variously named its foundation, substructure, or supporting system. It may be defined as *'the part of the total structure inserted in or placed on the seabed to support the wind energy converter permanently'*. It therefore may include anchorage piles placed in the seabed for sustaining the soil reactions to the loads or a gravity foundation, which is a large mass, laid on the seabed so its dimensions and weight together resist the wind and sea forces. The substructure is the portion submerged in the sea. The whole foundation is connected to the relatively standard tower, which in turn holds the nacelle and the rotor. Wind turbines on land have no substructure, but only a foundation and tower.

G*ravity foundations*, which apply almost entirely to shallow-water conditions and so are not analysed significantly in this chapter, three types of configurations will be discussed; (i) *monopile* support structures, (ii) *tripod* foundations, and (iii) a *three- or four-legged jacket*, as used for oil and gas extraction platforms. The designs and numerical analyses are based on general data, corresponding to an 'ideal' site in the Mediterranean or in Northern European Seas. This allows us to select the aerodynamic and hydrodynamic forces (winds, waves, currents, tides, etc.), the soil and subsoil composition and the other variables necessary to dimension the structure. Considering the number of windfarms installed in the Northern Seas of Europe, the reference site for this exemplar study is in the Adriatic Sea, which is characterised by soil properties similar to those of northern offshore sites. We realise of course, that any particular site has its own environmental characteristic that may differ considerably from the 'ideal' and will always require specific analysis.

The analysis is carried out on two main variables, the seabed depth and the wind turbine power. The selection of the substructure type and form may be possible through these two aspects. For this approach, it is important to determine the static loading to be applied to the structure. This phase of a project corresponds to a preliminary design, which is essential and necessary to dimension all the parts of the foundation and the substructure.

Constructions fixed to the seabed can also be divided into two other main groups, (1) those which have the *main vibrational frequency* of the structure (the *eigenfrequency*) *greater than the dominant wave frequency*, and (2) structures with an *eigenfrequency less than the dominant wave frequency*. In addition, the *stiffness* between the turbine and the foundation is a key factor for the complex design of an offshore wind turbine. Too large a stiffness with the foundation structure could constrain the wind generator operation, while too small a stiffness (i.e. too much elasticity), could reduce the natural frequency with the risk of dangerous resonant oscillation from waves.

2. STRUCTURAL SOLUTIONS

2.1 Selecting the Foundation

2.1.1 Substructure general characteristics

Initially, the following technical tasks should be considered for the structural analysis:

- *Seabed characteristics*: stratigraphic description, bearing capacity of the soil, depth and slope of the seabed, etc.;
- *Wind-sea actions*: wind speed and direction at specific heights; wave heights and wavelengths in the open sea, at the site and splash zone; tidal and other marine currents, etc;
- *Effects produced by the wind turbine* on the supporting structure: bending moment, shear action, vertical strength, impressed frequencies, etc.;
- *Calculation criteria*
    1. *Environmental forces + dead weight*
    2. *Operational forces + dead weight*
    3. *The worst case between 1-2 + environmental forces*
- *Structural materials*: physical/mechanical characteristics and allowable stress/strain, behaviour in low and elevated temperatures, transition temperatures, chemical reactions with salt water, etc.;
- *Surface protection* of metallic structure against erosion and corrosion;

Fatigue analysis over an operational life of 20 to 30 years;

- *Penetration length of the foundation piles* (if necessary), as functions of the pile diameter and thickness, which affect the impact force of the pile driver and the shear reaction of the soil;
- *Effects of the soil behaviour*, e.g. reaction to the loads transmitted by foundation piles, scouring effects due to water currents, etc.

For an initial outline design of the structural parameters, only limited data are needed. The general dimensions and the resulting estimate of dead weight should be determined first. The resulting approximate cost of the structure is then combined with the cost of installation and of the wind turbine to obtain a total cost estimate per turbine. This factor is based on a schematised time programme for the entire offshore windfarm (not discussed in this chapter). Having obtained estimates for the fixed and capital costs, the next step is to derive the variable costs for average operation of the windfarm.

2.1.2 Foundations as a function of wind turbine power and seabed depth

The supporting structure types are listed in the section 2.4 and represented in Figs 1 to 4. For a 2 MW wind turbine, three types of metallic substructures have been considered in our detailed study, i.e. monopile, tripod and three-legged jacket, Fig 1. The gravity foundation (Fig. 6) received less attention, since it becomes less attractive as the turbine power and seabed depth increase. Some details on geometry, feasibility and stability of each solution are presented in the next paragraphs as functions of seabed depth; increasing depth dramatically reduces the spectrum of available substructures.

For very large 4 to 6 MW wind turbines in deep water, the number of possible foundation types reduces to two, namely the three- and four-legged jacket. This is because, in practice, the monopile and tripod types cannot sustain the loading generated by sea and wind. The four-legged jacket is a unique way to solve the structural problems, especially for deeper waters.

Below are expanded descriptions of the structures, together with various observations for design, fabrication, installation and ultimate removal. We assume that the structures are delivered ready-to-be-installed from the dockyard and can be transported directly from the dockyard to the site.

2.2 Offshore Wind Turbine (OWT) and Deep Foundation Data

2.2.1 Wind turbine data

The wind actions on rotor and tower are transmitted to the wind turbine base and thence to the foundation. The data are evaluated for three powers of turbine in Table 1, i.e. 2 MW, 4 MW and 6 MW; this gives a range of turbines to determine their influence on the substructure.

Table 1 Data/loads for 2/4/6 MW wind turbines

	2 MW	4 MW	6 MW
Schematised bending moment [kN m]	20,000 – 40,000	80,000 – 140,000	140,000 – 250,000
Vertical load [kN]	2,000	≤ 5,000	≤ 8,000
Rotor axis height [m] above mean sea level	50 – 80	80 – 110	90 – 140
Idealised swept area [m²]	5,000 – 7,000	7,000 – 10,000	10,000 – 14,000

We need to determine the characteristics that influence the design of the associated supporting structure, and hence its size and weight. Therefore, we must analyse and estimate the possible technical and economic limits for each substructure model.

The principal purpose of the support structure is to sustain a 2, 4 or 6 MW wind turbine under the imposed aerodynamic and hydrodynamic forces associated with both operational and extreme event conditions. Obviously, the parameters depend on the regional characteristics of the site. For instance, for sites in the Adriatic and Baltic seas, the design data can be derived from bibliographic sources. Thus, the analysis of the different types of foundations has been carried out, imposing a rational variation of the seabed depth (see par. 2.2.3).

2.2.2 Seabed soil conditions

The pile stability has to be verified, considering, (i) the ultimate strength in tension, which is equal to the friction resistance on the pile surface when it is subjected to traction, and (ii) the lateral resistance, which depends on the pile-soil interaction and the p-y curves, as in the RINA code (see para. 3.1.1). The ultimate resistance (p) is proportional to the pile diameter and is a function of the undrained shear strength. The characteristic displacement (y) is proportional to the pile diameter and the plastic deformation of the soil, when it is subjected to a compression equal to half the ultimate compression.

For our example, soil conditions have been assessed on data communicated by the designers of platforms installed in sectors of the Adriatic and Baltic seas. Usually soil consists of fine to medium sands, typically medium to low density, with thin interspersed layers of silt and clay. The seabed may be horizontal or slightly sloped. The soft soil characteristics are modelled using p-y curves to evaluate the anchorage pile length below seabed, where necessary.

An important characteristic for the offshore foundation is the seabed instability. Four types of soil instability have to be considered:

(1) instability from the natural slope of the bed, which could be relevant in clay/sandy soils or in a seismic zone;
(2) hydraulic instability caused by water fluxes through a porous bed or by draining-canals, due to turbulence of currents around the structure;
(3) instability caused by scouring;
(4) the sand upper layers could be subject to possible liquefaction under dynamic effects, e.g. as originated by an earthquake.

At sites with sandy soils and significant tidal currents, *scouring* is a common problem for large offshore foundations, with the production of shallow and wide depressions and, perhaps, a cavity around the structure. Probably some scour protection would be required, but it is a costly procedure. This could be requested, especially for monopile and gravity structures, as dependent on their relevant diametrical sizes, which could be around 6 to 8 m in the case of a monopile, and much more for a gravity foundation. Scour effects are identified by the influence of the structure on the flow pattern and by overall seabed movement. Sand waves occur in soil of loose material, which is easily transported by sea currents. Uncertainty about the supporting function of the seabed is a fundamental problem for a monopile substructure. Particular effects include in a reduction of soil stability, a variation (mainly a decrease) of natural frequency and an increase in bending moments.

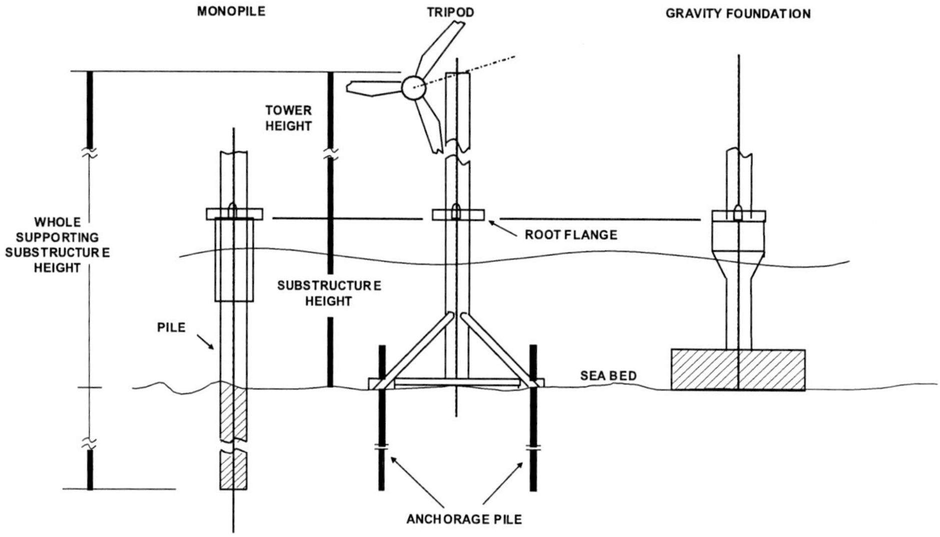

Figure 1. Schematic representation of the different substructures

For the sites in the Northern Adriatic Sea, scouring is not a significant aspect affecting the substructure, especially for three- and four-legged jackets. This favourable situation is justified by a combination of the sea and seabed conditions and, mainly, the external diameter of the legs, which is always less than that of a monopile.

2.2.3 Environmental data

The water depth to *lowest astronomical tide* (LAT) at a site may range from 25 to 200 m; allowing for this large range is effectively a sensitivity analysis on foundation forms and their economics for windfarms.

Typical extreme (100-years return period) and operating (1-year return period) *wave data* for the North Adriatic Sea are listed below.

Operating design wave conditions	
Maximum wave height	*10.3 m*
Associated mean period	*9.7 s*
Surface current speed	*0.86 m/s*
Current speed at -50 m depth	*0.30 m/s*
Wave length	*135 m*

Storm design wave conditions	
Maximum wave height	*12.6 m*
Associated period	*10.8 s*
Surface current	*0.99 m/s*
Current speed at -50 m depth	*0.30 m/s*
Wave length	*161 m*

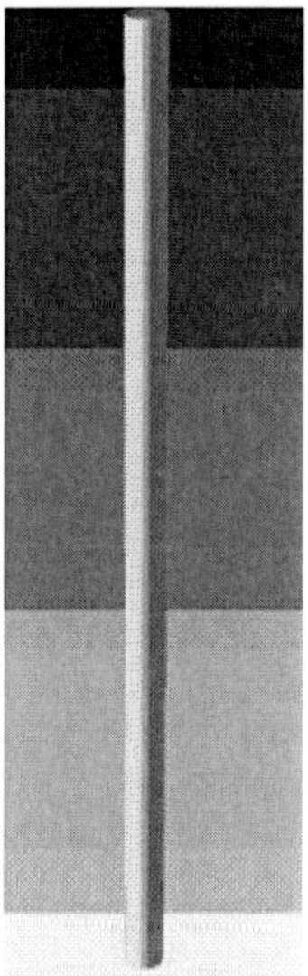

Figure 2. Monopile

Figure 3. Tripod

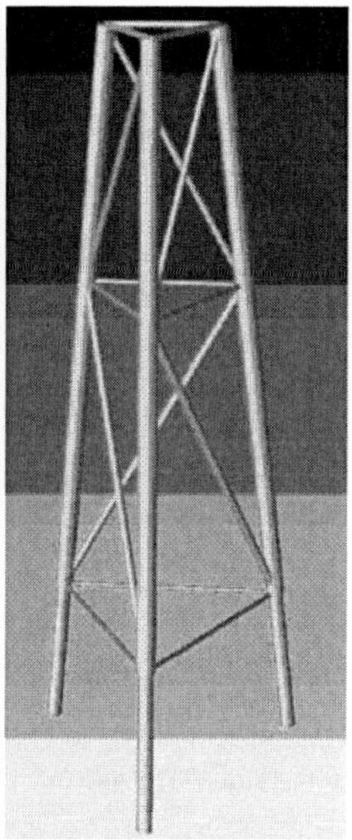

Figure 4. Three-legged jacket

Figure 5. Four-legged jacket

The above wave and current conditions are assumed omnidirectional; they may change if the depth increases or decreases. At different depths, the speed of the current also varies. From the surface, it decreases quasi-linearly to 0.3 m/s. This corresponds to the estimated speed at 50 m depth or at the seabed level considered in the calculations. In our analysis, the depth of the seabed varies between 25 m and 50 m or more. The mean wave height of the 50-year return sea state is probably less than the maximum value listed (12.6 m). The operating maximum wave height is used for all the structural analyses on the foundation. In all the cases, the seabed soil is always soft, and the surface current speed and wave data are the same.

This non-integral design approach is only applied for the preliminary design, necessary to estimate the substructure weight and cost. On these two variables (weight and cost), we can compare the different solutions for the offshore supporting systems.

Considering rotor sizes in Mediterranean seas, the bending moment due to the wind action is calculated for maximum hub heights and large swept areas. The purpose is to increase the wind speed at hub height and so collect maximum wind energy. In regions with larger wind speeds, hub heights and swept areas need not be so large to obtain the same turbine rated power.

2.3 Metallic Substructures for Medium- to High-Power Wind Turbines

2.3.1 Concept evaluation

The three main options for seabed-fixed offshore substructures are *driven pile, gravity-based and skirted*. Of these, the driven-pile is the most common form of offshore foundation (Figs 1 and 2). The standard method to anchor the foundation to the seabed is to fix the piles by a diesel or hydraulically driven hammer. In contrast, the gravity foundation avoids tensile loads between the supporting structure and the soil. This is achieved by applying large dead loads over the relatively large base to stabilise the seabed soil structure. The third type is similar in appearance to the gravity foundation, being in the form of an overturned bucket, but its adoption is questionable owing to the sizeable static component of wind loading.

Figure 6. Gravity substructures to be installed in windfarm site

In comparison with the piled solution, the gravity foundation (Figs 1 & 6) avoids tensile loads between the supporting structure and the soil. This is achieved by applying strong dead loads over the relatively large base to stabilise the structure.

In this chapter for deep-water foundations, we will particularly consider the following metal structures:

1) *Monopile foundation or piled monotower* (Fig. 2);
2) *Tripod foundation or monopod* (Fig. 3);
3) *Three legged jacket* (Fig. 4);
4) *Four legged jacket* (Fig. 5).

The main difference in practice is the number of piles required. For the *monopile,* the tube used for piling may remain in place as the foundation tube itself.

Compared to the monopile, the *tripod* consists of three smaller and shorter piled legs connected to the central column supporting the tower. Visually, it seems to combine the concept of the monopile, through the central column, with that of the jacket, through the three legs; these legs are partially submerged in the seabed and, in position, act as guides for the piles. More information on the characteristics of the different types is in tables 2a & b "Schematic list of requirements".

The optimised solution depends not only on the stability, dead weight and relative costs of the substructure, but also on other aspects, such as design, construction, erection, installation and eventual removal.

2.3.2 Installation

The characteristics and development of offshore wind turbines require a variety of different methods of installation, achievable by *lifting or floating the component parts*. Other factors relate to the regional location, including availability of ports and boats (such as barges, pontoons, tugs), distance from the construction dockyard, soil resistance (e.g. need for possible use of a jack-up, need to drill the piles instead of driving and hammering them in), meteorological conditions, etc. The solutions to these problems may be:

1. Foundation *construction in situ* and subsequent wind turbine erection on the substructure;
2. *Transportation of the foundation*, (Figs. 6, 14 & 15), its placement in situ and anchorage to the seabed with the subsequent wind turbine installation;

Table 2 a Schematic list of requirements for monopile and gravity foundations

Substructure	Design		Construction		Installation		Removal	
	Advantages	**Disadvantages**	**Advantages**	**Disadvantages**	**Advantages**	**Disadvantages**	**Advantages**	**Disadvantages**
Monopile	Simple structural configuration	Relatively sensitive to scour for high diameters	Minimal construction site investment	Excessive slenderness in deep waters	No seabed preparation without scouring	Piling equipment required portion only	Removal of under-water removable	Pile in soil portion not
	Dynamic response to tune structural flexibility	Performance determined by soil characteristics	Rolling facilities widely suitable for not thick tubes	-	- in boulders, very stiff clays, etc.	Drilling equipment pile lifting after sectioning	Removal by in water	Sectioning
	Suitable to seabed depth of 3 m to 20 m	Not suitable for depths of 20/25m or more	Ease, simple & economic for tubes not thick	Cost sensitive to high thickness & diameter	-	Field splice & adjustment to inclination	Not heavy structures to be removed	Difficult cut for thicker pile
	No anchorage pile requested	Poorly suitable for weak soil	Ease transportation to site	-	Ease transportation by floating	Field welding for high pile	-	-
	Steel as unique structural material	Not suitable for rocky soil	No particular protection against ice	-	- in exposed location	Scour protection	-	-
Gravity Foundation	Structure requiring yard on sea shore	Poorly suitable under scouring	Large use of concrete	Large construction site investment	No anchorage pile requested	Seabed preparation required	Complete removal at end of life	Ballast costly elimination
	Performance insensitive to soil parameters	Stiff dynamic response	Construction in yard	Yard close to windfarm site	Whole installation as single unit	Costly ballast placement	Possible reposition-ing after removal	Big lifting required
	Anchorage pile not required	Configuration often complex	Yard with ease access to sea	Yard equipped by strong lifting	Transport to offshore site by barge	Barge with big lifting capacity	By means of crane barge	Heavy body on barge
	Suitable to seabed depths 3m to 15 m	Not suitable for depth over 25m	Mainly concrete for substructure	Appreciable space at construction yard	Vertical positioning on crane barge	Tonnage barge adequate	-	-
	Good resistance to ice in water	Poorly suitable in weak soil	-	Sizes dependent by OWT power	-	Large protection against scour	-	-
	Compact structural configuration	Heave forces during the wave passage	-	Sizes depending strongly by depth	-	-	-	-

Table 2b Schematic list of requirements for jacket and tripod substructures

Substructure	Design		Construction		Installation		Removal	
	Advantages	Disadvantages	Advantages	Disadvantages	Advantages	Disadvantages	Advantages	Disadvantages
Three/four legs jacket	Relative insensitive to scour	Complex, unsuitable configuration with ice in water	Configuration used in oil/gas extraction	Specific technology to be used	No seabed preparation without scouring	Piling equipment required	Removal of under-water portion only	Pile in soil portion not removable
	Dynamic response to tune structural flexibility	Anchorage pile in each leg	All weldable piled lattice in steel	Construction site not close to offshore site	Small tubes for legs & anchorage piles	Drilling equipment in boulders, very stiff clays, etc.	Removal by lifting after sectioning	Sectioning in water
	Suitable to seabed depth of 8m to 40 m	Not suitable for depths of 8m⇢h⇢40m	Easy applicable to different seabed heights	In yard unsuitable construction	Possible use of crane barge or juke-up	Difficult transport by floating	Not heavy structures to be removed	Difficult cut for thicker pile
	Small displacements & rotation at jacket top	Anchorage pile in each leg	Ease transportation to site	Appreciable space required at yard	Small hammer for piling	Field welding for pile connections	-	-
	Suitable to different OWT powers	Needs protection against corrosion	No limits for OWT types	Very long anchorage piles	No juke-up needed for installation	Unsuitable for rocky soil without drilling	-	-
	Insensible to wave passage heave forces	Complex & not compact configurat.	Easy adaptable to resistance needs	Large use of weldings	-	-	-	-
Three/four legs jacket	Relative insensitive to scour	Complex, unsuitable configuration with ice in water	Configuration used in oil/gas extraction	Specific technology to be used	No seabed preparation without scouring	Piling equipment required	Removal of under-water portion only	Pile in soil portion not removable
	Dynamic response to tune structural flexibility	Anchorage pile in each leg	All weldable piled lattice in steel	Construction site not close to offshore site	Small tubes for legs & anchorage piles	Drilling equipment in boulders, very stiff clays, etc.	Removal by lifting after sectioning	Sectioning in water
	Suitable to seabed depth of 8m to 30 m	Not suitable for depths of 3m>h>30m	Easy applicable to different seabed heights	In yard unsuitable construction	Possible use of crane barge or juke-up	Difficult transport by floating	Not heavy structures to be removed	Difficult cut for thicker pile
	Small displacements & rotation at jacket top	Anchorage pile in each leg	Ease transportation to site	Appreciable space required at yard	Small hammer for piling	Field welding for pile connections	-	-
	Suitable to different OWT powers	Needs protection against corrosion	Sufficiently short anchorage piles	–	No juke-up needed for installation	Unsuitable for rocky soil without drilling	-	-

3. Transportation of the *foundation with the tower already erected on it*, to the correct position in the windfarm and subsequent installation of nacelle/rotor;
4. Transportation of the *foundation already equipped with the wind* turbine, complete with nacelle and bladed rotor.

Option 1 is typical of certain forms of gravity foundations. Option 2 is the usual method adopted for offshore turbines. Options 3 and 4 apply to floatable foundations, 'sunk' on site. Each design must include the installation of all the components of the foundation and turbine, and the transportation of each part by means of barges and the lifting by cranes from barge to site.

Figure 7. Transition zone between monopile foundation and tower (Elsam)

2.4. Supporting Structure for 2 MW Wind Energy Converters

2.4.1. Monopile

The installation of very large diameter piles has become common for offshore windfarms. As piling equipment has been developed and theories for laterally loaded piles have been established, mono piles have been used. They are common as substructure for offshore wind turbines located in shallow or medium depth waters.

The balance among the following different aspects of monopiles could be considered as favourable or unfavourable, depending on the seabed depth, i.e.:

- *simple structural configuration*, not susceptible to scouring;
- *flexible foundation* offering opportunities to tune the structural dynamics, but sensitive to choice of soil parameters and to resistance of soil (poorly suited for weak soils);
- *suitable for water depth* of 3 m to about 20 m, but not more than 25 m;
- *simple construction* (minimal site investment, very widely available rolling techniques, cost not expensive for small thickness, possible to procure the pile directly from the steelworks to the site, etc.);
- *easy connection* with the tower through a conical transition (Fig. 7);
- *no seabed preparation*, but usually susceptible to scouring, so scour protection may be required in exposed locations;
- *piling equipment* always required, but also drilling devices for very stiff clay, mudstones, etc. (not suited for geotechnical locations with large boulders);
- *decommissioning* at end-of-life, pile cut at seabed and removed, so remaining embedded part not removed.

The monopile usually consists of two sections. For example, the *lower section* may be a circular hollow pile (e.g. with diameter of 2.5 m and a wall thickness of 40 to 60 mm, driven about 20 m into the seabed). The *upper section* may be conical, allowing connection between the top of the driven pile and the base of the turbine tower. Adjustment is possible to ensure the tower is accurately vertical, and after that, the upper section is welded to the driven pile. The loads acting on the foundation are mainly due to wind forces. The pile length is selected to have only elastic deformation of the soil when exposed to fatigue loading.

In our case-study, the monopile solution is not applicable for the 2 MW wind turbine. Flexing of the monopile in 25 m deep water in normal winds leads to a displacement amplitude at the tower base of 136 cm, which at 85 m nacelle height corresponds to 482 cm. The rotation amounts to 2.82°. These figures increase strongly in storm conditions, rendering the monopile completely unacceptable.

2.4.2 Tripod

The tripod concept, Figs 3 & 8, is sometimes misleadingly called a 'monopod'. It is the minimum structure to avoid bending, by having three steel piles located at a distance from central column and angled to connect to the tower base. The design has been used for marginal platforms in the oil and gas industries. It is considered lightweight and cost-effective; however, it has yet to be verified that these characteristics are valid for the offshore wind energy industry.

The following characteristics of tripods have to be considered and balanced against each other:

- *adaptable* to increased water depths;
- *minimum seabed preparation* required prior to installation;
- *specialised fabrication* methods, similar to legged jacket construction;
- not suitable for geotechnical locations with large boulders;
- not fully suitable for shallow water (< 10 to 15 m).

Figure 8. Tripod substructure

The construction consists of a steel space frame made by tubes and arranged with a triangular base. This structure transfers the sectional forces from the wind turbine tower primarily to tension and compression loads in three hollow steel piles driven into the seabed. From below the tie-in flange, connecting the turbine tower to the substructure, a large diameter tubular component extends centrally downwards. The upper end could have a conical transition section (as in Fig. 7) to reduce the diameter and to increase the wall thickness. Each leg consists of a pile, a pile sleeve and three large braces (in Fig 3, five in the first floor and three in the second floor). After pile driving, the annulus between the pile sleeve and pile itself is filled with grout, to form a rigid complex. From initial analysis of the pyramidal form, it is evident that the number of elements increases the total weight of the structure as compared with a monopile. To give sufficient stability and to support the wind turbine actions, the diagonal elements have been raised to a maximum elevation and the column wall has been adequately thickened in its upper part, obtaining acceptable displacements at the top of the substructure.

Figure 9. Four legged jacket with a different number of floors

The *total weight* (structure plus piles) for a 25 m depth is very large. If this weight is not a limit for transportation by barges, it leads to a consistent cost limitation. In our case study, because of the outline results of the preliminary investigation, it is decided not to consider the tripod solution for 4 MW and larger wind turbines.

2.4.3 Three-legged jacket

This jacket, Fig 4, is similar to the conventional structure for oil platforms. It consists of three columns, each slightly inclined from the vertical. Horizontal connections at a different heights tied by diagonal tubes stiffen the structure. The number and size of the planes and diagonals assure the expected stiffening. The internal surface of each leg guides its anchorage pile, which, for our study for a 2 MW wind turbine, has to be driven up to 55 m into the seabed at 25 m sea depth. The structural configuration, which assures good stability, has acceptable weight and therefore cost. In this arrangement, the structure appears to be an economic and profitable solution for a multimegawatt wind turbine.

The balance amongst the following different aspects could be positive, in contrast the same comparison for monopiles:

- conceptual formulation and detailed design of the *characteristics of traditional jackets for oil extraction platforms*;
- *complex structural configuration*, which could be weak in strong ice conditions (absolutely absent in the Mediterranean Sea); suited to water depth from 8 to perhaps 40 m and considerably more convenient than other alternatives;
- *stiff foundation* response, limiting the opportunities to tune the structural dynamic coupling between tower/nacelle and foundation;
- *specific requirements for construction* (specialised technology, workshop possible close to the windfarm site, appreciable space for storing/mounting the substructure, etc.);
- *acceptable cost and good stability*;
- *minimum seabed preparation* and defence against scouring, if any;
- *easily transported to site* by floating or by crane barge, which can also be used for its installation;
- *easy pile insertion* requiring smaller hammer and conversely longer time of hammering (repetition of this activity due to the number of legs, considerable length of foundation piles, time consumption for pile spools welding, probable field splice between different sections of tower, and of tower and foundation, etc.);
- *decommissioning at end of life*, piles not completely removed, requiring cutting at seabed level (or preferably a little below the mudline).

The structure is sufficiently rigid, allowing acceptable displacements and rotations at the substructure top in storm conditions (10 cm. and 0.53 degrees for a 25 m. sea depth). It may be necessary to improve these figures and to improve safety, by increasing the numbers of floors and diagonals, or by having a larger inclination of the legs (even if this alternative could complicate the anchorage pile insertion).

To determine the applicability limits using the same structural configuration, some other depths were considered, namely 50, 100 and 200 m. At 50 m depth, our calculations show that no supplementary stiffening of the structural configuration should be required; storm condition displacements and rotations at the top are increased, but not seriously (135 mm. and 0.63 degrees). To reduce the displacement and rotation, the stated improvements are easily made, demonstrating that the technical aspects do not introduce any limitation. The real limits relate to the dead weight and, consequently, the cost.

2.5. Supporting Structures for 4 and 6 MW Offshore Wind Turbines

We first analyse a three-legged jacket, as before, but for 4 and 6 MW turbines instead of 2 MW. Only some technical specifications (mainly diameter and thickness of tubes) were adjusted for the increased forces and bending.

2.5.1 Three-legged jacket

For the larger turbines at 25 m depth and without any other modifications on the geometrical model, the intermediate and base sections of the legs have to be thickened and the diagonals strongly crossed for support in storm conditions. The strains evaluated at the foundation top are a little too large (9.5 cm of displacement and 2.9 degrees of rotation), so reducing the ability of this solution to a safe response under static loads. Probably a three-legged jacket could be applicable for depths of about 20 to 25 m, but no more.

2.5.2 Four-legged jacket

Fig 10 shows an example of a four-legged platform used in the oil industry, where the dominant loads are the very strong vertical forces. This jacket is very similar to the three-legged jacket, with the main difference being the number of legs and the redistribution of static loads into four points of anchorage. The considerations on the balance on favourable versus unfavourable technical aspects of the model for wind turbines are the same as for the three-legged jacket.

Since the foundation has to be both strong and rigid, the dead weight is increased throughout, even for each leg and its pile (total length 135 m). In this arrangement at the tower base, both the displacements and the rotations are within an acceptable range. Further stiffening could be arranged by increasing the leg slope, which is limited by the spatial and verticality limitations of the pile hammer.

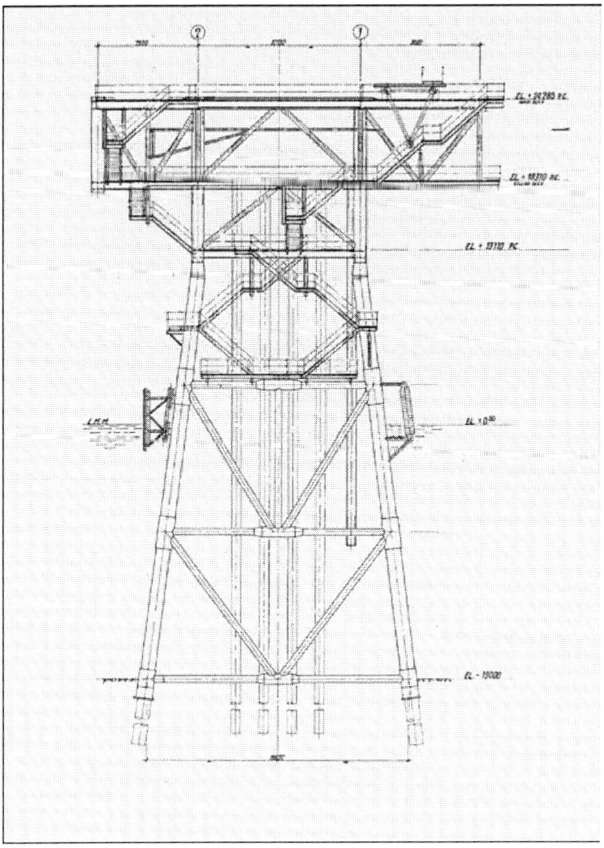

Figure 10. Four legged platform used in petroleum extraction (ENI)

As for the effects of deeper seabed on the structure, it is not sufficient to recalculate the characteristics for 50 m depth, since at this depth some change on the geometry is necessary and, in particular, the introduction of a third floor (Fig. 9). The configuration has to be slightly modified for a depth of seawater more than 50 to 70 m. For depths reaching 100 m, another braced floor has to be added and the sizing calculation has to be repeated for preliminary stress analysis.

If the foundation pile-length into the seabed is more than 100 m, an optimisation between its ratio of diameter to thickness and the length is required.(see para 2.2.2). The problems of static and dynamic stability of the structure could affect substantially the technical reliability and the economic availability of this model. At the top of the foundation, the lateral displacements and the rotations become at least 260 mm. and 0.7°, which are too large. The unity check (an important method to test the structural strength) is excessively high because of the modelled slenderness. This situation could have a dangerous influence in storm conditions. The evaluation of the weights and the corresponding costs are presented and discussed later in Section 6.

2.6. Floating and Non-Floating Substructures

Some theoretical studies have been conducted to compare wind turbines on fixed and non-fixed platforms (floating and semi-submerged) as alternatives to the typical offshore foundations. For fixed platforms at deep offshore sites at more than 25 m depth, we studied the use of oil and gas offshore platforms as substructures of wind turbines. An example of this kind is the offshore platform 22 km off the coast of Sicily, named VEGA; we assumed this is equipped with a 4 to 5 MW wind turbine, producing 12 GWh per year, founded on a 120 m seabed.

Figure 11. Semi-submersible design concept

Floating and semi-submerged structures (i.e. non-fixed structures) do not require fixed foundations, which improve their environmental impact The structures are made of two main parts, (i) the floating platform or semi-submerged structure with its anchorage, and (ii) the wind turbines. Fig 12a is an example of a floating device with dynamic motion compensation obtained by pumping ballast water. Fig. 12b shows a schematic view of tensioned platforms to be used in both the oil extraction and wind energy industries.

Design studies have applied the concepts to a single turbine, believing that a multi-turbines scheme on a single platform would too difficult and too expensive; an example is the tensioned leg platform (i.e. a submerged platform connected to the seabed by tensioned moorings).

Main technical aspects to consider for the design of these submarine floating sub-structures are:

- *Buoyancy* is maintained by water-tight chambers in the floating 'foundation' structure;
- *Placing the sub-structure below the sea-level*, reduces its wave loads and movements;
- *Materials* to be used for the sub-structure are concrete or steel; concrete is cheaper and heavier, but weaker in tension;

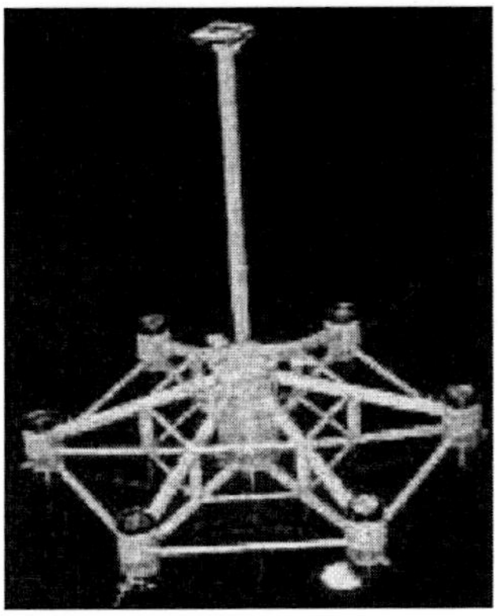

Figure 12a. Floating device for 5/10 MW wind mill (Hitachi)

Figure 12b. Deep water platform with tensioned anchorage for oil extraction and wind mill (NREL)

- *Moorings* may be of two different kinds, catenary moorings which act through their weight, and tensioned moorings which prevent, especially, pitch, roll and heave movements, so reducing fatigue loadings on the wind-turbine rotor (Fig. 13);
- *Solutions* for single or multiple turbines structures.

The technological challenges for future development are:

- *Dimensions:* multiple turbine vessels may require large amounts of materials;
- *Wave loads:* large structural surfaces may be exposed to waves, requiring significant materials-bearing capacity;
- *Stability:* the structure is subject to large overturning moments from the turbine rotor drag;
- *Motion response:* to reduce fatigue damage, it is necessary to minimize movements;
- *Anchorage:* tensioned moorings are expensive, and catenary moorings perform poorly in relatively shallow waters.

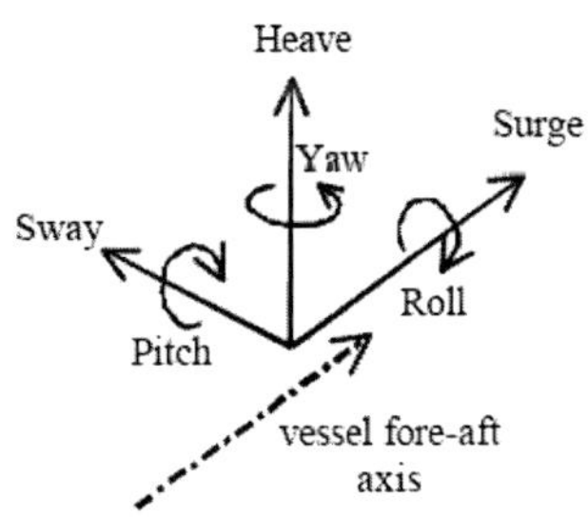

Figure 13. Rigid motions degree of freedom

To date, no structures of these kinds exist, although several studies have been conducted. These studies did not report good technical and economical results: An important feasibility project from the University of Delft and other partners, considered semi-submerged structures near the coast for a 50 m deep sandy seabed in the North Sea. Their results were: (1) for a 9 m/s wind speed at 10 m height above sea level, and installing 100 x 5 MW turbines, the generated electricity cost would be 0.068 Euro/MWh; (2) for a similar site 100 km off the coast, the electricity cost would be 0.074 Euro/MWh.

3. SUBSTRUCTURE DESIGN

The substructure needs to be designed for the functional requirements of the wind turbine. The design must also ensure that strength and stability are rigorously satisfied during the various stages of load-out, barge transportation, lift installation and in-service operations.

3.1 Design Data

The technique must satisfy the substructure design and reference codes that typical for designing other offshore structures. The main requirements to develop the structural analysis of the substructure are summarised in the following paragraphs.

3.1.1 Codes to be applied

For example, the main internationally recognised Codes for an offshore project in the Adriatic Sea, are

- API RP 2A: American Petroleum Institute – Recommended Practice for Planning, Designing and Constructing Fixed Offshore Platforms.
- AISC: American Institute of Steel Construction – Manual of Steel Construction.
- RINA: Registro Italiano Navale (Italian Navy Register) – Rules for the Classification of Steel fixed Offshore Platforms.
- GERMANISCHER LLOYD: Wind Turbines, Part 1, Rules for Regulation IV, Non-marine Technology, Part 2

3.1.2 Functional requirements

The requirement for the functionality of the wind turbines is a limitation of the displacements and rotations at the interface between the substructure and the turbine tower. Typically, rotation of the substructure interface of 0.5° is considered the maximum acceptable rotation for operating wind turbines. It is therefore essential that the environmental conditions which produce substructure rotations that exceed the allowable value, be of very limited occurrence. In principle, a reasonable design criteria limit is the 1-year return period for the environmental condition.

Table 3. Maximum wave height

Wave Approach Direction	1-year		100-year	
	THmax (s)	Hmax (m)	THmax (s)	Hmax (m)
North	8.7	8.1	108	12.6
East	7.0	5.2	10.3	11.5
South	5.6	3.4	7.6	6.3
West	5.8	3.6	7.4	5.9

3.1.3 Environmental data

The environmental data necessary for developing the substructure design are listed below. They are essentially

- *Water depths*
  Water depth referred to LAT (Lowest Astronomical Tide) and the elevation associated with the maximum astronomical tide and with the maximum surge.

- *Wave data for extreme and operational conditions*
  The *extreme* environmental conditions are defined as those of one occurrence in 50 and 100 years. The *operating* environmental conditions could be those with once in a year occurrence.

The two-dimensional wave kinematics may be determined from an appropriate wave theory, depending on the wave height and period and on water depth. For the envisaged sites in the Northern Adriatic Sea, the Stokes 5th order wave theory is applicable.

Table 3 (for a typical North Adriatic offshore site) indicates the maximum wave heights for the main approach direction and for the 100-years and 1-year occurrence.

Table 4. Sea current speeds, m/s.

depth	Surface	5 m	25 m	Bottom (40 m)
100 year storm	0.81	0.81	0.62	0.51
1 year storm	0.66	0.66	0.53	0.41

- *Sea-current data for extreme and operational conditions*
  The sea-current parameters are the speed profile at the sea surface, where the speed is a maximum with respect to that at the seabed, or at specified depths. For instance, for a typical 40 m depth site in the Adriatic Sea, the sea-current are listed in Table 4.

- *scatter diagrams of wave heights for fatigue analysis*
  The cyclic loads, like the wave loads, can cause a significant reduction of the welded connections of the substructure due to fatigue, e.g. Table 5..

Table 5. Scatter values of waves by significant wave height and period

HS (m) TZ (s) →	0.25s	0.75s	1.25s	1.75s	2.25s	2.75s	3.25s	3.75s	4.25s	4.75s	5.25s	5.75s	Total Occurrences
3.5m	28,255	2,611	26	0	0	0	0	0	0	0	0	0	30,892
4.5m	5,984	9,127	1,766	84	1	0	0	0	0	0	0	0	16,962
5.5m	270	2,270	3,354	1,177	168	13	0	0	0	0	0	0	7,252
6.5m	4	90	614	999	589	182	36	5	0	0	0	0	2,519
7.5m	0	1	20	109	203	180	96	37	11	2	0	0	659
8.5m	0	0	0	2	11	24	28	23	13	6	2	1	110
9.5m	0	0	0	0	0	0	1	2	2	1	1	0	7
TOTAL	34,513	14,099	5,780	2,371	972	399	161	67	26	9	3	1	58,401

Fatigue analysis is therefore mandatory for a correct design of an offshore structure. The fatigue verification is generally performed following a stochastic approach for reliable results. Therefore wave scatter diagrams are needed, which give, for each direction, the distribution of wave periods versus significant wave height, expressed in number of expected events of 3 hour duration in the service life of the structure. Table 5 shows the omnidirectional scatter diagrams for a site in a northern area of the Adriatic Sea and for 20 years.

- *wind data associated with extreme and operational conditions*
  For the substructure structural design, the wind data may be limited to the 1-minute sustained wind speed for the 100-year and 1-year return period.
  Typical wind data at elevation +10m for a North Adriatic Sea location are listed in the Table 6.

Table 6. Wind speed data

Return Period	1' Wind Speed (m/s)
1 year	37.0
100 years	52.0

From these values, the wind profile can be derived using the following standard formula

$$V_z = V_{10}\,(1+0.097\ln(z/10))$$

- *marine growth*
  The loads generated by waves and currents are affected by the marine growth that, with different thickness, covers submerged elements almost completely. Consequently, the rough surface of the marine growth causes a significant increase in drag forces.

  The marine growth thickness is taken as follows:

From	+2.00 m LAT	to	-20.0 m LAT	12 cm
From	-20.00 m LAT	to	sea bottom	5 cm

- *bottom morphology*
  Knowledge of the sea bottom morphology is essential for evaluating the stability of the substructure during installation and before (i) the permanent piled foundations are connected, and (ii) for identifying possible levelling problems.

3.1.4. Geotechnical data

Many of the proposed substructure configurations depend on steel piles, driven to a penetration sufficient to guarantee the correct anchoring. It is therefore necessary to know the seabed soil characteristics and all the necessary geotechnical parameters. This information can be obtained by a geotechnical survey before the design activity starts.

3.1.5. Seismic data

All the offshore devices to be installed in the Adriatic Sea must be designed to survive earthquakes, which are common events. Typically, the seismic analysis should quantify the Strength Level Earthquake (1 in 200-year occurrence) and the Rare Intense Earthquake (1 in 2,000-year occurrence). If not yet available for the selected site, a seismology study must identify the design spectra and the peak ground acceleration as basic data.

3.2. Design Analyses

The substructure design requires structural analysis so the substructure can withstand the worst combination of the design loads for all phases of the project.

The verifications include:

- *ice analysis;*
- *dynamic and spectral fatigue analysis;*
- *seismic analysis;*
- *load out analysis;*
- *lift analysis;*
- *barge transportation analysis;*

- *on-bottom stability analysis;*
- *free floating and upending analysis;*
- *pile driveability analysis.*

All the above analyses are performed using software specifically dedicated to the offshore structure design. Therefore, it is necessary to build-up a complete three-dimensional model of the substructure. In some cases, an experimental physical simulation is advisable.

Special care needs to be given to modelling the pile-soil complex. The behaviour of the pile subject to lateral force is non-linear and therefore the correct evaluation of the foundation response can be obtained only with an iterative process, which is included in most advanced software packages

3.2.1. In-service analysis

The static analysis of the installed substructure should consider:

- *operating conditions:* wind turbine loads + 1 year storm environmental loads;
- *storm conditions:* wind turbine loads + 100 years storm environmental loads.

For each of the design conditions, a minimum of eight wave approach directions for the environmental loads will be considered to generate the maximum forces acting on the structure. For the static installed analysis, the hydrodynamic loads for the design load conditions should be calculated for the wave actions, which provides the maximum global overturning moment and/or the maximum shear force on the substructure.

The extreme storm and operating environmental wave and sea-current forces should be amplified by Dynamic Amplification Factors (DAFs) evaluated on the basis of platform natural periods.

3.2.2. Dynamic and fatigue analysis

Dynamic analysis should establish the natural vibrational period and the eigenvectors of a sufficient number of vibrational modes so 90% to 95 % mass participation is achieved. Spectral fatigue analysis evaluates the fatigue lives at the welded structural connections of the substructure.

The full analysis should include:

- *Selection of Wave parameters to define the structural response adequately;*
- *Environmental Loading;*
- *Dynamic Structural Response;*
- *Stress Range Transfer Function;*
- *Stress Response Spectrum;*
- *Fatigue Damage Assessment.*

The dynamic structural response of the structure to the loading spectra generated by the waves should be calculated for a critical damping ratio of 2%, for all the modes of vibration considered. The dynamic response stress recovery should be based on the equivalent static load method. This approach develops equivalent static loading used to represent the forces on the structure due to fluid motion, including (i) *the relative motion* between the structure and the fluid, and (ii) *their inertia*. The method combines the inertia loads from modes with significant dynamic response to the actual hydrodynamic loads. The stress ranges transfer functions based upon member end-forces should be evaluated at each structural connection. The method of determining stress concentration factors, for use in the calculation of hot-spot stress range, should be appropriate to the type of connection under investigation. The calculation and summation of damage at each connection should be performed. The evaluation of fatigue damage and the calculation of joint fatigue life should be carried out, at each connection, following the Miner-Palmgren model.

3.2.3. *Seismic analysis*
Seismic analysis should be performed based on the given ground accelerations and spectrum data. The design horizontal spectrum should be applied equally in the two horizontal orthogonal directions, with an acceleration spectrum of 50% applied in the vertical direction. All three spectra should be applied simultaneously and the modal responses combined by the Complete Quadratic Combination method, (CQC). The directional responses along the horizontal and vertical axes should be combined separately by the square root of the sum of the squares method (SRSS). In the calculation of member strength, the stresses due to the static dead loads should be combined with the spectral earthquake analysis results.

3.2.4. *Load-out, lifting and transportation analyses*
The substructure should be analysed for all stages experienced before the final installation at site. The main stages are:

- *Load-out;* the transference of the fabricated structure from the fabrication dockyard quay onto the transportation barges;
- *Barge transportation*; once loaded on the barge, the substructure is secured with appropriate sea-fastening and the barge is then towed to the site. In this situation the structure is subject to the gravity loads and to the inertia loads generated by the barge motion during towing;
- *Lifting*; the substructure is unloaded from the cargo barge and placed in its final position through a lifting operation. The structural integrity during this operation should be assessed by a dedicated analysis.

3.2.5. *On-bottom stability analysis*
Before the foundation piles are installed, the substructure (with the exception of monopiles) remains supported on the seabed by suitable shallow foundations (mud-mats). On-bottom (or on-piled) stability analysis should be performed to estimate the range of loading conditions appropriate for the design of the mud mats and to establish the limiting installation sea-state in which the on-piled substructure will maintain a stable, free standing condition.

Figure 15. Lifting of a four-legged jacket (Micoperi)

3.2.6. *Free floating and upending analysis*
For all substructures transported to the site in a horizontal position, the installation requires the following steps

- *Lifting the structure* off the cargo barge and lowering it horizontally onto the sea. The structure must be designed to remain self-floating.

- *Releasing the lifting slings* and connecting the up-ending slings to the substructure head.

- *Up-ending the structure* by the combined effect of lifting the head and ballasting the legs.

This is the typical installation sequence for a three- or four-legged jacket. The sequence should be simulated with suitable software to assess that all the required phases are feasible and safe.

3.2.7. Pile driveability analysis

In all the proposed substructure configurations, the foundation piles need to driven to the design penetration using driving hammers. The selection of the appropriate driving tool is a very important step to assure that the target penetration is achieved and, in the meantime, to avoid overstress in the pile during driving. Therefore a pile driving analysis, using the wave equation technique, is performed. The analysis is done for hammers of different energy and derives the soil resistance to driving from the soil data.

Figure 16. Wind turbine installation on four-legged jacket (REpower)

4. CONSTRUCTION AND TRANSPORTATION

4.1. Construction

The construction of all the steel substructures, described in the above Section 2, will take place in a yard with good experience in the construction of offshore steel structures and provided with a loading quay. The yard should be located not too far from the offshore windfarm to minimize the transportation cost. Assembly equipment is as normally available in a yard dedicated to assembly work of this scale, e.g. cranes, moving devices, welding machines, trailers, etc

If many substructures are needed, the construction could be spread among various yards to meet the project time-schedule. The size of the project could justify the construction of a dedicated onshore assembly yard located conveniently for access to the offshore installation site. In this year, the structure would be assembled from components prefabricated in several manufacturing workshops. Once completed, the substructures will be loaded on suitable cargo barges for the transportation to the installation site. The load-out of the main structures onto the cargo barge will usually be performed using hydraulic trailers. The yard cranes will be used for the load-out of the piles and of the minor appurtenances.

Figure 17. Hammer to be applied to monopile

4.2. Transportation

The transportation of the structures from the fabrication yard will usually require *standard flat deck cargo barges* (Fig. 14). An exception could be transportation of the monopiles, which can be sealed at both ends with rubber diaphragms to make them buoyant. The sealed monopiles can then be lowered into the water at the fabrication yard and towed to the site with a tug, thus avoiding the use of a cargo barge. This solution is certainly convenient for the transportation of a single monopile, but becomes much less attractive for many monopiles.

All the other structures, i.e. tripods and 3-leg or 4-leg jackets, to be installed in limited water depth, say up to 30 to 35m, may preferably be transported in a vertical position, since this generally simplifies the offshore installation. For tripods, the vertical transport is mandatory. The jackets to be installed in deeper water should be assembled, loaded-out and transported horizontally. The number of structures loaded on the same barge depends on the size of the barge, on the distance of the yard from the installation site and on the type of transport arrangement (vertical or horizontal array).

Figure 18. Crane barge used for wind turbine installation (Elsam)

Note that, when a structure is ready to be installed at site, it is essential to have the foundation piles available at site also. For this reason, the foundation piles are nearly always loaded on the same barge as the structure. However, for a large number of substructures for the windfarm, it is probably more convenient for pile transportation to use a dedicated barge.

5. INSTALLATION AND DISMANTLING

5.1. Monopile

A monopile is a substructure foundation suitable for shallow water installation of wind turbines. Transportation of the monopiles to the installation site may occur with all the monopiles of the windfarm on the same barge.

The typical operation required for the installation of one monopile can be summarized as

- Up-ending from the horizontal transport position.
- *Lowering* onto the seabed.
- *Driving* to the design penetration.
- *Installing* the transition structure onto the embedded pile, to support the wind turbine towers.

All these operations can be performed either by using a crane barge or a self-elevating platform (jack-up). Due to the relatively light weights to be handled for offshore wind turbines compared with those of the offshore oil and gas industry (100t should be an upper limit), the required lifts are in the low-range for experienced operators. For example, a suitable jack-up could be a simple self-elevating platform provided with a revolving crane of sufficient capacity (definitely smaller than a typical drilling jack-up rig).

Figure 19. Jack up operation (Micoperi)

The choice between the jack-up (Fig 19) and the crane barge needs to be evaluated technically and economically, case by case. In principle, the self-elevating platform (jack-up) is much less affected by the sea state conditions than a crane on a floating barge, and has the advantage of reducing 'weather standby'. On the other hand, the crane barge requires less time for relocation from one turbine site to the next. In either case, the critical operation is lifting the monopile from the barge and its up-ending, especially if the selected cargo barge is small.

To guarantee verticality, the monopile should be lowered onto seabed through an installation guidance framework protruding outboard of the barge or jack-up hull. With the monopile resting on the seabed, the crane can be disconnected and used for the hammer handling. A diesel hammer could be the most appropriate driving equipment for monopile installation. Due to the anticipated large diameter of the monopile, the hammer should be provided with a special bell (see Fig 17) or a short 'follower', with a conical reduction, will be required.

An alternative to the conventional driving hammer is the vibro-hammer. With this equipment, pile penetration is achieved by vibrations applied to the pile top. Vibro-hammers are generally lighter than driving hammers, which could be an advantage.

5.2. Tripod

The tripod is a type of substructure that can be considered appropriate for water depth up to 30 to 35 m. This means that the maximum overall height of the tripod is in the order of 40-45m. Usually, a tripod would be transported on the cargo barge in a vertical position.

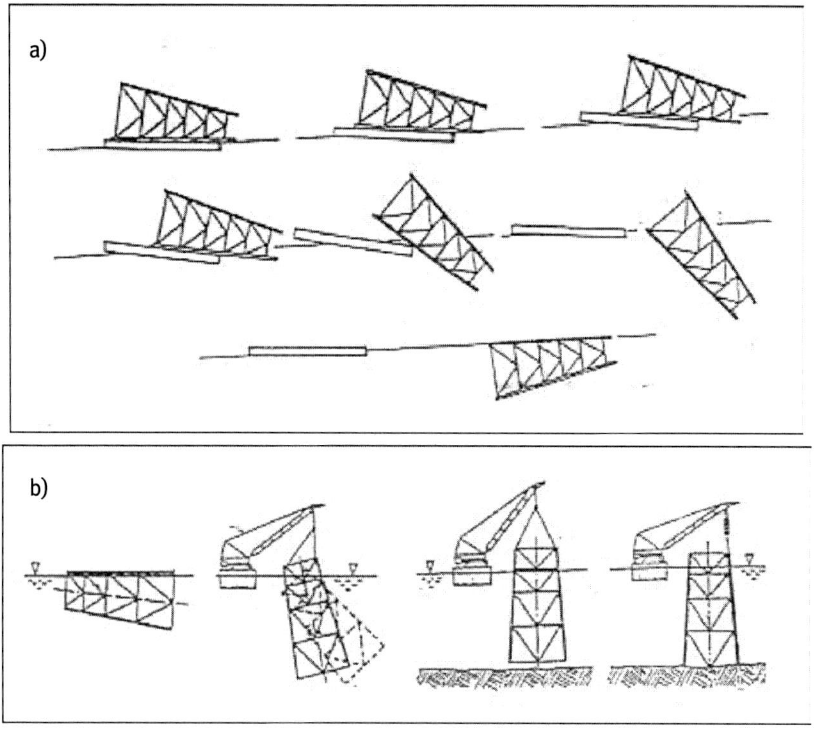

Figure 20 a/b. Launching (a) and positioning (b) of jacket on seabed

Once at site, the operations required for the installation are:

- *Lifting the structure from the cargo barge*; the operation can be done connecting the lifting sling to the centre column head.
- *Lowering the tripod to the seabed*; the tripod will have three skirt pile sleeves of suitable mud mats for stability until the piles are installed.
- *Lifting the foundation piles* from the cargo barge and positioning the piles in the sleeves.
- *Driving the piles* to the target penetration using an appropriate hammer, which is usually a hydraulic underwater hammer.
- Once the pile driving is completed, it is necessary *connect the structure* to the piles by grouting the annulus between the pile and the sleeve. The grouting operation is probably more convenient with a dedicate vessel instead of using the main crane barge.

The transition structure, required to support the turbine tower, can be either preassembled in the fabrication yard, thus reducing the duration of the offshore operation, or installed at site.

For the installation of tripods, the most appropriate equipment is a barge provided with a revolving crane having the required height and lifting capacity. The use of a jack-up (Fig 19) could be also considered, but, due to the size of the substructure, it becomes very difficult to find a jack-up with the required characteristics.

Figure 21. End of life recovery of jackets in the Gulf of Mexico (Bisso-Gulf)

5.3. Three- or Four-Legged Jacket

A three or four-legged tubular substructure can be transported from the fabrication yard to the site in a vertical or horizontal position, depending on the characteristics of the crane barge and on the height of the substructure. If the jacket height does not exceed 40 to 45m the vertical transport can be considered feasible and more convenient for time and economic evaluation. For greater height, the only possibility is transporting the structures horizontally.

The lifting and positioning onto seabed of a vertically transported jacket is a very simple and fast operation, as described above for the tripod. However, if the jacket is transported horizontally, the installation becomes more complicated and requires the following main steps:

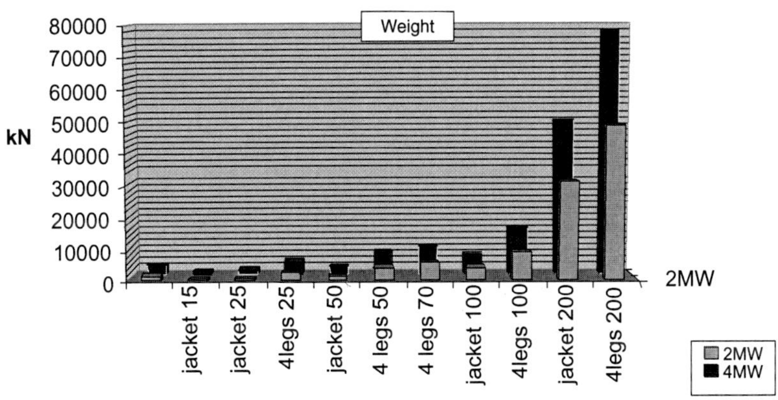

Figure 22 . Weight of the foundation – no piles

- *Lift* the jacket from the cargo barge and lower it into the water. The jacket must be designed to remain self-floating and stable, even after the lifting slings are disconnected. The minimum buoyancy-reserve of the floating structure should be not less than 10 to 12% of the weight, which is obtained generally by sealing the legs with rubber diaphragms. Note, that in case of three legs, the jacket, when lowered into the sea, rotates until reaching a stable position, which allows a single operation for lowering onto the sea surface.
- *Disconnect* the lifting slings and connecting the up-ending slings, pre-installed on the jacket head, to the crane hook. This operation is done while the jacket is free floating.
- *Overturn* the jacket into vertical position. The jacket rotation (up-ending) is achieved by lifting the jacket head with the barge crane and at the same time flooding the jacket legs in a controlled manner. The operation must be carefully simulated beforehand in order to identify the best and most safe sequences during these stages. To allow the controlled flooding of the jacket legs, the jacket has to be equipped with a ballasting system that includes, as a minimum, one flood valve and one vent valve on each leg.
- *Lower* the jacket onto seabed and fully flooding the jacket legs.

For a high jacket, an alternative method to that described above is transportation horizontally and subsequent launching. This is shown in Fig. 20a. When the substructure is in the water, it is possible to follow the preceding procedure, as in Fig. 20b.

Once the jacket is lowered onto seabed and the crane slings completely disconnected, the piling operation can start. The jacket foundation is achieved by driving the piles into the seabed through the legs from the top with an appropriate hammer. The anchorage piles cannot be installed in one section only, because their length is such that the handling is impossible for manoeuvring. The piles are therefore prefabricated and transported in sections, to be assembled and welded offshore. The result is that the pile installation takes the longest time of all stages of installation, so becoming the most critical operation offshore.

Table 7. Weights [kN] of the foundations without piles

Turbine Power [MW]	Depth [m]	Mono-pile	Tripod	3 legged -jacket	4 legged -jacket
2	15			931	
2	25	1,940	4,186	1,093	2,764
2	50			1,688	4,517
2	70				5,851
2	100			4,301	9,309
2	200			31,277	49,173
4	15			1,518	
4	25	3,130	6,658	1,777	4,447
4	50			2,728	7,249
4	70				9,380
4	100			6,903	14,906
4	200			50,011	78,610

After the piles are all driven to the target penetration, the jacket legs are permanently connected to the piles by welding suitable steel wedges at leg top.

In general, the installation of a conventional jacket (even of very small size) requires the presence of the crane barge for a longer duration than for the installation of a tripod. This is one of the key points to be considered for the selection of the substructure configuration. As an alternative to the use of crane barge or of more crane barges, a jack-up equipped with an adequate crane (Fig. 19) should be considered. Its dimensions could be less than of a typical oil or gas industry jack-up, but it must give better stability for the installation in difficult sea conditions. The jack-up could be also used for the installation of wind turbine foundations. In this case, the same equipment could be used for the whole unit assembly and remained in position during the time necessary to complete all the installations.

5.4. Wind Turbine

The wind turbine consists of a supporting tower, a nacelle and the blades. The first component to be installed is, of course, the tower. Depending on the height and on the available installation equipment, the tower can be prefabricated and installed in one single piece or in more than one section to be joined offshore. The tower will be lifted and arranged to an accurate vertical position by the crane and installed on the transition frame previously fitted to the top of the substructure. Once the tower is in place and is connected to the substructure, the installation can be completed by lifting and fixing the nacelle, usually with two of more of the blades.

In practice, the vertical distance from sea surface to the rotor axis ranges between about 50m to over 100m. Therefore, the crane must be equipped with a very long sloping boom, able to reach the required elevation. However, note that the weights of the wind turbine components to be installed at the higher elevation are not very large. Therefore, the use of a crawler crane fastened on the cargo barge could be also considered, even if the downtime due to unfavourable weather condition will significantly increase. The typical operation is shown in Fig. 19.

Table 8. Costs of the foundation substructures (no piles) / thousand euro.

Cost [€/dN]	Power [MW]	Depth [m]	Monopile	Tripod	3 legs jacket	4 legs jacket
2,94 €/dN	2	15			€ 285	
2,94 €/dN	2	25	€ 594	€ 1,281	€ 334	€ 846
2,94 €/dN	2	50			€ 516	€ 1,382
2,94 €/dN	2	70				€ 1,796
2,94 €/dN	2	100			€ 1,316	€ 2,855
2,94 €/dN	2	200			€ 9,574	€ 15,058
2,94 €/dN	4	15			€ 464	
2,94 €/dN	4	25	€ 958	€ 2,057	€ 544	€ 1,361
2,94 €/dN	4	50			€ 835	€ 2,219
2,94 €/dN	4	70				€ 2,871
2,94 €/dN	4	100			€ 2,104	€ 4,563
2,94 €/dN	4	200			€ 15,300	€ 24,064
3,43 €/dN	2	15			€ 332	
3,43 €/dN	2	25	€ 693	€ 1.495	€ 390	€ 987
3,43 €/dN	2	50			€ 603	€ 1,613
3,43 €/dN	2	70				€ 2,096
3,43 €/dN	2	100			€ 1,536	€ 3,331
3,43 €/dN	2	200			€ 11,170	€ 17,568
3,43 €/dN	4	15			€ 542	
3,43 €/dN	4	25	€ 1,118	€ 2,399	€ 634	€ 1,588
3,43 €/dN	4	50			€ 974	€ 2,588
3,43 €/dN	4	70				€ 3,350
3,43 €/dN	4	100			€ 2,454	€ 5,323
3,43 €/dN	4	200			€ 17.850.758	€ 28.075.448

5.5. Decommissioning at End of Life

Substructure removal is not difficult in shallow to medium seabed depths, but may become critical for larger depths. The removal consists in cutting the foundation piles just below mud-line. The pile cutting can be executed by lowering the cutting equipment inside the pile by remote operation. After the foundation piles have been severed, the substructure can be retrieved, placed on a cargo barge (Fig 21) and transported to land for final dismantling and re-cycling of materials.

First phase: preliminary actions to statutory bodies

- decommissioning plan, including time schedule, means for onshore transportation to be requested, dismantling/disposal area, etc.;
- licences and permissions/consensus to be obtained;
- issue to be notified to the statutory bodies;
- project implementation/management for all the dismantling activities;
- mobilisation schemes/actions as harbour facilities, onshore disposal/recovery area/devices, etc.;

Second phase: wind turbine disassembly

- windfarm de-energizing and isolation from grid;
- removal of electrical components;
- removal/recovery of rotor/blades;
- removal/recovery of nacelle;
- dismantling of tower and its internals;
- lead onto crane barges and transportation to dismantling/disposal area;

Third phase: foundation substructure recovery

- cut off the connection piece and the upper part of monopile or the anchorage piles of legged jacket;
- lifting pile top and global jacket onto crane barge;
- transportation to dismantling/disposal onshore area;
- seabed reinstatement/sweeping;
- survey to verify/prove experimentally the clean state of seabed;

Fourth phase: offshore infrastructure elimination

- marine cable excavation, if its recovery is required by the decommissioning plan;
- its cut/rewinding onto cable recovery vessel;
- its transportation to dismantling/disposal area;
- seabed reinstatement, as performed after the foundation recovery;
- survey to verify/prove experimentally the clean state of seabed.

A similar programme has to be managed also for the onshore infrastructures.

6. WEIGHT AND COST EVALUATION

6.1. Comparison of weights

The first index we consider is the total weight of the substructure and the foundation piles. This quantity affects both the capital cost of the total foundation (which is directly proportional to the weight) and the costs for its construction and installation. The weights of the structures are described in the following tables.

Data, in bold italic type, are determined by static dimensioning calculations, while the others are obtained through data extrapolation. The weights of the 4 MW substructures for depths more than 25 m, have been derived by the relationship among weights of the three-legged jackets placed at 25 m depth, equipped with 2 and 4 MW turbines respectively. The weights of foundations, for 70 to 200 m depth range, can be found from the extrapolation of a parabolic curve. This has been obtained by interpolating the weights of jackets for 2 MW wind turbines as a function of the depth (15 m, 25 m, 50 m) and for 4 MW wind turbines placed at different water depths (25 m, 50 m, 100 m). The chosen interpolation function is an exponential curve, which allows a better correlation than parabolic (weights are expressed in kN).

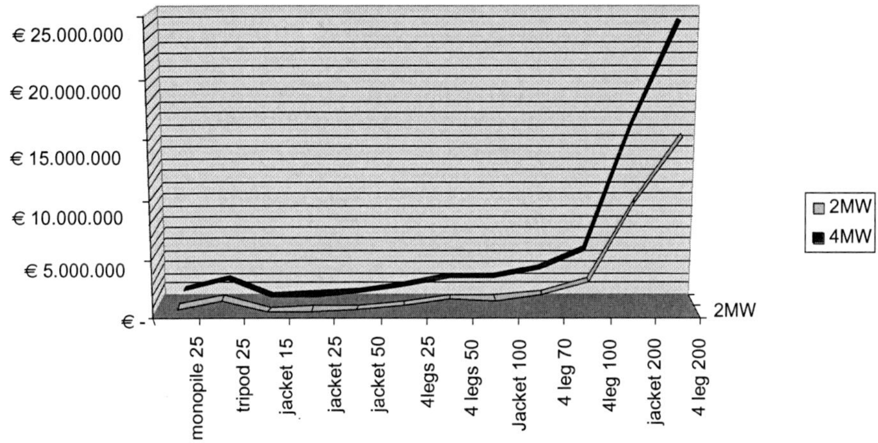

Figure 23. Costs of foundations (no piles)

Information contained in Table 7 summarizes the evaluations and considerations of the preceding paragraphs. Results of a separate analysis of data extrapolated for 2 MW and 4 MW turbines are reported in Fig. 22, which shows the different dead-weights of the structure, excluding the corresponding foundation piles for increased seabed depth.

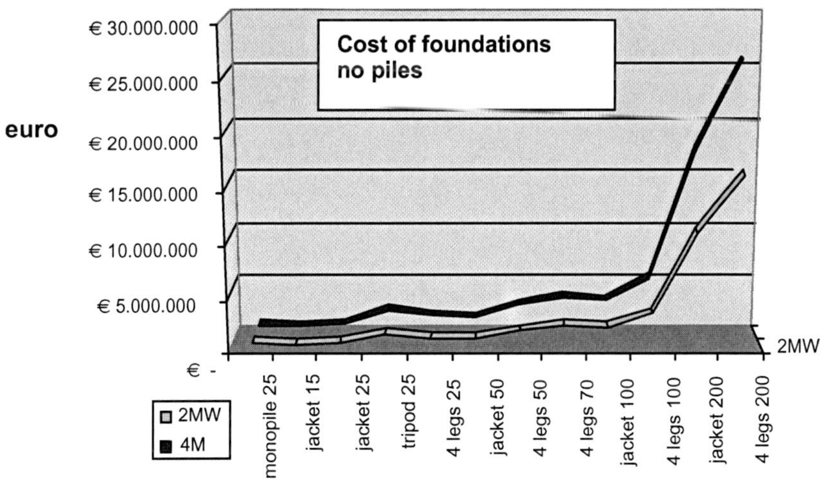

Figure 24. Costs of foundations (piles included)

A different treatment of the data has been made for substructures for 6 MW wind generators. Due to a consistent inaccuracy affecting the weight interpolation process from 2 MW units to 6 MW ones, a direct calculation was necessary. Preliminary analysis has been performed on all the design quantities (height, diameter and thickness of structure, supporting nacelle and blades, mean wind speed, shear forces and bending moment at tower root, rotor diameter, etc.; all evaluated at their minimum range) to obtain an acceptable sizing, even if the calculation is approximate and hypothetical. The substructure type consists of a standard four-legged jacket with double diagonal tubular beams between two adjacent legs and two planes of stiffening tubular rings.

For a 25 m water depth, the leg diameter is 48" (1219 mm). For seabed depth of 50 m, the same diameter is increased to 54" (1371 mm) or probably even more than 60" (1524 mm). For water depth of 25 m and 50 m, the structure weight, considering only legs and stiffeners without piles, is 2,800 to 4,500 kN and that with the piles 5,000 to 8,000 kN. For a 4 MW turbine placed on a four-legged platform, the structure weight without piles is respectively 4,500 to 7,300 kN for 25 m and 50 m of water depth.

Table 9. Costs of foundations (piles included)/ (1000 euro)

Cost [€/dN]	Power [MW]	Depth [m]	Monopile	Tripod	3 legged jacket	4 legged jacket
2,94 €/dN	2	15			€ 474	
2,94 €/dN	2	25	€ 594	€ 1,484	€ 628	€ 1,152
2,94 €/dN	2	50			€ 1,006	€ 1,791
2,94 €/dN	2	70				€ 2,296
2,94 €/dN	2	100			€ 2,042	€ 3,461
2,94 €/dN	2	200			€ 11,026	€ 16,270
2,94 €/dN	4	15			€ 768	
2,94 €/dN	4	25	€ 958	€ 2,380	€ 1,012	€ 1,851
2,94 €/dN	4	50			€ 1,618	€ 2,872
2,94 €/dN	4	70				€ 3,670
2,94 €/dN	4	100			€ 3,264	€ 5,531
2,94 €/dN	4	200			€ 17,620	€ 26,000
3,43 €/dN	2	15			€ 554	
3,43 €/dN	2	25	€ 693	€ 1,731	€ 732	€ 1,344
3,43 €/dN	2	50			€ 1,174	€ 2.090
3,43 €/dN	2	70				€ 2,679
3,43 €/dN	2	100			€ 2,383	€ 4,038
3,43 €/dN	2	200			€ 12,864	€ 18,982
3,43 €/dN	4	15			€ 896	
3,43 €/dN	4	25	€ 1,118	€ 2,777	€ 1,181	€ 2,159
3,43 €/dN	4	50			€ 1,887	€ 3,351
3,43 €/dN	4	70				€ 4,281
3,43 €/dN	4	100			€ 3,808	€ 6,452
3,43 €/dN	4	200			€ 20,557	€ 30,334

Considering the data of Table 7 for a 2 MW offshore wind turbine at a depth of 25 m, the substructure weight varies from 1,894 kN with a tree-legged jacket to 1,940 kN with a monopile, 2,700 kN with a four-legged platform and 4,200 kN with a tripod.

6.2. Comparison of Costs

The costs of the foundation structures are directly proportional to their weights. In fact, to calculate total costs it is possible to estimated a construction and supply cost on the basis of 2.94 to 3.43 €/dN

(Table 8). This cost must be added of the cost for foundation piles, that is about 1.078 €/dN for different geometric pile typology of 60" (1.5 m) diameter for the tripod, and 42" (1.0 m) diameter for all the other foundations. Fig. 23 shows the estimated costs of the foundation structures, excluding piles. By adding these costs to those of piles, the total cost of different types of foundation substructure can be determined, as in the Table 9 and Fig.24.

If we consider weights of the structures for 100 m or 200 m depth, we can determine the scale of these constructions costs. In Table 10 and Fig. 25, the costs for the 3 legged-jacket (without piles) are reported for a 2 MW turbine, while; likewise in Table 11 and Fig. 26 for the costs for a three-legged foundation for a 4 MW turbine.

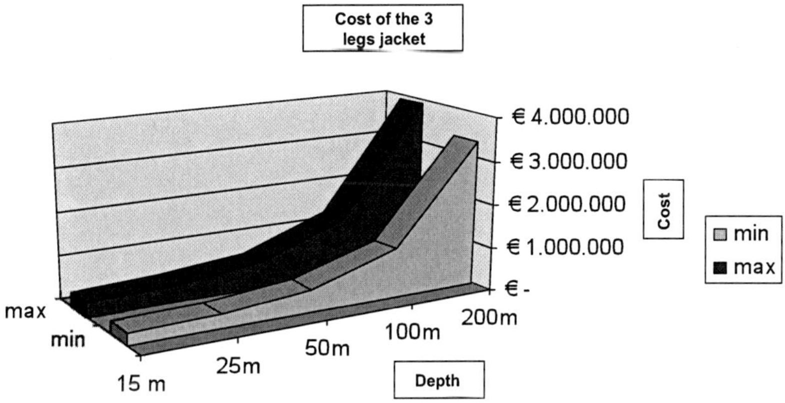

Figure 25. Costs for the three-legged jacket for a 2 MW wind turbine (no piles)

Costs are reported in Table 12 and in Fig. 27. for the 4 legged-jacket (piles included) and for the 4 MW turbine,

7. CONCLUDING REMARKS

To evaluate the costs of offshore foundations accurately, other factors have to be analysed, which could influence significantly the cost schemes presented above which are based only on weights. These other factors include the time and the requirements for fabrication and transport from workshop to warehouse and from the surface storage to the site, the procedures of placement in situ and of driving the piles into the soil and so on. Such additional factors may modify substantially the costs determined only through the weight evaluation. Nevertheless, such costs do not change *the trend* of our cost calculations for each solution as functions of the sea depth. With all the other factors constant (e.g. seabed composition and resistance, operative and extreme conditions, etc.), our comparison of costs based on component weight are a first step for rational and reasonable selection of offshore foundation structures and methods.

Examining Fig. 22, representing the variability of weights (with and without anchorage piles) as a function of the seabed depth, and comparing the cost trends of Figs 23 and 24, approximate guidance can be suggested for each type of supporting system.

Table 10. Costs for the three-legged jacket for a 2 MW wind turbine (no piles)/ 1000 euro

	15 m	25m	50m	100m	200m
Min	€ 285	€ 335	€ 517	€ 1,129	€ 3,347
max	€ 333	€ 391	€ 603	€ 1,317	€ 3,905

For the class of 2 MW wind turbines and for seabed depth more than 8 to 10 m, the trend shows a clear preference for *monopile foundations* (Fig. 2). For less depth, the choice has often been for gravity foundations (Figs. 1 & 6). The range for monopiles may extend to 15 to 20 m, but at 25 m, the technical characteristics of monopiles put the concept beyond the margin of acceptability.

Table 11. Costs for the three-legged jacket for a 4 MW wind turbine (no piles)/ 1000 euro

	15 m	25m	50m	100m	200m
Min	€ 465	€ 544	€ 835	€ 2.104	€ 15,301
Max	€ 542	€ 635	€ 974	€ 2,455	€ 17,851

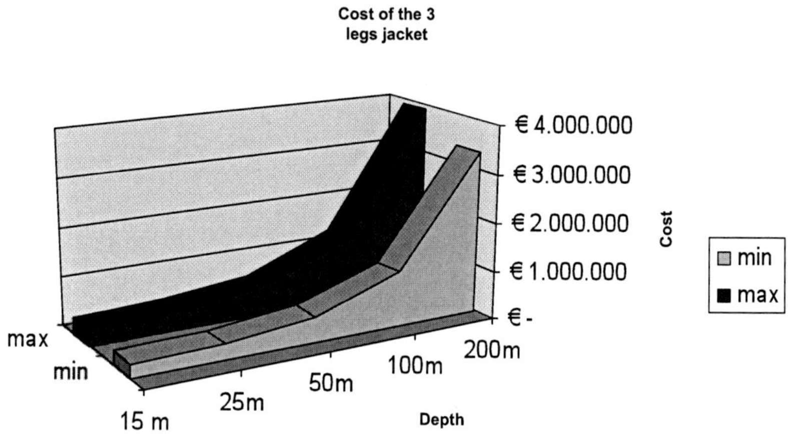

Figure 26. Costs for the three-legged jacket with a 4 MW wind turbine (no piles)

In the study, a model equipped by a *tripod foundation* (Fig 3) has been introduced. The structure is found to be technically compatible with deeper waters of more than 20-25 m, but not with shallow depths. There are two reasons for this: (1) the requirement for a sufficient depth of water above wall parts to allow service vessels to approach the supporting structure without colliding with it, and (2), where applicable, to keep the strong diagonal elements well below sea level to avoid interlocking and packing of ice.

In very shallow waters, say 5 to 6 m, the distances between upper and lower braces of the tripod substructure are insufficient to transfer all the forces for any reasonable cross section, so the structure is not suitable. Even for 25 m depth the technical behaviour is acceptable for displacements and rotations at the top level, but the dead weight, and hence cost, is too large. This means that the solution is too expensive for wind turbine foundation as the depth increases.

In the range of 15 to 20m, the *three-legged jackets* are also competitive with the monopile and tripod. This does not imply that piled solutions are not feasible for shallower waters, but a correct balance among different technical factors have to be compared with other solutions, due to their strength and stability. Their margins of technical and economical advantage could be valid and promising almost to depths of 30 to 40 m. Probably their applicability could be extended to 50 m, even if some economic doubts are evident.

Table 12. Costs for the four legged jacket with a 4 MW wind turbine (piles included)/ 1000 euro

	4 legs 25	4 legs 50	4 leg 70	4 legs 100	4 legs 200
Min	€ 1,361	€ 2,219	€ 2,871	€ 4,562	€ 24,064
Max	€ 1,588	€ 2,589	€ 3,350	€ 5,323	€ 28,075

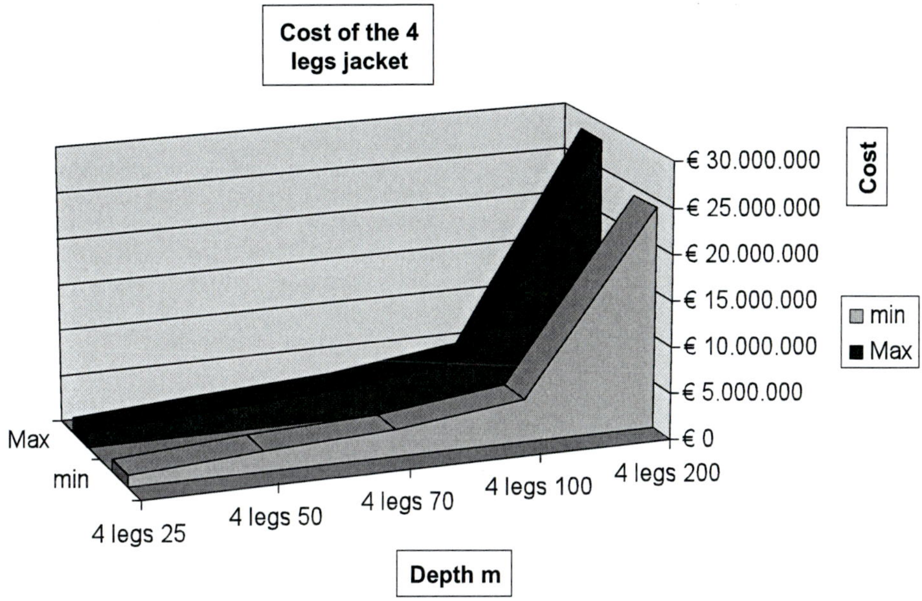

Figure 27. Costs for the four-legged jacket with a 4 MW turbine (piles included)

The *four-legged jackets*, even if technically advantageous, are not economically competitive with the three-legged jackets, except for very definite need for strength and stability required by particular wind-sea static/dynamic conditions.

For the class of 4 MW wind turbines, the main considerations, as presented and discussed before for the monopile and tripod supporting systems, remain substantially the same. The braced jacket structures have the properties and capability to be economic solutions to all the technical problems.

If the cost of a three-legged jacket increases non-linearly with the sea depth, its margin of technical convenience progressively decreases. The same applies for a four-legged jacket. Undoubtedly, it is not credible to suggest a three-legged jacket for depths more than about 50 m. The four-legged jacket should be applicable for deeper waters of more than 70 m.

The analysis has been extended to 100 m or more, just to demonstrate the difficulty and the impracticability of such proposals. In the oil and gas market, similar supporting structures at such depths are well known and of frequent application; however these industries are incommensurably richer than any windfarm, even equipped by the largest wind turbines.

8. REFERENCES

L. Pirazzi, R. Vigotti, "Le vie del vento. Tecnica, economie e prospettive del mercato dell'energia eolica", Franco Muzzio Editore, 2008 Roma (It).

F.G. Cesari et al., "Selection, configuration and cost evaluation for an offshore windfarm in the Adriatic Sea", *International Wind Conference,* 2008 Berlin (De).

AA.VV., *"OWEMES 94 (Offshore Wind Energy in Mediterranean and Other European Seas)", Proceedings*, 1994 Roma (It).

AA.VV. "The Horns Rev Windfarm project, Denmark, Status", *OWEMES 03 (Offshore Wind Energy in Mediterranean and Other European Seas), Proceedings*, pp.459-468, Napoli (It) 2003.

F. Cesari, "Ricerche per aerogeneratore da collocare in sito fuori costa: criteri e problematiche strutturali in appoggio a studio tecnico-economico di fondazione" LIN 1510, Bologna (It), Novembre 1990.

M.B. Zaaijer, J van der Tempel, "Scour Protection: a necessity or a waste of money ?", *43rd IEA Topical Expert Meeting – Critical Issues Regarding Offshore Technology and Deployment*, pp.43-51, 2004 Skaerbaek (Dk).

AA.VV., "Comparison of monopode, tripod, suction bucket and gravity base design for a 6 MW turbine", *OWEMES 03 (Offshore Wind Energy in Mediterranean and Other European Seas), Proceeding*, pp.255-269, 2003 Napoli (It).

F. Cesari, G. Gaudiosi, P. Battistella, F. Taraborrelli, S. Carrara, "Wind turbine on floating platforms for desalination plants", *European Wind Energy Conference*, pp.982, 1999 Nice (Fr).

P. Bertacchi, A. Di Monaco, M. Gerloni, G. Ferranti, "Eolomar, a moored platform for wind turbines", *OWEMES Conference (Offshore Wind Energy in Mediterranean and Other European Seas)*, 1994 Rome (It), II session.

K.G. Tong, "Technical and economic aspects of a floating offshore windfarm", *OWEMES Conference (Offshore Wind Energy in Mediterranean and Other European Seas)*, 1994 Rome (It), III session.

AA.VV., "Floating windfarms for shallow waters offshore sites", *OWEMES 03 (Offshore Wind Energy in Mediterranean and Other European Seas), Proceedings*, pp.433-446, 2003 Napoli (It).

North Adriatic Offshore, IDA C PLATFORM, "Design Premises for Monopod & Deck Structures", 2004 Ravenna (It),.

F.Cesari, P.Battistella, F.Taraborrelli, "Feasibility study for an offshore windfarm off the coast in the northern Adriatic sea", *Ecomondo International Conference,* 2006 Rimini (It).

F. Cesari, F.Taraborrelli, "Offshore wind plants", National Energy Conference, 2006 Bologna (It).

AA.VV. "Critical issues regarding offshore technology and deployment", *43rd IEA Topical expert meeting*, 2004 Sk_rb_k (Dk).

"Implementation of the 'Integrated Coastal Zone Management' (ICZM) Guidelines at Provincial Scale Study Area: 'Ferrara Coast'", 2008 Ravenna (It).

A.R. Henderson, R. Leutz, T. Fujii, "Potential for floating offshore wind energy in Japanese waters", *International offshore and polar engineering conference*, 2002 Kitakyhushu (Ja).

J. van der Tempel, "Differentiating Integrated Design", *43rd IEA Topical Meeting*, 2004 Sk_rb_k (Dk).

North Adriatic Offshore, IDA C PLATFORM, "Design Premises for Monopod & Deck Structures", 2004, Ravenna (It).

Chapter 13:

Offshore Wind Turbine Materials

L. Battisti and A. Brighenti,

1. INTRODUCTION

Offshore wind energy exploitation for electricity generation is a relatively new engineering development, driven by the huge potential of sea winds. In early developments, poor design, incorrect use of materials and inexperienced operation, have all contributed to failures at offshore wind turbines. Technological developments are still progressing, although much basic information is taken from the established offshore oil & gas industry, which has safe and cost-effective structural designs, material specifications and maintenance procedures. The marine environment is characterised by the presence of seawater and moisture, by large temperature excursions, by loading from wind, waves, sea currents and sea ice, and by special requirements for transportation. Therefore, special solutions are required for the *materials* of Offshore Wind Energy Conversion Systems (OWECS).

Modern offshore wind turbines are designed for the specific environment where they are placed, although the main design philosophy is still based on typical onshore large multimegawatt turbines. However, the maintenance and the high reliability requirements for offshore installations demand continued improvement in design and manufacturing.

Therefore the current approach is to consider the typical materials used in conventional turbines, and to analyse their behaviour in the marine environment by means of a criticality analysis. Table 1.1 shows the type and percentage (by weight) of materials used in current conventional megawatt and multi-megawatt turbine components.

Table 1.1. Type and percentage by weight of materials used in current wind turbine components [1.1].

Component material (% by weight)	Steel	Aluminium	Copper	Glass fibre reinforced resins (GRP)	Carbon filament reinforced plastics	Permanent magnetic materials	Pre-stressed concrete
Rotor							
Hub	100						
Blades	5			93	2		
Nacelle	80	3-4	14	1			
gearbox	98	2	2				
generator	65		35			7	
Frame, machinery and shell	85	9	4	3			
Tower	**98**						**2**

Each of the components listed in Table 1.1, will interact with the marine environment by three mechanisms represented by:

1. *forces*: considered as dynamic and static loads on the fixed and moving parts, as affected by both wind and waves;
2. *heating and cooling fluxes*, due to thermal gradients between the turbine components and the liquid and atmospheric environments, as affected by parts having different thermal properties (e.g. blades and hub);
3. *physical-chemical interaction* (erosion, corrosion, oxidation).

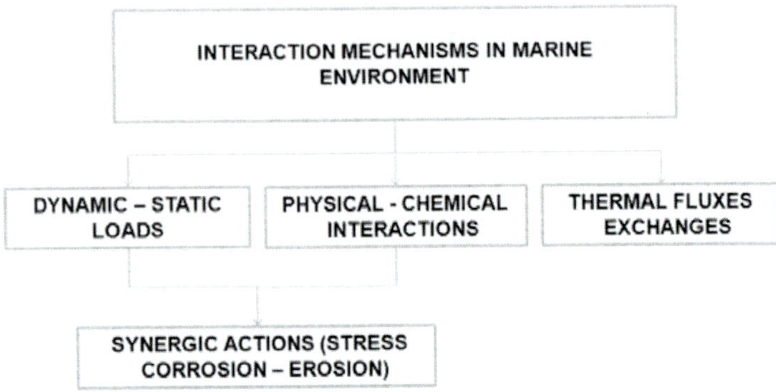

Figure 1.1. Schematic of the main interacting mechanism in offshore locations.

The *fatigue loading* in the offshore environment is one of the most severe in engineering practice: as many as 10^7-10^8 cycles in a 20 years life with complex loading patterns in a number of directions. The large number of cycles at low stress- levels lies in the extreme end of fatigue testing as replicated in a controlled environment.

The *synergic interaction* between loads and physical-chemical processes will modify the behaviour of most metals by emphasizing and accelerating the effects of fatigue loads and, for some metals, by eliminating the knee of the equivalent load-cycle curves. Cracking due to *stress corrosion and corrosion fatigue* will occur in saline environments combined with a rapid rate of wear of erosion, thus shortening the lifetime of some metal components significantly, as indicated in Figure 1.2 for typical mild steel.

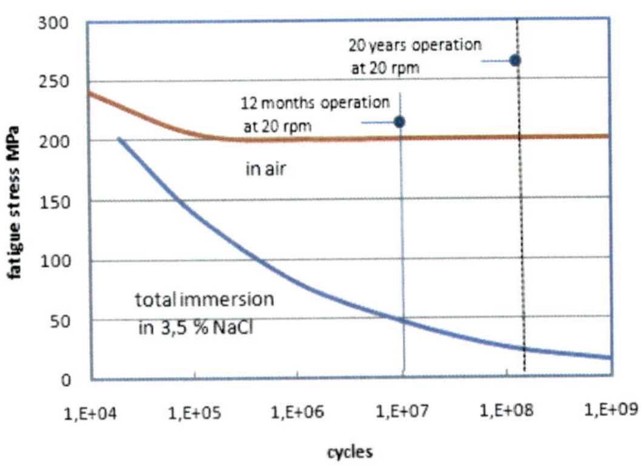

Figure 1.2. Fatigue and corrosion fatigue in pure air and in a 3,5% NaCl solution for a mild steel (0,17% C) with full stress reversal at 15 °C.

However, the offshore atmospheric environment has very aggressive salt sprays and saline fogs. Combined with the ambient oxygen, the above-water situation can be much more harmful than simple seawater immersion. Therefore, actual fatigue stress for components of offshore structures may be worse than indicated in Figure 1.2. This indicates that protection systems have always to be considered for steels and all metals used in marine environment. All these factors are most important at *junctions and contacts between materials*, which may be satisfactory at inland sites, but need to be analysed afresh for offshore machines. The differential heating and cooling of components, perhaps induced by the temperature difference between the seawater and the top of the turbine exposed to sunshine, may induce severe thermal gradients, with consequent additional loads on the structure. Therefore, both the turbine configuration and the types of material influence the degradation of the components.

As mentioned before, offshore structures must be corrosion protected to reduce damage from the combined actions of the aggressive marine environment. Therefore, assessing the environmental conditions is important. The choice of metal materials, welded joints, bolts, adhesion for the structures and corrosion protection strategies have special importance in the design phase to ensure durability and safety of all aspects of Offshore Wind Energy Conversion Systems.

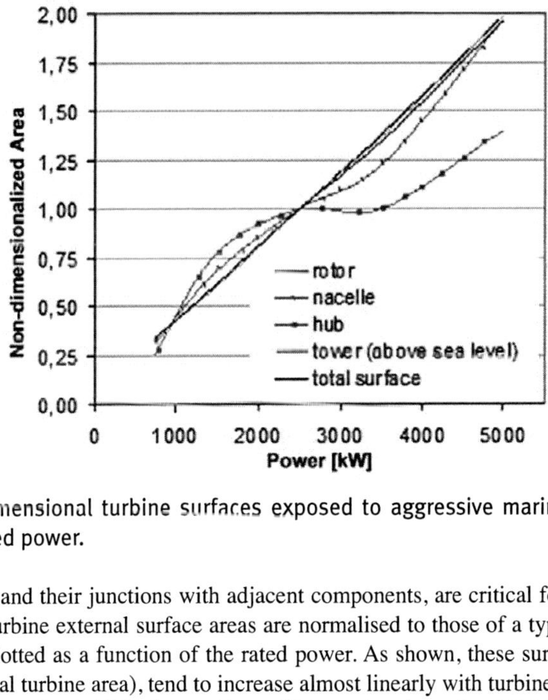

Figure 1.3. Non-dimensional turbine surfaces exposed to aggressive marine environment, as a function of the rated power.

Tower surfaces, and their junctions with adjacent components, are critical for corrosion problems. In Figure 1.3, the turbine external surface areas are normalised to those of a typical 2.5 MW offshore wind turbine and plotted as a function of the rated power. As shown, these surfaces (which are more than 50 % of the total turbine area), tend to increase almost linearly with turbine size. The surface area exposed to corrosion of a 3 MW size turbine will be twice that of a 1.5 MW size turbine. Thus, corrosion problems steadily increase with turbine size, together with their mitigation costs, including additional wall thickness, sacrificial anodes and coatings.

The selection of materials and shapes of structural parts is made considering the requirements of withstanding extreme and fatigue loads. At rotor heights, wind speeds are generally greater offshore than at the coast, but wind turbulence is generally less. However, offshore wind turbines operate in clusters, so wakes and turbulence is thereby increased. In addition, wave loads, wind/wave interactions and, in some cases sea ice and blade icing, produce additional combinations of load.

For steel materials, beside of the requirements on the strength of materials to withstand extreme and fatigue loads, the materials selection criteria include the thickness of products (with respect to welding and corrosion endurance), the design temperatures and the durability of the structural members, regarding limitations due to chemical composition.

For concrete structures, the major consideration is the protection of the reinforcement steel against corrosion; this requires adequate concrete composition and cover, and the limitation of the maximum crack width. For a wind turbine design to be successful, the single loads and vibrations they induce in the structure need to be analysed in terms of individual components and taking into account the merit of various competing materials.

The following sections review the state-of-the-art of offshore wind turbine materials, their properties and limitation, and the current manufacturing techniques. The wind turbine is divided into three major component assemblies (illustrated in the schematic view of Figure 1.4); these are:

- rotor components;
- nacelle components;
- tower components.

Section 2 discusses the rotor materials (blades, bolt connections, pitch drive, and hub and enclosure). Section 3 addresses the nacelle materials (main shaft, gearbox, generator, yaw drive, bearing, couplings, mechanical brakes, nacelle cover). Section 4 considers the tower and foundation materials, and their corrosion protection.

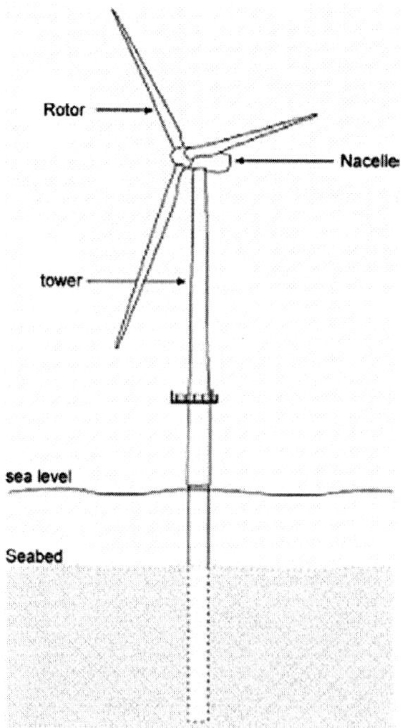

Figure 1.4. Offshore wind turbine : major components.

Section 5 outlines the special material requirements for cold climate offshore installations and the behaviour of the materials at low temperatures. Section 6 presents a design methodology for blades and tower, with regard to materials selection criteria, based on appropriate selection charts. Section 7 discusses briefly the importance of laboratory testing and finally, section 8 lists the main references and literature on the topic.

2. ROTOR COMPONENT MATERIALS

The rotor typically includes:

- *blades*;
- *extenders*, that attach the blades to the central hub;
- *pitch drives* to control the angle of the blades;
- *hub*, houses the control system for the pitch drive and serves as a base for the blades and the extenders.

2.1 Blades

Turbine blades are fatigue-critical structures. Blades must ensure several orders of magnitude more cycles than, for example, any aircraft wing. Moreover, wind turbine blades have to be made of lightweight materials to minimize the loads from the rotating mass. At present, *fibre reinforced plastics* (FRPs) have become the predominant blade material. High strength and stiffness to weight ratio, fatigue and corrosion resistant, electrical insulation and the ability to tailor the material to the loads, have led to their widespread use as a blade material. Both offshore and onshore wind turbine blades are usually made of a fibreglass mats impregnated with a polyester resin, known as GRP (*glass-fibre reinforced polyester*). Epoxy resin is sometimes used instead of polyester resin. The glass fibre is partially substituted by carbon fibre in large rotor blades.

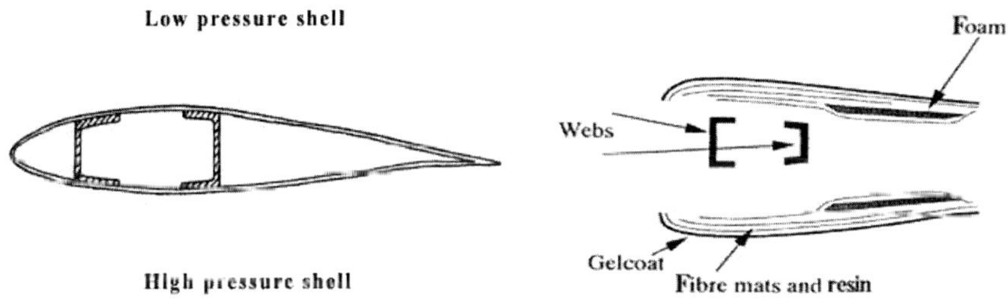

Figure 2.1. Example of a section of an actual blade structure (left drawing) and its primary elements (right drawing).

A typical blade is composed of two primary components (the section is shown in Figure 2.1):

- external shells;
- internal shear webs/spars bonded between the two shells.

Lightning protection (carries a lightning hitting the blade tip to the root) and aerodynamic brake (typically a tip turning on a shaft) are also included.

The external shells form the aerodynamic shape and carry part of the bending load. To stiffen the trailing edge, foam panels can be glued inside both shells. The internal longitudinal webs/spars,

(i) carry shear load and part of the bending load (making the blade stronger and stiffer in the flapwise direction), and
(ii) restrain the cross section against deformation and the shells against buckling. FRP is the favoured group of materials for external panels and internal spars.

FRP is a composite material, with fibres for transferring loads and a polymeric resin to distribute loads between the fibres and to restrain the fibres against local displacements. It is often used on both sides of a cellular core in a sandwich structures. The core is used for increasing the local bending stiffness of the shell or web. A thin film of gelcoat covers the whole blade external surface, giving a smooth, usually white coloured, finish to the blades. Gelcoats are used to protect the FRP against abrasion, UV radiation and moisture.

2.1.1 Fibre reinforcements

The role of the reinforcement in a composite material is fundamentally to increase the mechanical properties of the resin system. Reinforcement characteristics (such as type of fibre, fibre orientation, volume fraction of fibre and interaction between matrix and fibres) play a major role in static and fatigue composite properties, as reported in the DOE/MSU database and in the FACT database [2.3] (the main fibreglass composite databases for wind turbine applications).

Fibreglass materials are the most used and are formed by spinning glass into long threads. The most common glass fibre is *E-glass*. It is a low-cost material, with reasonably good tensile strength. E-glass is electrical grade with low alkali content. It exhibits better electrical insulation and strongly resists attack by water. More than 50% of the glass fibres used for reinforcement is E-glass.

Usually, glass fibres are not used directly, but first combined into other forms (known as "performs" or "pre-fabricated"). For instance, fibres may be woven or knitted into cloth, as shown in Figure 2.2. In the first case, fibres form a *continuous strand mat*, where long fibre stands are arranged in bundles with little or no twist. In the second case, fibres, prepared as chopped fibres, are supplied in *Chopped Strand Mat* (CSM), where the orientation of the relatively short fibres is random. Mats are used where high-strength is required, [2.1]. The first row of pictures in Figure 2.2 shows some biaxial weaves used in wind turbine blades.

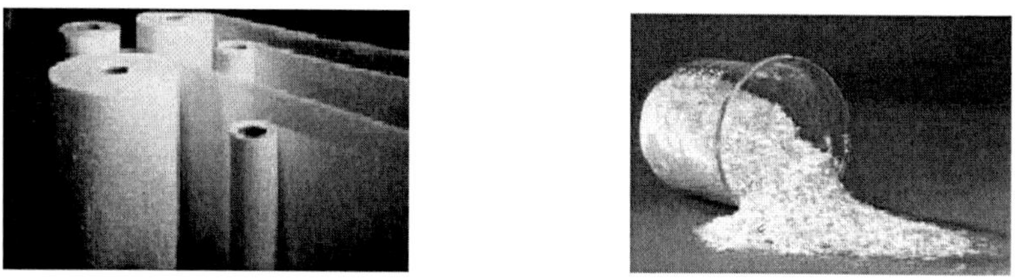

Figure 2.2. Woven (left) and chopped (right) glass fibres used in wind turbine applications.

For larger blades, as used in offshore applications, stiffness is important. E-fibreglass systems may not be able to meet these requirements within reasonable design constraints on weight. The obvious answer to this design issue is to introduce stiffer fibres into the composite.

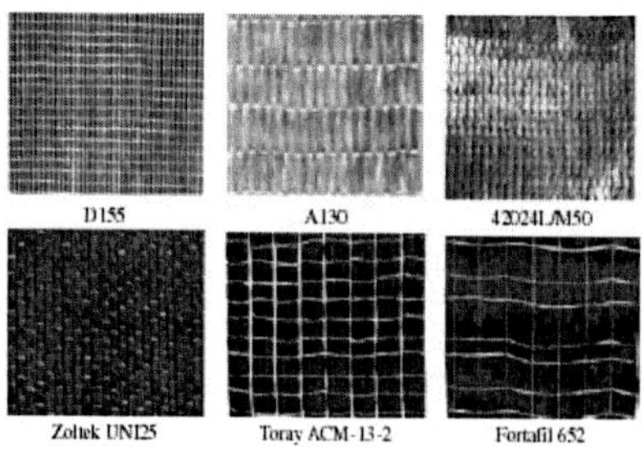

Figure 2.3. Some examples of glass (top) and carbon (bottom) biaxial weaves [2.2].

Carbon fibres have several advantages over E-glass in blade applications: larger modulus (by a factor of three), smaller density (by a factor of two-thirds), larger tensile strength and reduced fatigue sensitivity [2.2]. The major disadvantage is the increased cost, by approximately a factor of 10-15

[2.4], but this disadvantage is mitigated somewhat by a reduction in blade weight. In addition, carbon fibres are semiconductors, so introduce lightning issues that are complex, but manageable, although adding significantly to blade costs. Moreover, early results [2.5, 2.6] showed that the strain-related properties of lower-cost varieties of carbon composites, particularly in compression, are significant poorer compared with glass fibres (as shown in Figure 2.4). Fibre waviness and misalignment (Figure 2.4) have long been recognized as having the potential to cause major reductions in compressive strength.

A large uncertainty of values for physical properties of carbon fibres occurs in the literature. Therefore, the safety factors to be adopted for the carbon composites will be much larger than those used for better characterized materials such as E-glass, leading to an uncontrolled increased in weight that can offset their potential advantages.

Carbon fibre is produced by the controlled oxidation, carbonisation and graphitisation of carbon-rich organic precursors which are already in fibre form. Variation of the graphitisation process produces either high-strength fibres (T ~2,600°C) or high-modulus fibres (T ~3,000°C), with other types in between. Once formed, the carbon fibre has a surface treatment applied to improve matrix bonding and chemical sizing, which serves to protect it during handling. Carbon fibres are usually grouped according to the modulus band in which their properties fall. These bands are commonly referred to as: high-strength (HS), intermediate-modulus (IM), high-modulus (HM) and ultra high-modulus (UHM).

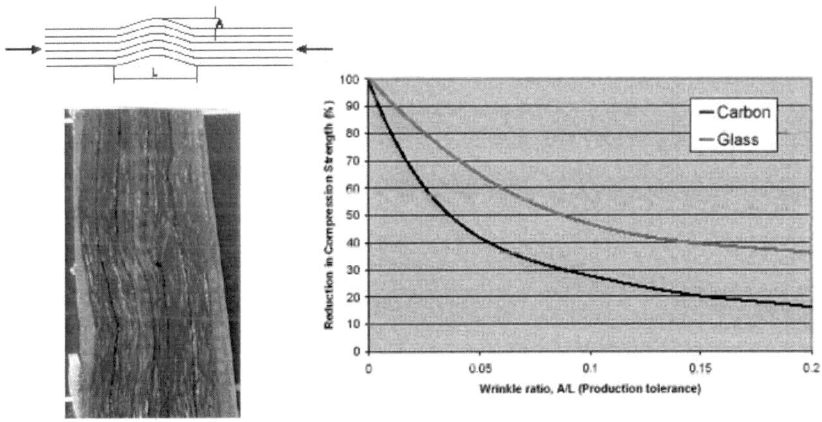

Figure 2.4. Wrinkles in carbon fibre laminates (picture on the left) and the consecutive reduction on compression strength (graphic on the right) [2.6].

The search for materials that lead to more efficient designs for large blades is forcing re-examination of established materials as well as the "new" materials (such as carbon). Two of the most promising are S-glass and carbon/wood hybrids [2.7].

S-glass fibre is used in applications where a combination of large tensile strength and large strain at failure are needed, such as for helicopter rotor blades. The S-glass fibre is a magnesium aluminosilicate with larger alumina content than the more commonly used E-glass. Originally designed for marine applications, it has significantly better properties than E-glass, but its higher cost has limited its use in turbine blades. However, as blade length is increased, the enhanced properties of S-glass offer an alternative to carbon fibres. It provides most of the benefit of carbon without the very high cost. Moreover, with the mass production of the fibres required by the turbine blade industry, the relative cost of S-glass should reduce significantly [2.5].

One way to utilise the advantages of carbon fibre, without paying the full cost, is to mix some carbon fibres with glass fibres in the overall composite (*carbon/glass hybrid*) for specific transitions sections, Figure 2.5. This is structurally efficient and cost-effective to manufacture. However, joining glass and carbon laminate adds weight at the transition region).

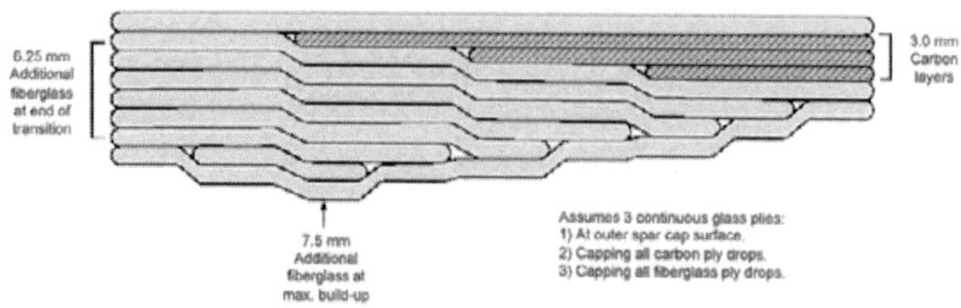

Figure 2.5. Example of a fibreglass-to-carbon spar transition region [2.8].

Carbon/wood hybrids also offer unique material characteristics suitable for very large turbine blades; however, the biggest problem is the supply of high-quality veneer. Joining to these hybrid materials is also a technical challenge [2.5].

Zebrawood (***carbon/wood/E-glass hybrid***) is a convenient name used for the inclusion of carbon fibre between the layers of epoxy laminated Douglas fir, the regular black lines suggesting the "zebra" term. This concept makes dual use of the epoxy, to both bond the veneer and wet the carbon, so the weight and cost increments are not much beyond that for just the carbon fabric. Fir and carbon were known to have similar working strains, so this material is studied as a cost-effective way for increased strength and stiffness wherever a wood/epoxy blade might need it. Further, it was thought that the fir could provide a measure of damage tolerance, and protection from external impact, which is often a concern for all carbon structures [2.5].

The basic mechanical properties of some fibres, compared to standard materials, are given in the Table 2.1.

Table 2.1. Basic properties of fibres and other engineering materials.

Material type	Tensile strength (MPa)	Tensile modulus (GPa)	Density (g/cm³)
Carbon HS	3500	160-270	1.8
Carbon IM	5300	270-325	1.8
Carbon HM	3500	325-440	1.8
Carbon UHM	2000	>440	2.0
Glass-E glass	2400	69	2.5
Glass-S glass	3450	86	2.5
Aluminium alloy (7020)	400	1069	26
Titanium	950	110	24
Mild steel (55 grade)	450	205	26
Stainless steel (A5-80)	800	196	25
HS steel (17/4 H900)	1241	197	25

The surface interaction of fibre and resin is controlled by the degree of bonding between the two. This is heavily influenced by the treatment given to the fibre surface. Surface finishes are nearly always applied to fibres, both to allow handling with minimum damage and to promote fibre/matrix interfacial bond strength. A particular sizing that has been developed for one type of application or one type of resin is not necessarily adequate for another application or another resin. It is important that the supplier of the fibre material provides adequate information about this.

The reinforcement fibres for FRP panels in wind-turbine blade manufacturing are supplied in fabrics. In polymeric composite terms, a "fabric" is defined as a manufactured assembly of long fibres of carbon or glass, or a combination of these, to produce a flat sheet of one or more layers of fibres. These layers are held together either by the mechanical interlocking of the fibres themselves or with a secondary material to bind these fibres together and hold them in place, giving the assembly sufficient integrity to be handled.

Fabric types are categorised by the orientation of the fibres, and by the various construction methods used to hold the fibres together. The four main fibre orientation categories are*: unidirectional* (UD), 0/90°, *multiaxial*, and *random*. Typically, the fabrics used in wind turbine applications have either *stitched* (combining two layers of unidirectional material into one fabric) or *woven strand* structures (produced by the interlacing of warp-0° fibres and weft-90° fibres in a regular pattern or weave style).

2.1.2 Resins

There are three types of resins commonly used in matrices of composites. They are *unsaturated polyesters*, *epoxies* and *vinyl esters*. Polyester is dominant, with epoxy being used in structurally more demanding applications. These resins all have the general property of being liquid during the lay-up of the composite, but solid when cured. As solids, all of the resins tend to be somewhat brittle. The choice of resins affects the overall properties of composite.

Polyester resins have been used frequently in the wind industry because they have a short cure time and low cost. However, shrinkage upon curing is relatively large [2.4]. Cure time is from a few hours to overnight at room temperature; but with the addition of an initiator, curing can be done at elevated temperatures in a few minutes.

Epoxy resins are stronger than polyester, have better chemical resistance, good adhesion and low-shrinkage upon curing, but they are also more expensive (almost twice as expensive as polyester) and have longer cure time than polyesters [2.4].

Vinyl ester resins have become more widely used in recent years. These resins have similar properties to epoxies (as shown in Figure 2.6), but are somewhat cheaper and have shorter cure time. They have good environmental stability and are widely used in marine applications [2.4].

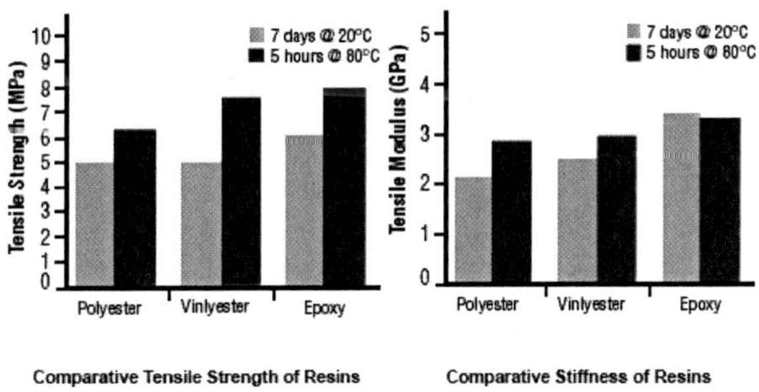

Figure 2.6. Comparative tensile strength (left) and stiffness (right) of resins, post-cured for 5 hrs & 176°F (80°C) or for 7 days & 68°F (20°C) [2.9].

The adhesive properties of a resin in the fibre reinforcement affect the mechanical properties of a composite. Of the three resin types discussed (polyester, vinyl ester and epoxy resin), polyester resins generally have the weakest adhesive properties of the three systems. Vinyl ester resin has improved adhesive properties over polyester, but epoxy systems offer the best performance of all, and are therefore frequently used in many high-strength adhesives. This is due to their chemical composition and the presence of polar hydroxyl and ether groups.

The strength of the bond between resin and fibre is not solely dependent on the adhesive properties of the resin system but is also affected by the surface coating on the reinforcement fibres. Also of importance to the composite designer and builder is the amount of *shrinkage* that occurs in a resin during and following its cure period. Shrinkage is due to the resin molecules rearranging and re-orientating themselves in the liquid and semi-gelled phase. Polyester and vinyl esters require considerable molecular rearrangement to reach their cured state and can show shrinkage of up to 8% [2.10]. However, the different nature of the epoxy reaction, leads to very little rearrangement and with no volatile by-products being evolved, typical shrinkage of an epoxy is reduced to around 2%. The absence of shrinkage is, in part, responsible for the improved mechanical properties of epoxies over polyester and vinylester (as shown in Figure 2.6), as shrinkage is associated with built-in stresses that can weaken the material.

An important property of any resin, particularly in a marine environment, is its ability to *withstand degradation from water ingress*. All resins will absorb some moisture, adding to a laminate's weight, but what is more significant is how the absorbed water affects the resin and resin/fibre bond in a laminate, leading to a gradual and long term loss in mechanical properties. This damage to the matrix has minimal effect on the tensile behaviour of the laminate because the fibres are the primary load carrier. However, the effects are significant in compression and shear, where the matrix is the primary load carrier. Both polyester and vinyl ester resins are prone to water degradation due to the presence of hydrolysable ester groups in their molecular structures. As a result, a thin polyester laminate can be expected to retain only 65% of its inter-laminar shear strength after immersion in water for a period of one year, whereas an epoxy laminate immersed for the same period will retain around 90% [2.9].

2.1.3 Core materials

The purpose of a core in a composite laminate is to increase the laminate's stiffness by effectively "thickening" it with a low-density core material. This can provide a dramatic increase in stiffness for very little additional weight.

Generally, the most common core materials are *structural foams* (Figure 2.7) and *wood products* (such a *balsa*). Foam is by far the most commonly used in large wind turbine blades. Foams are based on thermoplastics, e.g. closed-cell polyvinyl chloride (PVC), and are delivered in a range of densities. Stiffness and strength increase with density (as shown in Table 2.2). Consequently, different types and grades of foam can have significantly different properties. The most important mechanical properties are the shear modulus, shear strength and ductility or yielding behaviour.

Core-sandwich adhesives are used to bond the sheets and pieces together. They are also used to fill the voids between sheets and pieces [2.10]. A good bond is necessary to maintain the shear capacity of the core. The bond should have at least the same shear strength and fracture elongation as the core material. Large ductility of the adhesive normally increases the strength of the joint.

Figure 2.7. Some commercial foams for wind turbine applications [Reproduced by permission of DIAB,www.diabgroup.com].

Table 2.2. Some basic properties of semi-rigid PVC foams provided for wind turbine market.

Material type	Density (g/cm³)	Tensile strength (MPa)	Shear strength yield (MPa)	Shear modulus (MPa)	Shear strain (%)
H45	48	1.3	0.485	18	10
H60	60	1.9	0.680	22	20
H80	80	2.2	0.950	31	18
H100	100	3.1	1.330	40	26

2.1.4 Manufacturing techniques

The final properties of composites are not only a function of the individual properties of the resin matrix and fibre (and in sandwich structures, the core as well), but depend how these materials are integrated into the component. They are also a function of how they are processed. This section reviews common wind-turbine blade production methods and the trend of current production.

Wet hand lay-up

A layer of resin is applied in a mould. The mould has previously been covered with a release agent and a layer of gelcoat. The first ply of reinforcements fabric is then applied in the wet resin. Subsequently the remaining plies are applied (according to the laminate schedule) alternating with application of resin. During the application of all plies, the fabrics are worked using metallic or bristle rollers (see Figure 2.8) to ensure a thorough wetting out of the fibres, to compact and consolidate the laminate and to make sure that all entrapped air is removed. In addition, superfluous resin is removed using an elastic scraper.

Since the quality of the material is dependent on the system of laying, any form of automation is beneficial for quality control. Systems involving automatic mixing of resins and catalysts require mixing pumps, wet-out machines (where cloth is weighed dipped in resin and squeezed through rollers, and then reweighed to verify resin ratio before being laid), gel time measuring machines, etc. will be of great benefit.

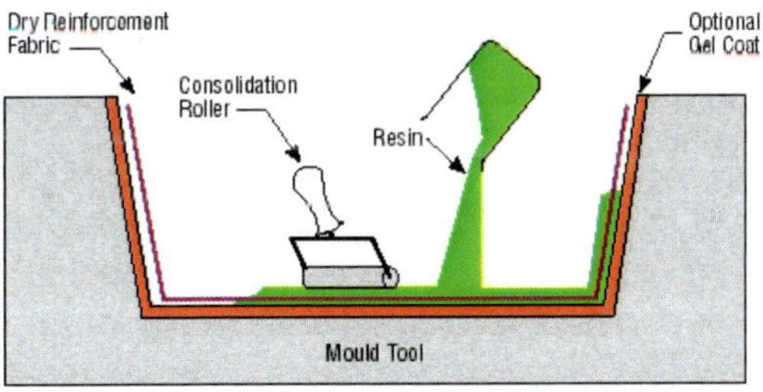

Figure 2.8. Wet hand lay-up technique [Reproduced by permission of Gurit, www.gurit.com].

If the preceding ply has cured completely the operation is no longer a wet-in-wet lamination and the new bond will be treated as a new bond. The surface preparation for a new bond is critical, since there will be no effective cross-linking between the laminates. In this case, the new applied resin will act as an adhesive.

This technique, widely used for many years, offers low cost tooling and simple principles to teach, together with the following disadvantages [2.9]:

(i) resin mixing, laminate resin contents, and laminate quality are very dependent on the skills of laminators;

(ii) the smaller molecular weights of hand lay-up resins generally means that they have the potential to be more harmful than larger molecular weight products,
(iii) the smaller viscosity of the resins also means that they have an increased tendency to penetrate clothing, etc;
(iv) resins need to be low in viscosity to be workable by hand, which generally compromises their mechanical/thermal properties due to the need for high diluents/styrene levels.

Vacuum-bagging is basically an extension of the wet lay-up process, described above, where pressure is applied to the laminate after being laid-up, in order to improve its consolidation. This is achieved by sealing a plastic film over the wet laid-up laminate and onto the tool (as shown in Figure 2.9). The air under the bag is extracted by a vacuum pump and thus up to one atmosphere of pressure can be applied to the laminate to consolidate it. The vacuum-begging technique is used to improve the compacting of the laminate, thus to increase the volume fraction of reinforcement and to improve the consistency of the finished laminates. Normally, a porous mat is applied on top of the laminate and under the membrane, to facilitate an even resin distribution and the escape of the air. Vent holes may be required through the core for the same purpose.

The vacuum bag reduces the amount of volatiles emitted during cure and the void content, as compared with the standard wet lay-up technique. Moreover, it allows better fibre wet-out, due to pressure and resin flow throughout the structural fibres. This technique requires greater skill by the operators. In addition, polyester and vinylester resins may give difficulty due to excessive extraction of styrene (for dilution) from the resin by the vacuum pump [2.9].

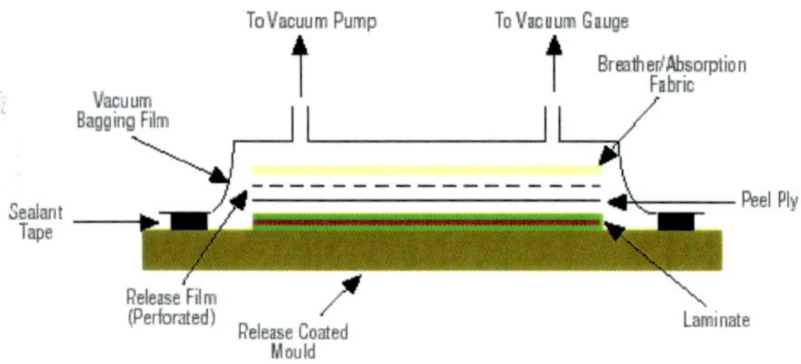

Figure 2.9. Vacuum bagging technique [Reproduced by permission of Gurit, www.gurit.com].

Pre-pregs are fabrics and fibres are pre-impregnated by the materials manufacturer, under heat and pressure or with solvent, with a pre-catalysed resin. The catalyst is largely latent at ambient temperatures giving the materials several weeks, or sometimes months, of useful life when defrosted. However to prolong storage-life the materials are stored frozen. The resin is usually nearly solid at ambient temperatures, and so the pre-impregnated materials (prepregs) have a light sticky feel to them, such as that of adhesive tape. Unidirectional materials take fibre direct from a creel, and are held together by the resin alone. The prepregs are laid up by hand or machine onto a mould surface, vacuum bagged and then heated to typically 120-180°C (as illustrated in Figure 2.10). This allows the resin to reflow initially and then eventually cure. Additional pressure for the moulding is usually provided by an autoclave (effectively a pressurised oven) which can apply pressures up to 5 atmospheres to the laminate.

This system has the following advantages [2.9]:

- *resin/catalyst levels and the resin content* in the fibre are accurately set by the materials manufacturer. High fibre contents can be safely achieved;
- the materials have excellent *health and safety characteristics* and are clean to work with;

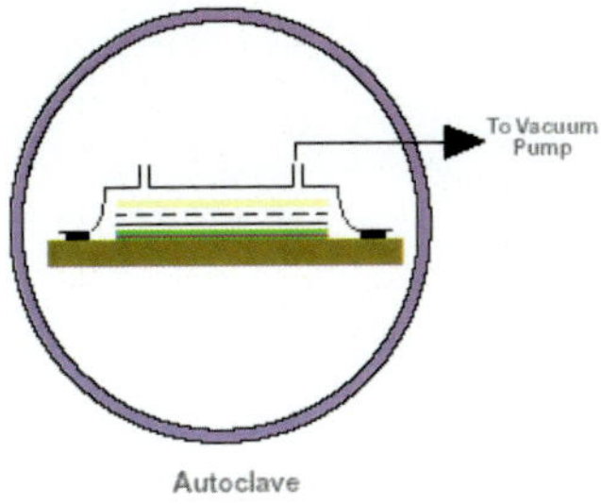

Figure 2.10. Pre-pregs technique [Reproduced by permission of Gurit, www.gurit.com].

- resin chemistry can be *optimised for mechanical and thermal performance*, with the high viscosity resins being impregnable due to the manufacturing process;
- the *extended working times* (of up to several months at room temperatures) means that structurally optimised, complex lay-ups can be readily achieved;
- potential for *automation* and labour saving.

Against, are the followings disadvantages:

- preimpregnated fabrics are *expensive*;
- *autoclaves* are usually required to cure the component; these are expensive, slow to operate and limited in size;
- *tooling* needs to be able to withstand the process temperatures involved;
- *core materials* need to be able to withstand the process temperatures and pressures.

Vacuum-assisted resin transfer moulding (VARTM) has similarities to vacuum-bagging and is illustrated in Figure 2.11. Fabrics are laid up as a dry stack of materials. These fabrics are sometimes pre-pressed to the mould shape, and held together by a binder. These "*performs*" are then more easily laid into the mould tool. The fibre stack is then covered with peel ply and a knitted type of non-structural fabric. The dry stack is then vacuum bagged and, when bag leaks have been eliminated, resin is allowed to flow into the laminate. The resin distribution over the whole laminate is aided by resin flowing easily through the non-structural fabric, and wetting the fabric from above.

The advantage of this technique is the good control of, and the relatively high, volume fraction of fibres one achieves and thus consistency in the properties of the laminates and a lower laminate weight [2.10]. Moreover, large components can be fabricated and cored structures can be produced in one operation. While some disadvantages are listed below [2.11]:

- *relatively complex process;*
- *resins must be very low in viscosity, thus comprising mechanical properties;*
- *unimpregnated areas can occur, resulting in very expensive scrap parts;*
- *some elements of this process are covered by patents (SCRIMP ™).*

Filament winding: Shafts for aerodynamic brakes and some internal spars are typically manufactured by filament winding. Spars may take on a variety of forms, but the purpose is to create a lightweight member which can resist the applied moments. The shape of a spar for large wind turbine blade may be that of a web, a box beam or a D (illustrated in Figure 2.12). Spars in fibreglass or in alternative composite material, are usually made by building up layers of fibre and resin around a mandrel, which is then later remove [2.4].

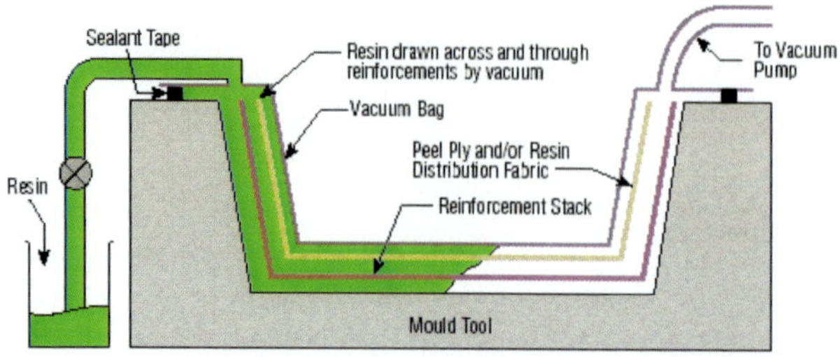

Figure 2.11. VART (Vacuum-Assisted Resin Transfer moulding) technique [Reproduced by permission of Gurit, www.gurit.com].

Figure 2.12. Large blade spars such as web (left), box beam (centre) and D spar (right) [2.11].

In the *filament winding process*, fibre tows are passed through a resin bath before being wound onto a mandrel in a variety of orientations, controlled by the fibre feeding mechanism, and rate of rotation of the mandrel (as shown in Figure 2.13). This can be a very fast and therefore an economic method of laying material down. Moreover, fibre cost is minimised, since there is no secondary process to convert fibre into fabric prior to use. Structural properties of laminates can be very good, since straight fibres can be laid in a complex pattern to match the applied loads. Main disadvantages are:

- *the process is limited to convex shaped components;*
- *fibre cannot easily be laid exactly along the length of a component;*
- *mandrel costs for large components can be high;*
- *low-viscosity resins usually need to be used with their attendant lower mechanical and health and safety properties [2.9].*

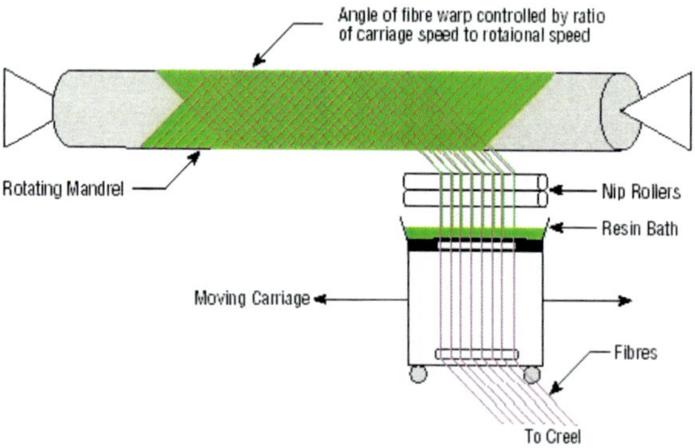

Figure 2.13. Filament winding technique [Reproduced by permission of Gurit, www.gurit.com].

Wind turbine manufacturers

Many wind turbine manufacturers now have in-house production of their own blades, and in doing so are using diverse materials and manufacturing methods. Hand lay-up was the most popular manufacturing method for wind turbine blades through the 1980s. For blade sizes to 30 m, the most common manufacturing approach has been open-mould wet layup. The most notable exception was VESTAS Wind Systems (Ringkøbing, Denmark), which has long used fibreglass/epoxy prepreg [2.5,2.8].

In current production, although several manufacturers still use open-mould wet lay-up processes, increasingly stringent environmental restrictions are likely to require processes with less emissions. Two methods are emerging as the most common replacement for traditional methods, *prepreg and vacuum-assisted resin transfer moulding* (VARTM). The majority of blade manufacturers presently use VARTM [2.5,2.8]. Nordex and GE Energy have built blades in the 40-50 m length range using hand lay-up of primarily fibreglass structure in open-mould, wet processes. NEG Micon (now in the Vestas group) builds 40 m blades with carbon-augmented wood-epoxy. TPI Composites manufactures 30 m blades using their SCRIMP™ vacuum-assisted resin transfer moulding (VARTM) process. Among the more novel approaches in current use for large blades is by Siemens Wind Power (that has taken over Bonus), where blades 30 m and longer are being produced in one piece from a dry preform with a single-shot infusion, eliminating the need for secondary bonding. This patented technology is named IntegralBlade® [2.5,2.8].

The company 'LM Glasfiber' is the major supplier of blades for wind turbines and has been developed the longest blade (61.5 m long). It uses VARTM technology for its production.

Another promising alternative is partially pre-impregnated fabric, marketed by Gurit Holding AG (ex SP Systems) under the name SPRINT®, and by Hexcel Composites as HexFIT™. When laid-up, the dry fabric regions provide paths for air to flow, and vacuum can be used to evacuate the part prior to heating. Under heat and pressure, the resin flows into the dry fabric regions to complete the impregnation. The use of automated preforming and automated lay-up technologies are also potential alternatives to hand lay up in the blade moulds. Benefits could include improved quality control in fibres/fabric placement and a decrease in both hand labour and production cycle times.

Figure 13 a. *Top pictures:* Blade manufacture at LM Glasfiber [Reproduced by permission of LM Glasfiber, www.lmglasfiber.com]). *Bottom pictures:* hardening of blade at Siemens (left, reproduced by permission of Siemens Power Generation, www.powergeneration.siemens.com) and mould building at TPI Composites.

For large blades, selection of materials and manufacturing processes are interdependent. Carbon's structural performance is sensitive to fibre alignment, particularly for compressive strength, and infusion of thick carbon/epoxy parts is difficult. This has prompted nearly exclusive use of prepreg for carbon spars in large blades. Again, there is a diversity of processes in use. The UK-based company DeWind (now subsidiary by Composite Technology Corporation) is using an innovative approach to produce 40 m carbon/fibreglass hybrid blades. In that process, the spar cap is produced using prepreg carbon. After curing, the spar caps are then placed into a preform and infused into the fibreglass blade skins. Vestas Wind Systems has testing several prototypes of its 44 m (V-90) turbine with carbon-fibre spars. GAMESA has announced that it will include carbon spars for 87 m and 90 m rotor diameters. NEG Micon (merged with Vestas) has built 40 m blades with carbon-augmented wood/epoxy.

The largest currently installed turbine (the 6 MW E-126 from the German manufacturer ENERCON) has fibreglass/epoxy blades. In contrast LM Glasfiber's reports "selective" use of carbon fibre in its new 61.5 m fibreglass/epoxy blade, scheduled for the 5.0 MW prototype under development by German manufacturer REpower.

Major manufacturers of blade wind turbine

Blade wind turbine manufacturers	Manufacturing technique	Location
Enercon	GmbH VARTM	DE (EU)
Gamesa	Pre-pregs	ES (EU)
Hexcel Composites	HexFITTM	CT (US)
LM Glasfiber A/S	VARTM	DK (EU)
TPI Composites Inc.	SCRIMPTM (a VARTM method)	RI (US)
Siemens Wind Power A/S	IntegralBlade®	DE (EU)
Gurit Holding AG	SPRINT®	CH (EU)
VESTAS Wind Systems A/S	Pre-pregs	DK (EU)

The blade manufacturer, and not the supplier of the raw materials, controls the final properties of the blade laminates in according with the existent standards [2.10, 2.11]. Properties to be considered (by testing or calculations) include, but are not limited to:

- *stiffness, ultimate and fatigue strength at relevant temperatures, stability (against buckling);*
- *eigenfrequencies (resonance frequencies);*
- *toughness (at low temperature if appropriate);*
- *creep;*
- *ageing characteristics (considering humidity and temperature);*
- *resistance of wood to rot and fungus.*

Figure 2.14. Bolt connections at the root of a blade.

Table 2.3. Average lifetime for hot dip galvanized bolts

Environment	Average lifetime [years]
Open agricultural areas	45
Small towns	30
Large towns	12
Heavy polluted industrial areas	5
Coastal and offshore areas	30

2.2 Bolt Connections

Stainless steel bolts support the rotor blades by securing them to the hub (see picture in Figure 2.14). Each blade connection is mounted to a four-point ball bearing (to allow the blade rotation), which is then mounted to the hub. Bolts are made from three different type of stainless steel:

- *austenitic steel;*
- *ferritic steel;*
- *martensitic steel.*

Normally austenitic steel is used. All stainless steel bolts are standardised according to ISO 3506 (corrosion-resistant stainless-steel fasteners specifications). Different surface treatments are available to achieve specific properties, such as corrosion protection and low-friction. The most commonly used bolts are black (no surface treatment) and then hot-dip galvanized. In this case, bolts offer an excellent protection against corrosion. The lifetime for hot-dip galvanized bolts depend on the environmental conditions (as shown in Table 2.3).

Some manufacturers of bolt connection for blade wind turbine.

Bolt wind turbine manufacturers	Location
The Dyson Corporation	OH (US)
Cooper & Turner Limited	UK (EU)
AH Bolte A/S	DK (EU)

2.3 Pitch Drive

Today's prevailing advanced offshore wind turbine technology uses three-bladed, variable-speed turbines. The rotor blades are adjusted around the longitudinal axis, to allow controlled pitch via motor drives on each blade. Thus, aerodynamic efficiency and reduced loads on the soft drive-train is assured, providing reduced maintenance and longer turbine life [2.4].

One way of controlling the pitch is shown in Figure 2.15, namely a *mechanical pitch drive*, typically using three electrical motors; one for each blade. The motors are placed inside the hub and generally are made in forged carbon steel. The teeth of these gears are hardened by carburizing or other forms of heat treatment.

When a *hydraulic pitch drive* is employed, each turbine blades is pivotally driven by a pair of hydraulic actuators placed inside the hub (Figure 2.16). Such an actuator is a hydraulic cylinder, usually powered nearby by a hydraulic power package, including a proportional valve, placed in the nacelle.

Pitch device wind turbine manufacturers	Type	Location
AVN Hydraulik A/S	Hydraulic system	DK (EU)
Parker Hannifin Corporate	Hydraulic system	OH (US)
Rexroth Bosch AG	Hydraulic and mechanical systems	DE (EU)
Maxwell Technologies Inc.	Mechanical system	CA (US)

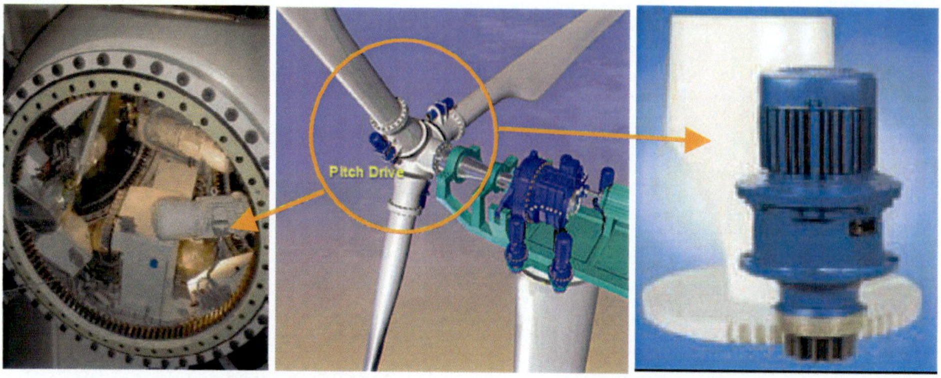

Figure 2.15. The Maxell mechanical pitch drive in the left picture [www.maxwell.com], and the Bosch mechanical pitch drive in the right picture [credit www.boschrexroth.com].

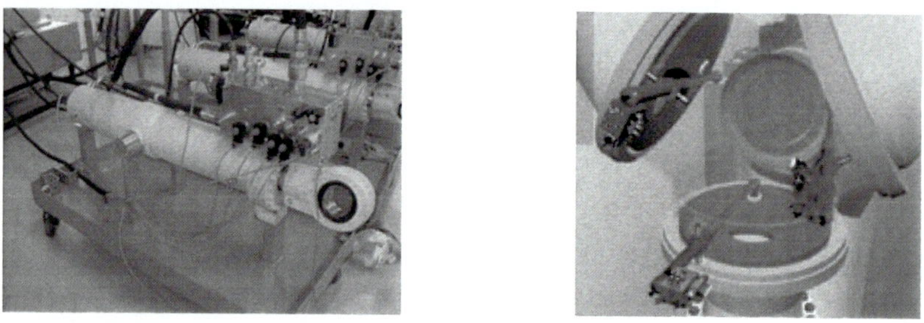

Figure 2.16. Hydraulic actuator of the pitch drive in the left picture [Reproduced by permission of AVN Energy, www.avn.dk] and whole pitch mechanism in the left picture [credit www.boschrexroth.com].

2.4 Hub and Hub Enclosure

The hub (the fixture for attaching the blades to the rotor shaft) usually consists of nodular cast iron components (also known as *spheroidal graphite cast iron*) for distribution of the blade loads to the tower. A major reason for using cast iron is the complex shape of the hub, which makes it hard to produce in any other way. It is important to consider whether the chosen structural material possesses the necessary ductility.

Cast iron is classified according to its mechanical properties, such as strength and hardness, as specified in the EN1563 European standard. Cast hubs are usually tested by non-destructive testing for verification of the mechanical properties and for detection of possible defects and internal discontinuities. The following tests are available:

- *ultrasonic*
- *magnetic particle*
- *visual inspection;*
- *hardness.*

Low ambient temperatures can be critical for cast hubs and the choice of hub material should relate to the temperatures of the surroundings [2.10].

The hub enclosure (Figure 2.17), sometimes called the 'nose cone', is usually made of glass fibre reinforced polyester (see paragraphs 2.1.1 and 2.1.2).

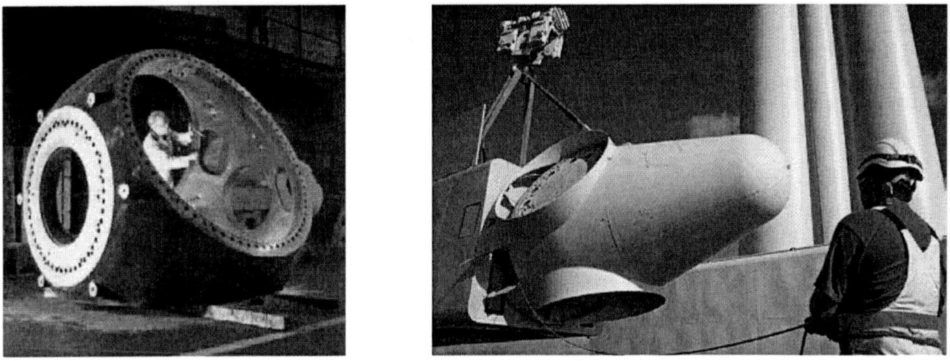

Figure 2.17. The 'hingeless' hub type (top picture) is the most common configuration for wind turbine hubs [credit, www.cwtaylor.co.uk]. The left picture shows the hub with the front enclosure (nose cone). [Reproduced by permission of Siemens AG, Energy Sector, www.siemens.com/energy].

Hub wind turbine manufacturers	Component	Location
Hodge Foundry Inc	Hub	PA (US)
CAB Inc.	Hub	GA (US)
K&M Machine-Fabricating, Inc.	Hub	MI (US)
Richter Maschinenfabrik AG	Hub	DE (EU)
Metso Foundries	Hub	SE (EU)

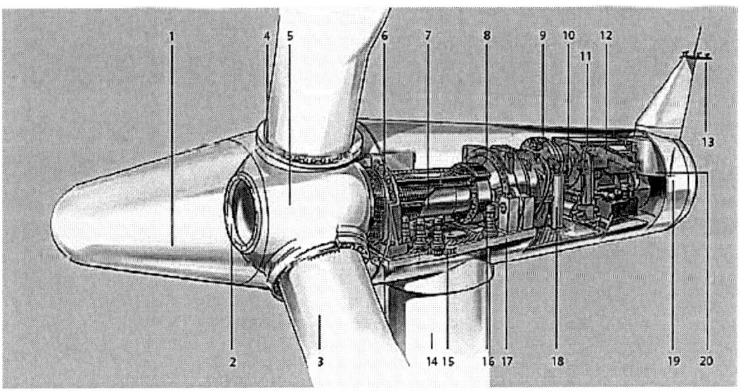

Figure 2.18. Typical drive train for a wind turbine with gearbox [Reproduced by permission of Siemens AG, Energy Sector, www.siemens.com/energy] and components: 1) Spinner, 2) Spinner bracket, 3) Blade, 4) Pitch bearing, 5) Rotor hub, 6) Main bearing, 7) Main shaft, 8) Gearbox, 9) Brake disc, 10) Coupling, 11) Generator, 12) Service crane, 13) Meteorological sensors, 14) Tower, 15) Yaw ring, 16) Yaw gear, 17) Nacelle bedplate, 18) Oil filter, 19) Canopy and 20) Generator fan.

3. NACELLE COMPONENT MATERIALS

The nacelle is the housing for the drive train (with the exception of the rotor), ancillary components and the main frame. Figure 2.18 illustrate typical nacelle components, including the *nacelle cover, main shaft, couplings, gearbox, brakes, generator, oil cooler, control panel, yaw system, flanges and heat exchanger.* These individual parts are usually purchased from sub-contractors and subsequent assembly occurs at the manufacturer's factories.

For offshore and coastal wind turbines, interior corrosion is generally avoided by maintaining the nacelle interior air at low-humidity with *de-humidifying devices*. The de-humidifying system maintains the interior relative humidity below the limit of any steel corrosion risk limit (RH 60%). For additional protection, the main electric components (generator, control systems, etc.) have standby heating systems, preventing condensation, even during sudden variations in ambient temperature.

3.1 Main Shaft

Every wind turbine has a main shaft at the rotor, best referred to as the *low-speed or rotor shaft*. The low-speed shaft connects the rotor hub to the gearbox (Figure 3.1). Their primary function is to transmit torque from the rotor to the rest of the drive train. The main shaft supports the weight of the rotor and is itself supported by bearings, which transfer reaction loads to the main frame of the turbine. Shafts are often subject to bending and the combined loading is time-varying, so fatigue is an important consideration.

For machines with *gearboxes*, and depending on the design, the low-speed shaft and/or the bearings may be integrated into the gearbox or they may be completely separate from it, connected by a coupling. On a MW-scale wind turbine, the rotor rotates relatively slowly, about 19 to 30 revolutions per minute (RPM) to maintain a blade-tip-speed about seven times the wind speed (i.e. a constant tip-speed ratio).

Materials used for shafts depend on the application. In most cases, non or low-alloyed machinery steels are used for forged shafts (carbon content of 0.3%-0.7%, with less than 5% alloy of other metals such as Mn, Cr, Mo, Ni and V). For the least severe conditions, hot-rolled plain carbon-steel is used. For greater strength applications, somewhat higher-carbon content steel may be used. After machining, shafts are often heat-treated to improve their yield-strength and hardness. These steels have different mechanical properties, ranging from non-alloyed steel with an ultimate strength of about 500 MPa and failure elongation of 15%, to low-alloyed steel with an ultimate strength of up tom 1500 MPa and failure elongation less than 10%. Under the most severe conditions, alloy steels are used for shafts [3.1].

In some cases, cast low-speed shafts are used. Casting provides a great degree of freedom for shaping, but poses limits in terms of relatively low ultimate-strength and failure elongation. Usually, nodular iron is used [3.2].

The *low-speed shaft* is one of the most critical components in a wind turbine, especially for very large rotors, which have to transmit very large torque. Thus, quality assurance is essential, so the necessary material quality is met and the manufacturing has not caused surface cracks or other imperfections. Hence, the material must be checked for imperfections by a suitable non-destructive testing procedure (such as ultrasonic testing) and delivered with acceptable certification.

Figure 3.1. The main shaft for a 2MW wind turbine [credit www.middelgrunden.dk]

Figure 3.2. An example of a planetary gearbox and its components.

3.2 Gearbox

Most wind turbine drive-trains include a gearbox to increase the speed of the low-speed shaft to input speed of the electrical generator (usually 1800 rpm or 1500 rpm). The gearbox is one of the most expensive and the single heaviest (its weight increases dramatically with power rating) component in a wind turbine [3.3]. Note that turbines with multipole synchronous generators and electronic power electricity couplings may not require a gearbox.

Gearboxes are normally designed and supplied by a different manufacturer than the one constructing the wind turbine. There are several gear types used in main gears for wind turbine, but *planetary gears* (Figure 3.2) are most common because they are relatively light and compact. The gearbox consists of a case, shafts, gears, bearings and seals.

Gears may be made from a wide variety of materials, but the most common material in wind turbine is forged-steel. High-strength and surface hardness in steel gear-teeth is often obtained by carburizing or other forms of heat treatment. Common heat treatments for steel include:

- *preheat treatments* (anneal, normalise, temper);
- *heat treatments*, such as through-hardening (anneal, normalise, normalise and temper, quench and temper), surface hardening (flame and induction harden, carburise, carbonitride, and nitride);
- *subsequent heat treatment* (stress relief).

Requirements for testing and inspecting gears depend on which type of heat treatment has been applied. Generally, certificates are required for chemical composition, mechanical properties (including Charpy-V) and ultrasonic testing.

Lubrication is a significant issue in gearbox operation. The gears are splash-lubricated to ensure that all gear meshes and bearings are sufficiently supplied with oil. Oils must be selected to minimize wear on the gear teeth and bearings, and to function properly under the external environmental conditions in which the turbine will operate. In some case, it may be necessary to provide filtering, or cooling or warming (in cold weather) of the oil. The choice of a lubricant and lubrication system should be the responsibility of the gearbox manufacturer. Wind turbine gears have relatively small pitchline velocity and large gear tooth loads; they operate in a continuously time-varying torque mode. These conditions require either synthetic or mineral oils [3.2]; recommended criteria for acceptance of gear lubricants are given in standards.

Gearbox manufacturers, examples for wind turbines	Location
Niebuhr Tandhjulsfabrik A/S	DK (EU)
Moventas	FIN (EU)
Rexroth Bosch AG	DE (EU)
Cincinnati Gear Company	OH (US)
Tecknatex Aps	DK (EU)

Oil filtration solution for wind turbine	Location
CJC™	DK (EU)
Hilliard Corporation	NY (US)

3.3 Generator

The generator transforms mechanical energy into electric power. There are two major types of generator: synchronous and asynchronous/induction (see chapter 5). In all asynchronous generators (for fixed speed and significant variable speed operations), the coupling of the power through the generator is 'soft' (i.e. speed-ratio variation occurs), with considerable benefit for mechanical components in the drive-train.

Traditionally, the active materials in a generic generator consist of magnetically-conducting iron and electrically-conducting copper coiled wire within insulation. Permanent magnets (made in NeBFe or in ferrite materials), instead of magnetic coils, are increasingly common, together with electrical components (such as temperature sensors), as integral parts of the generator. Mechanical stability of NeBFe magnets can be protected against corrosion by coating with tin, zinc or similar metals [3.2].

Figure 3.3. A multipole generator component [credit www.middelgrunden.dk].

Generator manufacturers, examples for wind turbines	Typology	Location
ABB A/S	Synchronous generators	DK (EU)
ELIN EBG Motoren	Induction & synchronous generators	AT (EU)
Hitachi Europe GmbH	Induction generators	DE (EU)
Motors & Controls Internationals Hazelton	Induction generators	PE (US)
GE Energy	Induction generators	GA (US)

The generator should be designed to be fully functional when the external air temperature is within the range [-20, to +30]°C, in a working environment so that damaging condensation and saline deposits cannot occur. These effects can be prevented by *auxiliary heating* when the generator is not running or when self-heating is insufficient. The encapsulation, the cooling/heating mechanisms and the insulation of the generator are all designed to withstand these impacts.

3.4 Yaw Drive

Yaw denotes the rotation of the nacelle and the rotor about the vertical tower axis for their orientation with respect to the wind direction. The main purpose is to maximise the power generation, but some very small microturbines may also use yaw as a means of power regulation. All commercial grid connected wind turbines have 'active yaw' as powered by motor drives under the nacelle.

An active yaw system employs a mechanism of hydraulic or electrically driven motors and gears to yaw the turbine (Figure 3.4). The *number of yaw drives* can be determined by the conditions on the site, but is normally 4. Together with yaw brakes (to keep the nacelle in position and to spare the gears) and yaw bearings, these components are most often delivered as standard components from a supplier and may be exposed to conditions which have not been taken into account by the supplier. The yaw drives and yaw brakes are generally made by forged-steel. The yaw drive gears have hardened teeth (by carburizing or other form of heat treatment) to resist the wear from extensive loads.

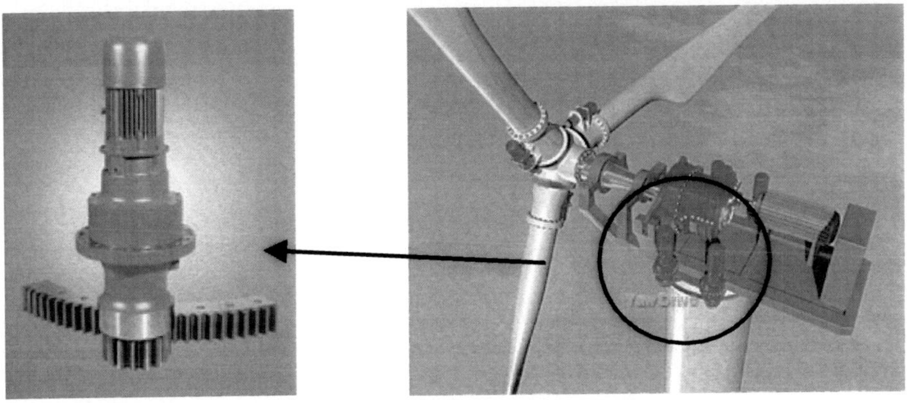

Figure 3.4. The Bosch yaw mechanism [credit www.boschrexroth.com].

Yaw drive manufacturers for wind turbines	Location
Parker Hannifin Corporate	OH (US)
Rexroth Bosch AG	DE (EU)

3.5 Bearings

Bearings are required for the shafts, gearbox, yaw and pitch mechanisms, generator, and other rotating parts. Larger wind turbines generally use *four-point contact ball bearings* (Figure 3.5) for the blade retention pitch-bearing and for the yaw bearing positions, and *spherical roller bearings* to support the main gear (Figure 3.5).

The bearing consists of two, ring carbon steel rolled forgings with carbon or stainless steel balls. The inner and outer rings are hardened to provide adequate yield and fatigue strength. The actual ball or roller path is induction heated and quenched to provide a hard case. The surface hardness of the race path is RC 58 minimum. The depth of the hardened case is defined as the depth to a hardness of RC 50. Most yaw and pitch bearings are supplied with plastic *spacers between each rolling element*. A four-point contact ball bearing has cylindrical spacers with hemispherical indented ends. The cross roller bearing has plastic saddle type spacers which conform to the two adjacent rollers with axes of rotation 90 degrees apart. The ball spacers for larger balls (5 cm diameter and over) often have a steel plate reinforcement cast into the plastic [3.4,3.5].

Proper lubrication are required to obtain a correct working of bearings. The main purpose of lubrication is to create a lubricant film between the rolling elements to prevent metal-to-metal contact. This is to avoid wear and premature rolling bearing fatigue. Additional functions may include protection against corrosion (marine environmental). When selecting lubrication, the operating temperature range have to be consider: low temperature (that may cause high viscosity of lubricant), and high temperature (that may cause low viscosity of lubricant) should affect the operating characteristic of bearings. Grease (a base of oil with thickeners and possibly additives added) is the most commonly used lubrication in the main shaft bearing. A possible choice of grease would be lithium soap-base grease (waterproof and usable at a wide of range temperatures –35°C to +130°C). Other possibility are sodium grease or calcium soap base grease [3.2, 3.5].

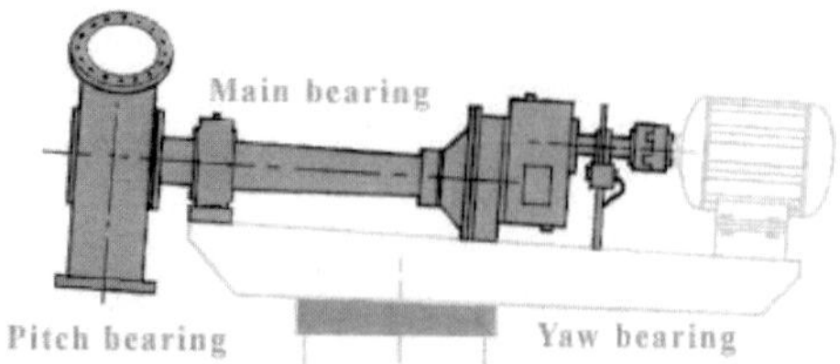

Figure 3.5. Bearing s in large wind turbines: four-point contact ball bearings in the pitch and yaw drives, and spherical roller bearings in the main shaft.

Bearing providers for wind turbines	Location
Kaydon Bearings Division	MI (US)
SKF	SE (EU)
TRI Transmission and Bearing Corp.	PA (US)

3.6 Couplings

Couplings connect two shafts together for transmitting torque. There are two locations in particular where large couplings are likely to be used in wind turbines: between the main shaft and the gearbox, and between the gearbox output shaft and the generator.

Couplings consist of two major pieces in steel, one of which is attached to each shaft. An example of a solid coupling is shown in Figure 3.6. The two pieces are in turn connected to each other by bolts, which are in general prestressed. Normally austenitic stainless-steel bolts are used and different surface treatments are required to achieve specific corrosion protection [3.2].

Figure 3.6. Typical solid coupling in wind turbine application [Credit www.mayr.de].

Coupling provider for wind turbines	Location
Flender GmbH	DE (EU)
Mayr GmbH	DE (EU)

3.7 Mechanical Brakes

Mechanical brakes are usually used as a backup emergency system for the aerodynamic braking system and/or as a parking brake, once the turbine is stopped, e.g. for service. Mechanical brakes are also used to lock the yaw position (paragraph 3.4), e.g. at maintenance. The pictures in Figure 3.7 show a yaw mechanical brake example.

In a mechanical brake, brake clippers, brake discs and brake pads form crucial parts. A hydraulic system is usually used for the actuation and release of the brake.

Brake discs and brake pads must be able to withstand temperature loading, since the friction during braking leads to dissipation of energy as heat, so causing local high temperatures. Brake pads may be made of different kinds of material and require temperature design calculations and temperature measurements. . For instance, ceramic pads lose their frictional resistance at temperatures in excess of 300 to 400°C. For high temperatures, brake pads are made of sinter bronze [3.2]. Otherwise, brake discs are generally made in carbon steel, with an appropriate thickness to absorb temperature loading.

Brake springs can be made in a variety of materials, but the most common is steel, as spring steel wire [3.3].

Figure 3.7: A typical yaw mechanical brake [credit www.hanning-kahl.de].

Mechanical brakes providers for wind turbines	Location
HANNING & KAHL GmbH & Co	DE (EU)
Hilliard Corporation	NY (US)
Parker Hannifin Corporate	OH (US)

3.8 Nacelle Cover

Nacelle casing are made of glass fibre, with steel reinforcements, and fixed to the main frame with steel supports through rubber dampers. The nacelle frame (Figure 3.8) is made in ductile casting iron. The nacelle enclosure must be designed for the wind loads, with an inside height to allow maintenance staff to stand when servicing components in bad weather. Often, however, the nacelle roof can be opened when the weather allows this.

Cooling and ventilation are controlled within the nacelle, particularly for the gearbox and generator. By control of the cooling air to and from the gearbox and generator, the nacelle temperature will generally be kept at least 7°C above outside temperature, thus *preventing condensation and subsequent corrosion.*

It is important that the nacelle enclosure can be opened to allow for removal of damaged components for replacement, for example, by a hoist through a hatch or by a helicopter through the roof.

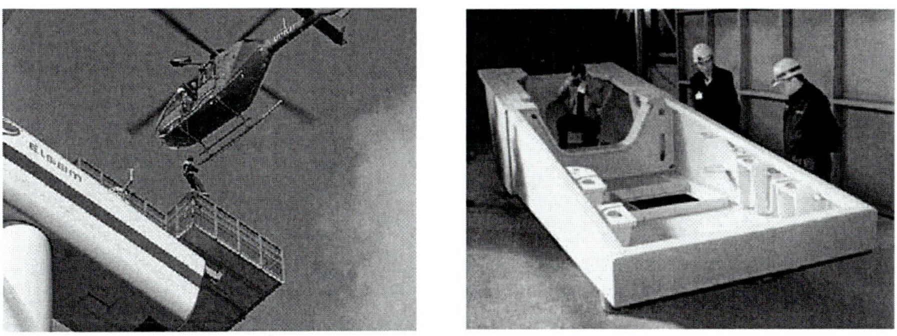

Figure 3.8. The nacelle frame of a wind turbine [Reproduced by permission of CAB Incorporated, www.cabinc.com]

Nacelle component wind turbine manufacturers	Component	Location
Bach Composite Industry A/S	Casing	DK (EU)
DIAB Inc.	Casing	TX (US)
TPI Composites Inc.	Casing	RI (US)
Hodge Foundry Inc	Frame	PA (US)
CAB Inc.	Frame	GA (US)
K&M Machine-Fabricating, Inc.	Frame	MI (US)

4. SUPPORTING STRUCTURE MATERIALS

4.1 Tower

For offshore wind turbines, the towers themselves and the connection pieces between the foundations and the towers, are made of *hot-rolled soft-carbon steel* (Figure 4.1). Reinforced concrete towers would allow a novel and cheaper approach, now being considered for offshore systems. Pre-stressed concrete is used in order to obtain the required tower stiffness, especially in near-shore [4.1, 4.2].

Tubular steel towers are conical (i.e. with their diameter increasing towards the base) in order to increase their strength and to save materials at the same time. These towers are bent and welded, and manufactured in sections of 20-30 m, with flanges at both ends to allow neighbouring sections to be bolted together on site. For flanged connection, *corrosion-resistant steel bolts* are generally used [4.3]. An alternative concept would be to weld sections together on site for a stronger connection. However, for all metal components, it is essential to choose suitable welded joints, bolts, and adhesion of metal to *avoid corrosion* when metals with different standard electric potential come into contact. Steel tower admissible stresses are determined by the quality of the welded unions, for which both the ultimate tensile stress and the admissible fatigue stresses are least [4.1].

Figure 4.1. A section of steel tower for large wind turbines.

The external and internal surfaces of the tower, the connection components between tower and foundation, and the access platforms have to be corrosion protected. This is usually by *special paints* that minimize corrosion and erosion caused by continual wave action, salt spray and wetness.

Low temperature conditions, as in the North Sea oil and gas exploitation, can occur (i.e. typically down to 5°C during day and often sub-zero cure overnight). Historically, the protection of offshore steelwork can be traced over 100 years to the shipping industry, with the acceleration in the offshore oil and gas industry over the last 40 years. Offshore windfarms have several similarities with these proven technologies, thus adopting their coating system to provide extended lifetimes to first maintenance. Coatings based on polyurethane or epoxy resin have long-term stability. Due to the dilute solid contents of these coatings, usually between five and seven coats are applied to obtain the required total dry film thickness. A significant improvement to these systems is to utilize zinc silicate or zinc epoxy as a primer (thus improving both under-film corrosion creep and long term corrosion resistance), a tie-coat, an intercoat epoxy-based paint and a polyurethane-based topcoat. Depending on the coating system used, the total dry film thickness may be from 275 micron to 335 micron thick [4.4, 4.5].

Tower manufacturers for wind turbines	Location
Beaird Company, Ltd.	LA (US)
Morrison Berkshire, Inc.	MA (US)
DS SM A/S	DK (EU)
Flanschenwerk Thal GmbH	DE (EU)

Steel Tower

Attack mechanism	*Protection solution*		
	Paint	*Additional cover protection*	*Material characteristics*
Saline air and/or frost corrosion	Polurethane or epoxy resins + zinc silicate or epoxy primer (~0,3 mm)	None	Soft steel (S275, S235, S355)
Severe seawater corrosion Seawater loads	Glass fibre reinforced epoxy paint (~1,5 mm)	Antifouling Corrosion allowance Cathodic protection systems	

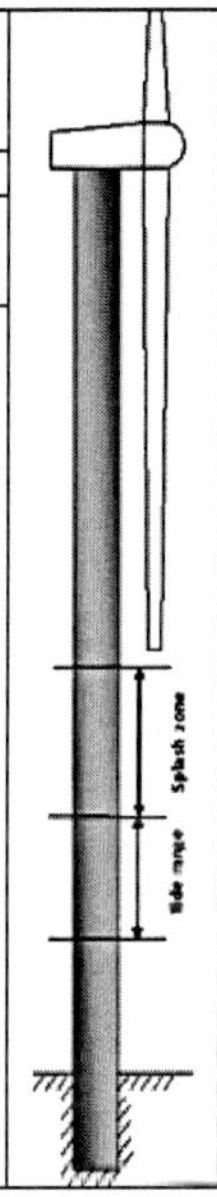

Reinforced Concrete Tower (class E DS411)

Attack mechanism	*Protection solution*				
	Paint	*Steel characteristics*		*Concrete characteristics*	
Frost damage		Concrete cover > 40 mm	40 mm prestressed steel reinforcement	Water/cement ratio < 0,4 (class E) Air pore additive (for freeze/thaw protection)	*Concrete density* 320-500 kg/m3
Lime leaching (dissolving calcium hydroxide from the concrete matrix) Seawater corrosion	Glass fibre reinforced epoxy paint	Concrete cover > 50 mm Concrete cover > 40 mm	50-100 mm prestressed steel reinforcement		*Concrete density* 400-500 kg/m3

Figure 4.2. Scheme representing environmental problems and currently used mitigation solutions for steel and concrete towers (modified from ref. [4.6]).

Figure 4.2 is a schematic table summarizing the mentioned environmental problems and mitigation solutions for steel and concrete towers. The tower is divided in four segments, that remaining permanently beneath the sea surface (permanently wetted), the tidal range area, the splash zone, and the atmospheric area. For each area, the type and the severity of the attack mechanism changes, thus requiring different protection solutions. The solutions vary markedly for steel and concrete towers. Composite towers, with rational use of concrete and steel in the same structure can lead to the optimum use of peculiar properties of each material.

4.2 Foundation

The foundations of offshore wind turbines have their upper structural parts within the wave splash zone, which is usually defined as the distance between (1) the normally occurring high water level *plus* the significant wave crest height, and (2) the normally occurring low water level *minus* significant wave trough depth. Maximum corrosion is located in this zone. Once a foundation concept has been selected, the corrosion protection has to be planned during the foundation design phase (see also Figure 4.2).

Offshore wind turbine foundation structures experience loads both from the turbine and from the sea current, waves and icing in the marine environment. These structures have to be corrosion protected so damage does not occur within the expected lifetime of the structure for which reductions in safety standards are not acceptable.

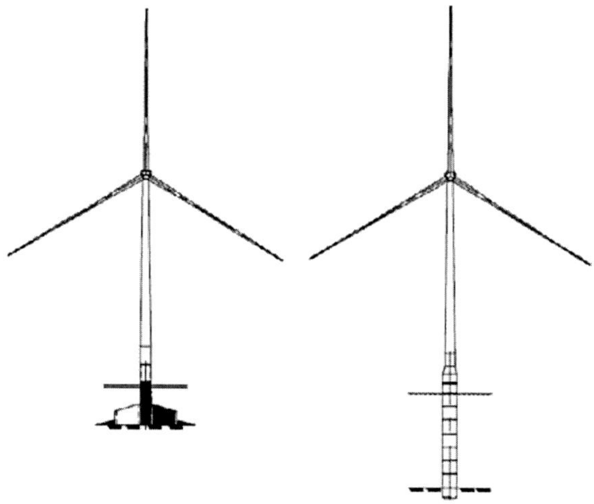

Figure 4.3. Gravity (left) and monopile foundation concepts (right) [4.7].

Two different foundation concepts currently used for offshore wind turbines, are illustrated in Figure 4.3:

- *gravity structures*, made in reinforced concrete material (in the past) or steel (cylindrical steel tube placed on a flat steel box);
- steel *monopile* structures.

Gravity base foundations have been developed for shallow sheltered waters. The gravity foundation has been utilized for several offshore European wind farms that have bottom characteristics that preclude the use of monopiles (e.g. ledge, rocks or soil characteristics unsuitable for piles). The Middelgrunden wind farm in Denmark represents the largest gravity foundation utilized is on a 2 MW wind turbine. Experience has shown that this solution is not suitable for larger offshore wind farms. They require a shipyard and dry dock near the site to construct and allow the massive foundation structures to be floated out to the site and sunk. The gravity structure design also presents a much greater environmental impact than a monopile due to the large diameter (approximately 18 m).

The ***monopile foundation*** is the most common technology in existing parks. The monopile is simply a large diameter pile (4-5 m), driven 15-30 m into seabed depending on the local load-bearing characteristics of subsurface marine sediments. It is open-ended, allowing sediment to be encased within the monopile to provide for additional structural support. As the water depths and loads increase, the pile diameter and thickness increase as well.

Because of the popularity of this type of structure, it will be the main subject for the development of the design philosophy. It is expected that the monopod will prove to be too soft when the top mass and water depth increase further, demanding the application of the stiffer *multipod foundation* (generally tripod type, Figure 4.4). This concept provides a very stiff foundation [4.5], and is governed by the fatigue loading and the high stress concentrations inherent at welded tubular joints.

The ***lattice structure*** is a similar concept (Figure 4.4). The recently erected offshore wind turbine REpower 5 MW (rotor diameter of 126 m) sits on a steel lattice-jacket structure, piled to the seabed at a depth of 44 meters in the North Sea, off Aberdeen, Scotland.

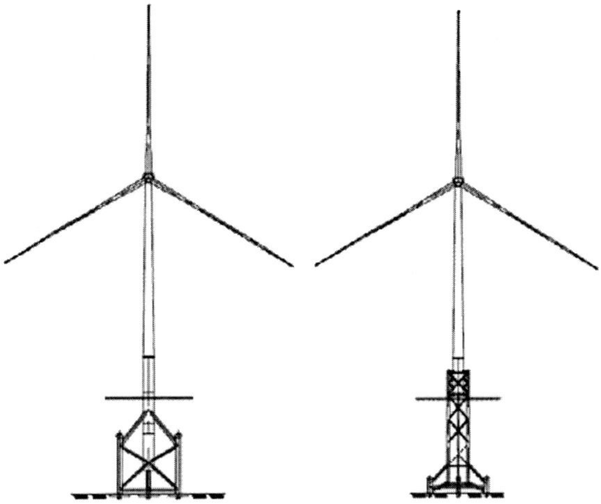

Figure 4.4. Tripod (left) and lattice structure foundation concepts (right) [4.7].

Alternative solutions, inspired by the oil platforms and therefore used to great depths, but still under study are ***floating foundations***. Blue H Technologies company has launched a test an 80 kW Submerged Deepwater Platform (SDP) off the coast of Puglia, Italy. The technology uses a tension-leg platform design (Figure 4.5) which relies on a S355 steel platform, cathodic protected (Zinc) that floats below the surface of the water. It is moored to the bottom with steel chains.

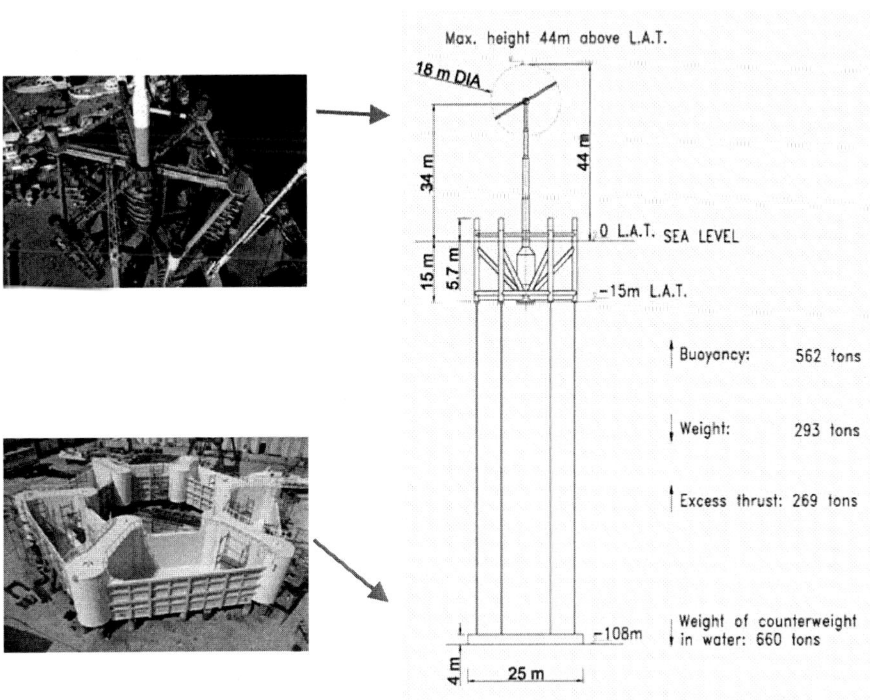

Figure 4.5. The large scale prototype Submerged Deepwater Platform (SDP) of Blue H [Reproduced by permission of Blue H, photo G: Radogna].

4.2.1 Concrete foundation

The *steel reinforcement* in reinforced concrete structures should be protected against corrosion by ensuring that there is at least the minimal required concrete cover thickness. In the splash zone, this minimum is 50 mm, and outside the splash zone 40 mm. The cover should form a dense concrete structure, with limited crack formation and crack widths. This can be ensured by an adequate concrete composition of environmental class E (high class supervision in DS411 specifications), with the following minimum requirements [4.3]:

- *the characteristic compression strength of the concrete,* f_{ck} *> 40 MPa;*
- *water/cement ratio for the concrete < 0.45;*
- *maximum aggregate size* d_{max} *< 32 mm or minimum distance between reinforcement bars;*
- *maximum distance between non-prestressed reinforcement bars is 150-200 mm;*
- *application of reinforcement with relatively insignificant reinforcement diameters (12-20 mm) to the extent possible.*

Additional corrosion protection should be considered, such as *cathodic protection* (by sacrificial anodes, plates made from magnesium or zinc, which corrode in preference to the steel), curing membrane on the reinforcement and application of stainless steel reinforcement on exposed parts. Furthermore, in the splash zone, glass-fibre-reinforced epoxy based paint should be considered for surface protection of the concrete. Alternatively, a corrosion-protection steel hood can be used around the foundation [4.3, 4.8, 4.9].

4.2.2 Steel foundation

For *unprotected steel*, generally assume that the rate of corrosion in the splash zone is in the range of 0.3-0.5 mm/year [4.9], and the corrosion rate will increase proportionally with the age of the structure [4.3]. Moreover, a corroded surface generates an area with very large stress concentrations and so cracks will propagate faster in corroded steel than in un-corroded steel. Considering these properties, reliable corrosion protection is necessary.

In the *splash zone*, recognized design practice involves the application of a '*corrosion allowance*' as the main method of corrosion protection, i.e. the wall thickness is increased to allow for corrosion. The particular corrosion allowance should be assessed in each particular situation.

In addition to the 'corrosion allowance', glass-fibre-reinforced epoxy paint can be used, such as 1.5 mm multi-layer glass flake epoxy or polyester, or 450 micron multi-layer high-build epoxy, or 200 micron thermally sprayed aluminium with a sealer coat [4.5]. It is normal practice to assume that the surface treatment reduces the rate of corrosion.

Below the "splash zone" of the foundation, steel surfaces having mechanical loads from seawater are protected primarily by *cathodic protection*, i.e. sacrificial anodes (Zn-Al-In type) and/or by impressed electrical current, supplemented by surface treatment. For larger structures, sacrificial anodes cannot deliver enough current economically to provide complete protection. Impressed Current Cathodic Protection (ICCP) systems use anodes (include high silicon cast iron, graphite, mixed metal oxide, platinum and niobium) connected to a DC power source.

All submerged zones must also be coated. In zones around the seabed, it is recommended that cathodic protection is combined with corrosion allowance. Depending on the type of sediment and seabed water currents, those parts of the foundation below the seabed (to 2-10 m) may need corrosion protection.

Wind turbine foundation manufacturers	Type	Location
DS SM A/S Steel	foundation	DK (EU)
Bladt Industries A/S Steel	foundation	DK (EU)
Densit A/S Concrete	foundation	DK (EU)

5. COLD CLIMATE ISSUES

Wind turbines in some location are exposed to:

- *sea ice* (flows, driving ice, land-fast ice);
- *atmospheric icing* (due to water in the air, as mists, fogs, rainfall and sea sprays) which may potentially lead to ice formation on turbine structures (Figure 5.1);
- *low temperatures*, that affect the various wind turbine materials (metals, composites and rubbers) and lubricants. Thus, all wind turbines have an operating and standstill (non-operating) minimum temperature limit. This limit is specified by the manufacturer. It is required to ensure the safe and reliable operation of the equipment with acceptable degradation of equipment life.

Cold climates are characterized by either icing events or temperatures that are lower than the operational limits of standard wind turbines. Cold climate issues could be a big concern for some planned offshore wind parks, that will be sited above the 52° latitude parallel [5.1].

Figure 5.1. Floating sea ice pushed from wind to shore in the northern part of the Baltic Sea (top picture) and heavy iced rotor blade (bottom picture) due to low temperature (below zero Celsius degrees) and high humid environment.

Typical *minimum operating temperature* of a commercial wind turbine, without any special temperature package, is -20°C, with a 'standstill temperature' of -30°C. It is possible to decrease the operating temperature limit through the purchase of a cold weather package, whereby the utility-scale wind turbine generator is typically capable of operating to -30°C and at standstill temperatures as low as -40°C (as shown in Table 5.1). With additional measures, it is possible to decrease the temperature limit further [5.2] Typical Cold Weather Packages (CWPs) generally include:

- *lubricant heaters (usually in gearbox);*
- *additional generator heaters;*
- *control cabinet heaters;*
- *nacelle space heaters;*
- *ice detectors;*
- *special alloy ductile iron for hub and machine frame*;
- *special alloy tower steel;*
- *improved nacelle sealing;*
- *low temperature lubricants;*
- *anemometer and weather vane equipped with heaters.*

Table 5.1. Manufacturers' specified minimum ambient operating/standstill temperatures of utility scale turbines with a Cold Weather Package [5.2].

Wind turbine with CWPs	Operating temperature	Standstill temperature
Vestas Turbine (1.8 MW, V80/V90)	30°C	40°C
GE Turbine (1.5 MW, 1.5S/SL)	30°C	40°C
NEGMicon Turbine (NM72/82,1.65 MW)	30°C	40°C
MHI (Mitsubishi) (1 MW, 1000/1000A)	40°C	N/A
Gamesa (G52/G80)	30°C	N/A

For many materials, short-period temperatures, such as daily minimum temperature, have too short a duration to pose any problems for the materials and their behaviour. Moreover, the duration can be too short to heat up the materials in response. Therefore, it is more apt to consider minimum values of the one-hour mean temperature.

During low temperatures, steel becomes more *brittle*, resulting in its energy absorbing capacity and deformation being reduced; this can result in failure. Additional stress on composite materials, i.e. blade materials, can result in *micro-cracking* in the material. Power applied to electrical equipment at cold temperatures can cause windings to suffer thermal shock and become damaged. The next section reviews these material issues.

5.1 Composite Materials

The performance of FRP (*fibre reinforced plastic*) composites can be adversely affected by their environment. *Subzero temperatures and freeze-thaw* conditions can cause matrix hardening, matrix micro-cracking and fibre-matrix bond degradation that affect mechanical properties. Moreover micro-cracking and 'brittling' of the polymer matrix when cooled, leads to very variable characteristics, as reported in the literature [5.3, 5.4, 5.5]. Such variability is a strong function of the quality of manufacture (for example, percentage of void and resin-rich areas), fibre fraction, orientation and distribution. As example, Figure 5.2 shows four types of fibreglass epoxy composites tested in tension at room temperature (23°C) and at low temperatures (-5°C, -20°C, -40°C). Such composite data are customarily related to fibre orientation [5.3].

Some composite changes at low temperatures can be explained by the stiffening of the matrix. Consequently, at low temperature, the stiffness of the FRP increases and, generally, the resulting failure modes are more related to brittleness than for specimens at room temperature [5.4]. This means that FRP at low temperature fails more violently, which should be considered for structural design.

Composite materials, due to unequal shrinkage of their fibre/matrix components, are subjected to residual stress during freeze-thaw. If this stress is sufficient, it can result in micro-cracking in the material. These micro-cracks reduce both the stiffness and the permeability of the material, which hasten deterioration [5.5]. One way to prevent this would be to use fibres and a matrix that have similar thermal expansion coefficients.

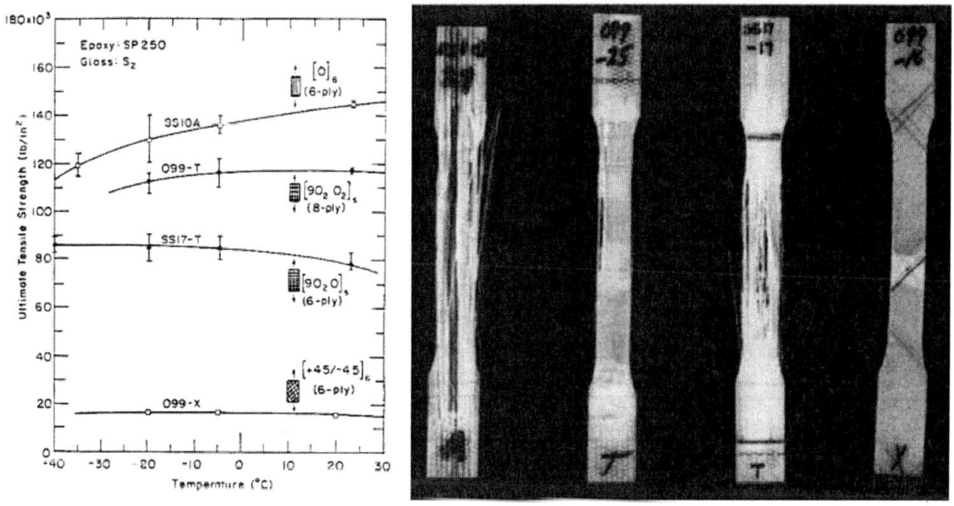

Figure 5.2. Temperature effects on strength of fibreglass-epoxy laminates (graph left) and the specimens tested (right photo) [5.1].

5.2 Metallic Materials

The effect of subzero temperatures must be considered in the selection of those structural and mechanical parts that are generally made in low and high carbon steel (tower, shaft, ball bearing), cast iron (hub, parts of gearbox, generator, main bearing, nacelle) and copper (generator parts). Low temperatures, high stress states, and fast loading rate may all cause a marked change in the fracture resistance of some metals is observed in service. The transition from ductile to brittle behaviour is called *ductile to brittle transition* (DBT). This is characteristic of metals with body-centred-cubic (bcc) crystal structures.

The transition of ferritic steels from ductile to brittle behaviour often occurs over a small temperature range, named as 'transition temperatures'. For this reason, at low-temperature *carbon steel* could become less ductile, as illustrated in Figure 5.3 and become brittle [5.7, 5.8]. Thus, its energy-absorbing capacity and its deformation prior to failure are both reduced. In the brittle state (the worst case) the strain is close to zero and steel breaks as if glass. A collision between an ice pack and a low-carbon steels turbine tower at temperatures below the standstill temperature might be disastrous. For the same reason, it is preferable to avoid erecting towers in extreme low temperatures [5.9, 5.10].

Metals have been used in low temperature conditions for many years. For instance, it is well documented that alloys, such as nickel and aluminium, improve the strength of steel at low temperatures. For applications involving exposure to temperatures below 0 °C, the ferritic nickel alloys usually are considered first, if they have sufficient corrosion resistance. The use of cold resistant steel in all structural members with welds does not increase the costs significantly. Standard hot-dip galvanized bolts have proven adequate in low temperatures [5.7].

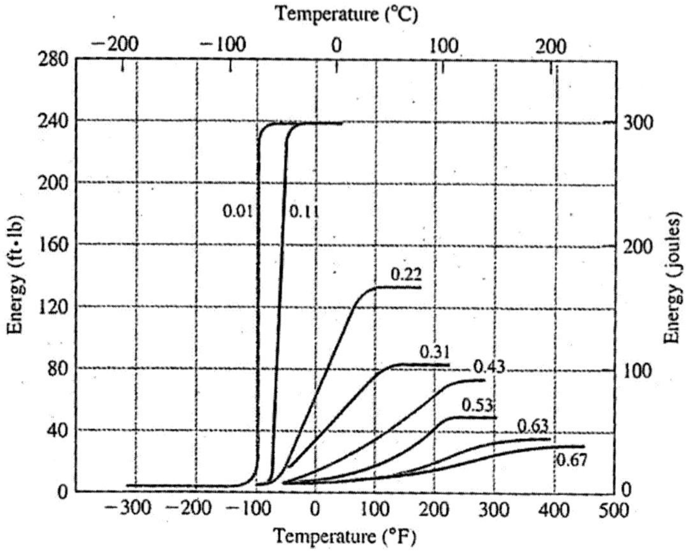

Figure 5.3. Effect of carbon content (0.01-0.67%w) on the impact energy versus temperature [5.11].

If the ductile-to-brittle transition corresponds to an increase in elastic modulus, variations in the natural oscillatory frequencies (ω_I) of components may occur. For example, if a wind turbine tower is considered as a tube, its first natural tower-bending frequency (rad/s) can be estimated by the following formula [5.12]:

$$\omega_I = 1.75\sqrt{\frac{EI}{h_H^3(m_{top} + m_T/4)}}$$

Here E is the elastic modulus (which may change dramatically at a temperature transition), I is the second moment of area, h_H is the hub height, m_{top} is the top mass and m_T the tower mass. Designers should verify that w_I of the tower does not ever coincide with the rotor and blade-passing frequencies, i.e. the 1P and 3P frequencies respectively, so allowing enforced oscillations to failure.

5.3 Other Materials

Gearboxes, hydraulic couplers and dampers may suffer from long exposure to cold weather. As the temperature decreases, the viscosity of the *lubricants and hydraulic fluids* increases. Heating and less viscous lubricants should be used to facilitate the normal operating phase [5.13].

At low temperatures, the *flexibility of rubber* (in seals, cushions, etc.) can be lost. High mobility of the molecular segments in a polymer is necessary for the rubbery state, i.e. high elastic elongation. The rubber's elongation is combined with a light cross-linking of the molecular chains. With decreasing temperature, the movements of the segments are reduced. At a certain temperature (the glass transition temperature, T_g), movement of the molecular segments is completely frozen and the material becomes stiff, brittle and plastic-like, with low elongation at breaking point. Selection of rubbers with an appropriate T_g (as phenylmethyl-dimethyl silicones rubbers) may insure that seals and other rubber parts retain their elasticity and prevent their brittleness at low temperatures [5.13].

6. OFFSHORE WIND TURBINES MATERIALS SELECTION PROCESS

Minimization of cost of energy production over wind turbine expected lifetime is achieved by rational design based on:

1. *reduced system complexity (reduced number of components);*
2. *reduced mass of components.*

These aspects affect manufacturing, handling and transportation costs, the latter being of utmost importance for offshore. Innovative materials offer an effective tool for the designer to perform this task.

Wind turbine complexity and mass (weight) can often be reduced by using higher strength materials and the innovative shape and functional capability offered by new manufacturing processes.

Effective forward steps need rational criteria for material selection. Beside of the material properties, the function of the component, the shape and the manufacturing process have to be taken into account. Among any possible material, the choice can be narrowed by applying primary constraints dictated from the design, and then by seeking the subset of materials which maximize the performance of the component. Primary constraints are given by non-negotiable characteristics of the design, as the loads the turbine must withstand, the dimensions, the typical characteristics of air and water of the marine environment, etc. With these specifications, only some subset of materials will result as applicable. Further narrowing will come from seeking the combination of physical and/or chemical properties which maximise a given performance.

Such combinations, called *performance indices*, can guide the designer to select the best materials for light stiff structural components, as function of dimension, loading requirement and environment characteristics. This procedure, introduced by M. F. Ashby [6.1, 6.2, 6.3, 6.4], provides a general method allowing the optimal subset of materials to be individuated. This procedure will be adopted to the design of typical offshore wind turbine tower and blade. The focus of the next sections will be on materials selection. Design and dimensioning procedure of the structures is beyond the scope of this chapter and will be introduced at a very simplified level to discuss the physical parameters involved the component performance maximization. In general, material design will be exposed for a specific load case, but the procedure can be extended by including more refined load models.

6.1 Tubular Tower Material Design

The main aspects involved in tower material selection are strength, buckling of the shell wall in compression and stiffness requirement to adjust the material frequency. Therefore, the tower can be modelled as a *loaded cantilever beam* as shown in the pictures of Figure 6.1 showing a monopile foundation. According to the model, the main external forces acting on the tower turbine structure are:

- *gravity forces* due to top (rotor and nacelle) mass ($F_{G,H}$) and tower mass ($F_{G,T}$),
- wind forces on the rotor ($F_{A,R}$) and on the tower ($F_{A,T}$),
- *wave and current forces on the tower (F_W),*
- *ice sheet force* on the tower base ($F_{ICE,T}$)
- additional *gravity forces* due to mass of ice accretion on the rotor and nacelle (if the case).

Stresses associated with rotor rotation and yaw motion need not to be considered in parked condition.

Since the tower is the heaviest component, tower mass minimization in a prior design target in offshore applications. This will reduce not only fabrication but also transport and handling costs. The following simplified equation can be utilized to calculate the tubular tower mass (m_T):

$$m_T = \pi\, D_T\, h_T\, t\, \rho_T \tag{6.1}$$

where D_T is the mean tower diameter, h_T the tower height, t the mean tower wall thickness, and $_T$ as the material density. By considering the taper of the tower, the mean diameter appearing in eq. (6.1) can be as following written:

$$D_T = \frac{D_{top}}{2}\left(\frac{D_{foot}}{D_{top}} + 1\right) = \frac{D_{top}}{2}\left(f_D + 1\right)$$

where f_D is the taper factor of the tower structure (very modest). Mass minimization of complex engineering structures as the tubular towers is a typical over-constrained design: statically speaking, strength, stability/bucking and stiffens have to be verified for light design. A multi-constraint analysis has therefore to be carried out to meet these requirements.

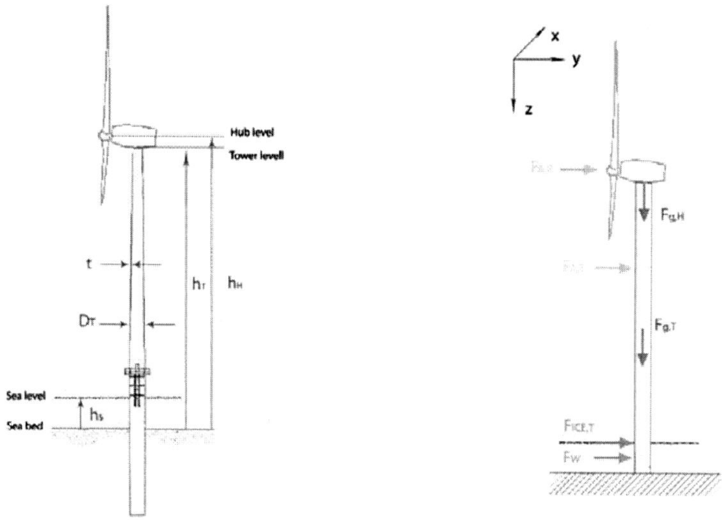

Figure 6.1. Geometric parameters of tubular tower simplified model (left) and the cantilever beam simplified model (right) of a parked wind turbine.

The first constrain is the condition on the axial yield stress:

$$\sigma_d \leq \frac{\sigma_f}{\gamma_f \gamma_m} \tag{6.2}$$

Maximum design axial stress (σ_d) in the tower section have to be less than material allowable stress (γ_f) reduced by proper loads (γ_f) and material (γ_m) partial safety factor. Referring to Figure 6.1 and assuming that the height of the tower h_T is approximately equal to the hub height h_H from the mudline, the equation (6.2) becomes:

$$\frac{F_{G,H} + F_{G,T}}{?D_T t} + \frac{F_{A,R} h_h + F_{A,T} 2/3\, h_h + F_{ICE,T} h_s + F_W 2/3\, h_s}{?\frac{D_T^2}{4} t} \leq \frac{\sigma_f}{\gamma_f \gamma_m} \tag{6.3}$$

Since the stress due to the top mass is negligible compared to stresses resulting from rotor axial forces, and the wind loading on the tower is negligible compared with the rotor loading, then the tower base moment is mainly due to the aerodynamic force on the rotor ($F_{A,R}$). The equation (6.3) simplifies to:

$$\frac{F_{A,R}h_H}{\pi\frac{D_T^2}{4}t} \leq \frac{\sigma_f}{\gamma_f\gamma_m} \tag{6.4}$$

A system of equation (6.1) and (6.4) gives the following expression for the mass:

$$m_T = \frac{\gamma_m}{D_T} 4F_{A,R}h_H^2 \frac{\rho_T}{\sigma_f} = \left(F_{A,R}\gamma_f\right)\left(\frac{h_H^2}{D_T}\right)\left(\frac{\rho_T}{\sigma_f}\gamma_f\right) = F\left(L,G,M\right) \tag{6.5}$$

where loading ($F_{A,R}$ safety factors), geometrical (D_T, h_H), and material parameters are grouped to give rise to a functional equation expressing the dependency of the mass on loading [L], geometrical [G] and material [M] indexes.

The hub height and aerodynamic force can be considered as proportional to rotor diameter [6.5] considering the follow scaling rules:

$$h_H = \frac{D_R}{2} + h_{wave,max} \tag{6.6}$$

$$F_{A,R} \propto D_R^{2\alpha+2} \tag{6.7}$$

where D_R is the rotor diameter, $h_{wave,max}$ the maximum wave height and α the wind shear exponent. Inserting the variables of eq.s (6.6) and (6.7) in the (6.5), the latter becomes:

$$m_T = 4\gamma_f\left(\frac{D_R}{2} + h_{wave,max}\right)^2 \frac{D_R^{2\alpha+2}}{D_T} \frac{\gamma_m}{\frac{\sigma_f}{\sigma_T}} \tag{6.8}$$

In particular, if we consider the loading and the geometrical indexes as not adjstable, the minimization of the tower mass can be achieved by the choice of:

1. a material that maximizes the first index parameter M_1 generated by the constraint on the axial stress verification:

$$M_1 = \frac{\sigma_f}{\rho_T} \tag{6.9}$$

2. a processes allowing assuming smaller safety factors which depends on the knowledge of the material and the fabrication process of the component. The presence of imperfections, in particular those introduced by welding, reduces substantially the tower wall compression resistance and each national code gives specific rules to take this into account.

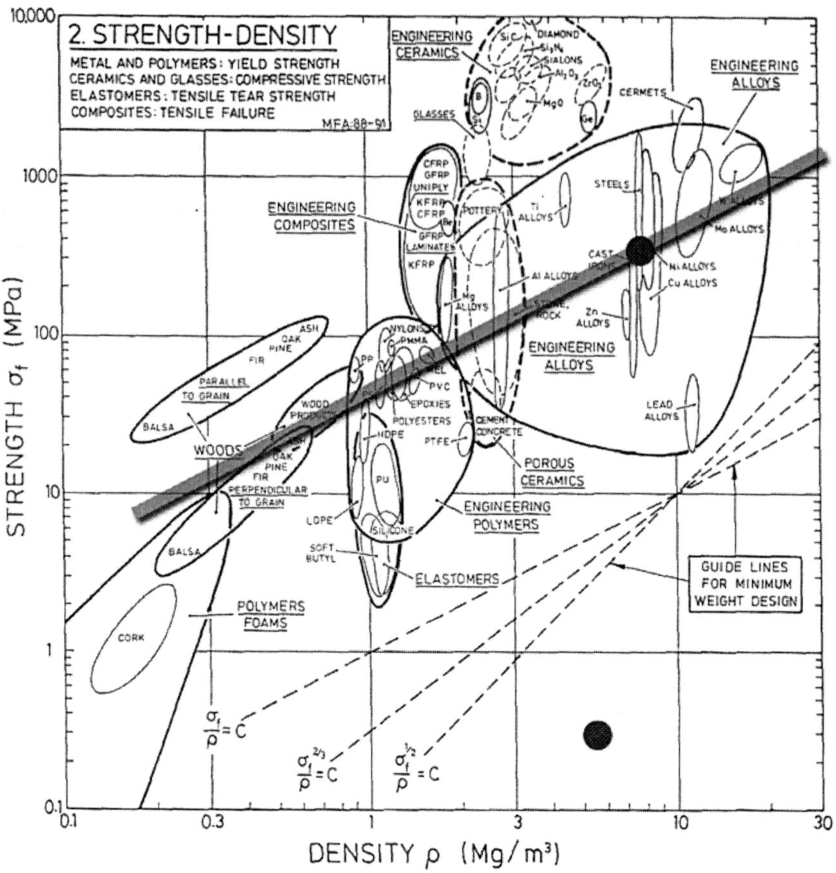

Figure 6.2. Ashby material selection chart with strength on abscissa axis and density on ordinate axis. The shaded blue band shows the primary constraints. [Reproduced by permission of Prof. M.F.Ashby].

Candidate materials for the optimization process can be obtained by the help of the Ashby charts, and in particular, the one shown in Figure 6.2, where failure strength (_f) is plotted vs density (_) for several materials on logarithmic scales. Specific material boundaries in Ashby charts are changing with technology advancements. New materials and applications (especially true for composites) appear in the market and the boundaries of the regions displaced in the charts are therefore changing with time.

By taking the logarithm of both left and right hand side of eq. (6.9) we can reproduce on the Ashby chart the family of straight parallel lines of slope 1:

$$\log(\sigma_f)=\log(\sigma_T)+\log M_1 \tag{6.10}$$

We shall refer to these lines as *design guidelines*. It is easy to read off the materials which optimally meet the primary constraint for the given loading and geometrical indexes. All the materials that lie on a line of constant σ_f/ρ_T will perform equally; those above such line will perform better, those below, worse.

Suppose now we have a satisfactory tower design made by an ordinary material, such as S355 steel. This steel is a common choice because it has good mechanical properties/weight ratio, it is easy to be tooled and is relatively cheap. For S355, $\sigma_f = 355$ MP and $\rho = 7.85$ Mg/m^2 and therefore $M_1 = 45$.

It is now possible to draw the primary constraints shown as a blue guideline and that goes through M_1=45 in Figure 6.2.

Stability/buckling analysis for the tower introduces a second constraint in the mass minimization problem. The strength of a cylindrical tower in axial compression is less than the yield strength, because the structure will buckle elastically when the Euler loading (F_{CR}) is exceeded by compressive loads:

$$F_{G,H} \leq \frac{F_{CR}}{\gamma_f \gamma_m} = \frac{EI}{\gamma_f \gamma_m} \frac{\pi^2}{4h_T^2} \tag{6.11}$$

where E is material elastic modulus and the second moment of area I is given by:

$$I = \pi \left(\frac{D_T}{2} \right)^3 t \tag{6.12}$$

Substituting (6.11) in the equation (6.1), the tower mass results:

$$m_T \propto (\gamma_f g) \left(\frac{h_T^3 D_R^{2.63}}{D_T^2} \right) \frac{\gamma_m}{\left(\frac{E}{\rho_T} \right)} \tag{6.13}$$

A second material performance index M_2, is now defined as:

$$M_2 = \frac{E}{\rho_T} \tag{6.14}$$

By assuming E = 205 GPa for S355 carbon steel, this yields: M_2 = 26.1.

When both M_1 and M_2 are shown in an appropriate chart (specific Young's modulus against specific strength) (see Figure 6.3), a new candidate subset of concurrent materials is identified for the minimum mass design:

- *cermets and some special engineering ceramics;*
- *Beryllium and Beryllium engineering alloys;*
- *unidirectional GFRP and CFRP composites;*
- *High Performance Concretes.*

Cermets and ceramics are too brittle and are weak in tension, therefore they are not suitable for wind turbine tower applications.

Beryllium is one of the lightest metals and has a high strength-to-weight ratio and stiffness-to-weight ratio. It has significant ductility, is readily machined and can be rolled, drawn, or extruded. It is used in several applications, such as cryogenic, aircraft and satellite structures. For that reason, Beryllium may be an attractive material, but it is rather expensive for a massive use, and it will not be considered further.

Characteristic properties of CFRP and GFRP include high tensile strength and high modulus of elasticity coupled with a relative low weight. The anisotropic variation of their mechanical properties (depending on the fibre amount and orientation) has to take in account in the design phase. In practice, they tend to have higher tensile strengths than compressive strengths.

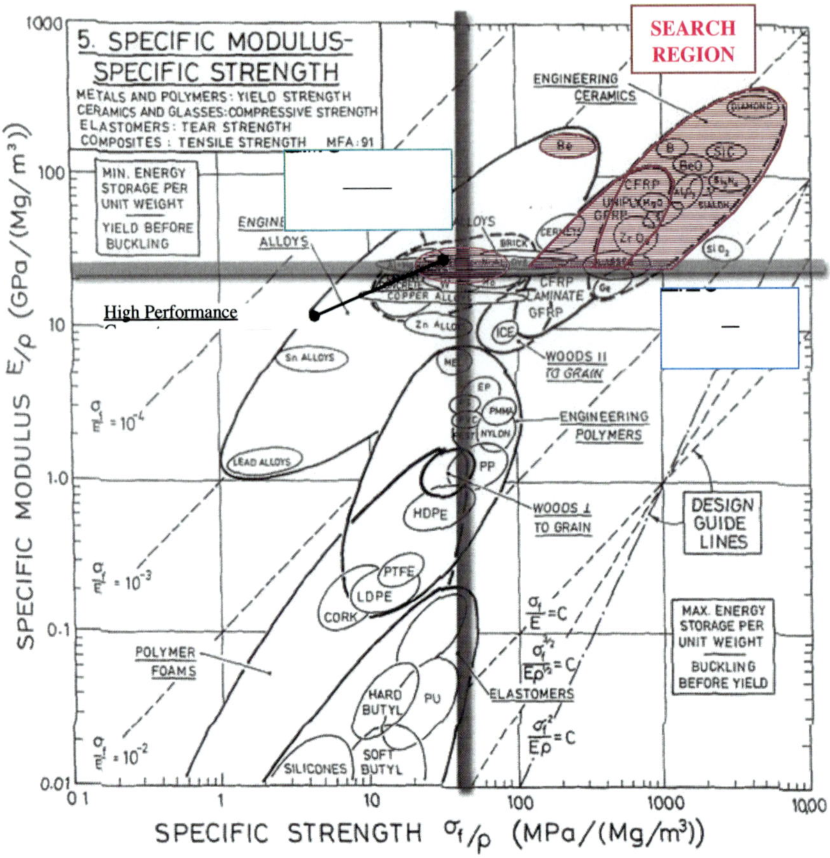

Figure 6.3. Ashby material selection chart with specific strength along the abscissa and specific modulus along the ordinate. The shaded green band shows the primary constraints. [Reproduced by permission of Prof. M. F. Ashby].

CFRP composites is an advanced composite materials that finds several applications today, but it still represents a costly solution, unlike GFRP. The principal benefits of glass fibre reinforced plastic composites are corrosion resistance, lightweight and ease of handling. Applications of glass fibre composite are increasing rapidly, along with the concurrent need for research in the area. Moreover, GFRPs have become attractive candidates for applications in offshore applications because they are corrosion resistance and exhibit good fatigue performance (there are several examples of successful and innovative applications of composite materials in offshore oil industry).

High Performance Concretes are very high compressive strength reinforced concretes [6.6, 6.7]. The incorporation of steel fibres (4-6%) in the extremely dense matrix (with quartz sand or bauxite) provides also ductility and low permeability (low chloride penetration there will be low corrosion of the reinforcing bars). High Performance Concretes are used in offshore industry and are being tested in offshore wind energy applications, where their dynamic behaviour is currently under investigation.

We now analyze the effect of this choice on the tower stiffness. The first natural bending frequency of a cantilevered tubular structure is given by [6.8]:

$$\omega_I = 1{,}75 \left[\frac{EI}{h_H^3 m_T \left(\frac{m_{top}}{m_T} + \frac{1}{4} \right)} \right]^{\frac{1}{2}} \tag{6.15}$$

where m_{top} is the nacelle and rotor mass. By introducing eq. (6.12) and eq. (6.1) within eq. (6.15), we obtain:

$$\omega_I \propto \frac{D_T}{H_h^2} \sqrt{\frac{E}{\rho_t}} \tag{6.16}$$

The material index governing the tower first natural frequency (M_3) is the root of the ratio of the elastic modulus to the material density:

$$M_3 = \sqrt{\frac{E}{\rho_T}} \tag{6.17}$$

According to the kind of wind turbine (constant or variable speed) and the design strategy, either a stiff or a soft tower may be required. We are not discussing here the design approach to the problem, but we want to stress the fact that a *change in the material will introduce some effects on the natural frequency,* that has to be taken into consideration for the discussion on tower mass minimization, and for the possible interferences with rotor frequency.

Table 6.1 presents a comparison based on the three material indexes M_1, M_2 and M_3 for five offshore wind turbine candidate materials: S533 steel, reinforced E concrete, High Ultra Performance Concrete (HUPC), High Strength Concrete CRC and the unidirectional GFRP composite. Steel S533 is taken here as baseline.

According to the constraint imposed by the maximum allowable tensile axial stresses (index M1) Concretes lead to lighter towers (with the same geometry and subjected to the same loadings) compared to steel towers. However, GFRP could lead, in theory, up to 80% savings of material, but the very high material safety coefficient adopted (7.4 for the composite analyzed), offsets the gain compared to steel.

Compressive strength and stability constraint (index M2), show the superior performance of steel compared to GFRP and Concretes. Low E/_ values and the unfavourable material safety factors, whose uncertainty is exacerbated in marine environments, lead to a huge mass increase for the GFRP based design (about 15 times) compared to steel. If we further consider the costs of the GFRP, being in an about 2:1 ratio with steel, the above figures will almost double.

High resistance concrete offers comparable axial stress performance compared to steel (both in compressive and tensile load), but the not satisfactory bucking verification requirements because of the 10% to 90% increased wall thickness, depending on the structure and fabrication process. Moving to M_3 index, thicknesses up to 90% are required to compensate for the reduction in the material frequency of the baseline steel design, depending on the reinforced concrete used. A somewhat smaller value of the mass increase (54%, but also a diameter increase has to be included, which bring the estimate to 62%), is reported in reference [6.11, 6.12]. From the same report, the manufacturing and erecting cost

Table 6.1. Effect of material indexes on the mass reduction for equal loading and geometrical indexes compared to the baseline steel case [UTS – ultimate tension strength, UCS – ultimate compression strength].

Material	Density [Mg/m³]	UTS (UCS) [MPa]	E [GPa]	M1	M2	M3	safey factors [g_m]	mass ratio (M1)	mass ratio (M2)	mass ratio (M3)
Steel [6.10]	7,9	355,0	210,0	45,2	26,8	5,2	1,1	1	1	1
Unidirectional GFRP Composite [6.9]	1,8	566 (313,0)	21,5	314 (174)	11,9	3,5	7,4	1,0 (1,8)	15,1	1,5
Reinforced Concrete (class E) [6.10]	2,5	550 (>40,0)	>38,5	220 (16,0)	15,4	3,9	1,2	0,22 (3,1)	1,9	1,3
High Ultra Performance Concrete (HUPC) [6.7]	2,3÷2,8	(110,0÷210,0)	35,0÷70,0	(48,9÷76,4)	15,6÷25,5	3,9÷5,0	1,2	(0,6÷1,0)	1,9÷1,1	1,3÷1,0
Compact Reinforced Concrete (CRC) [6.6]	2,6÷3,0	(100,0÷400,0)	40,0÷80,0	(38,5÷133,3)	15,4÷26,7	3,9÷5,2	1,2	(0,4÷1,3)	1,1÷1,9	1,0÷1,9

of the CRC design is comparable to that of steel tower, but maintenance has to be considered. By looking at the M_3 index, the tower's fundamental frequency is less that that of the steel tower for the listed materials.

6.2 Blade Material Design

Blades are an expensive part of a wind turbine and their mass can drive the design and costs of other parts of the system. Reduction of blade mass has been the focus of much design effort. The general design principle so far adopted is of a stressed member in the form of a box or a spar cross-section with an aerodynamically shaped outer shell. From a structural point of view, simple beam theory is applied here for the structural analysis. It is assumed that blade loads are carried by a hollow circular spar with diameter equal to the thickness of the airfoil section [6.13]. Spars and aerofoil sections should be designed to offer adequate fatigue strength and sufficient structural stiffness to ensure that the frequencies are far from that of the rotor. Compliant systems are also considered in a comprehensive design strategy, but will not treated here. It is assumed that the blade mass is lumped with the spar. This mass drives also the additional blade shell mass. We can write:

$$m_B = \int_0^{\frac{D_R}{2}} \frac{\pi d^2}{4} \rho_S dr = 1.25\pi^3 \rho_S \frac{R^3}{\lambda^4 C_{L,R}^2} \left(\frac{t_B}{c}\right)^2 K \tag{6.18}$$

where ρ_S is the spar material density, λ is the tip speed/wind speed ratio, $C_{L,R}$ is the lift coefficient of rotor, c is the mean chord of blade, and t_B is the mean thickness of airfoil (=diameter of spar). K is a constant grouping the linear assumed variation of the lift coefficient, and the thickness to chord ratio.

A number of equivalent forces and moments are experienced by the blade of a typical horizontal-axis wind turbine during normal operation, such as:

- *aerodynamic loads;*
- *gravity loads;*
- *centrifugal forces and Coriolis forces due to rotation;*
- *gyroscopic loads due to yawing.*

The aerodynamic loads can cause torsion of the blades because the aerodynamic centre cannot be coincident with the structural axis. This fact presents a difficulty in the design with composite construction. The gravitational loading causes either direct 'compression/tension' or 'bending in plane' depending upon the blade azimuthal position. Inertial loads will also cause tension stats in the blades.

In this section a simplified method to determine blade mass based on bending criteria will be presented.

To model the aerodynamic loads, the wind turbine blade can be considered a cantilevered beam uniformly loaded by the wind. The wind flow near the wind turbine rotor is complex, but a static approach can be used for a preliminary analysis of the problem of mass minimization. Thrust force on the rotor can be calculated from elementary momentum theory as:

$$F_T = C_T \frac{1}{8} \rho_a \pi V^2 D_R^2 = \frac{\pi}{9} \rho_a V^2 D_R^2 \tag{6.19}$$

where the thrust coefficient (C_T) should be approximately equal to 8/9 for optimum work harvest. Because of the thrust force, blades tend to deflect out of the rotor plane and the deriving flapwise bending moment can be expressed by:

$$M_T = \frac{F_T}{B}\frac{5}{8}R_R = \frac{5\pi}{144}\frac{1}{B}\rho_a V^2 D_R^3 \tag{6.20}$$

where B is number of blades, F_T/B expresses the thrust force per blade and 5/8R gives the centre of aerodynamic pressure for the whole blade [6.14].

The edgewise bending moment is given by the aerodynamic torque and gravitational loads. The average torque M_Q can be calculated as the power output P_{out} divided by rotational speed W

$$M_Q = \frac{P_{out}}{\Omega} = \frac{1}{4}\rho A_R V^2 D_R \frac{C_P}{\lambda} = \frac{\pi}{16}\rho V^2 D_R^3 \frac{C_P}{\lambda} \tag{6.21}$$

where C_P is the turbine power coefficient. The edgewise bending under gravity is given by:

$$M_G = F_{G,B} r_G = m_B g r_G \tag{6.22}$$

where g stands for the gravitational acceleration and r_G is the distance of the blade centre of mass from the rotational axis.

In terms of blade strength, torque and gravitational bending moments are generally less important than their flapwise counterparts [6.14, 6.15, 6.16]. Therefore, blade mass will be designed on the flapwise bending loads. The maximum bending stress (σ_b) is:

$$\sigma_b = \frac{M_T}{I}\left(\frac{t}{2}\right) = \frac{M_T}{ct_B^3/12}\left(\frac{t_B}{2}\right) = \frac{5?}{24}\sigma_a V^2 \frac{D_R^3}{ct_B^2 B} \tag{6.23}$$

The rotor solidity is related to the tip speed ratio and the lift coefficient via the following formula [6.12]:

$$S_R = \frac{Bc}{\pi R} = \frac{5}{6\lambda^2 C_{L,R}} \tag{6.24}$$

Following the procedure outlined in the tower rotor paragraph, we can write the limit strength condition:

$$\rho_b \leq \frac{\sigma_{lim}}{\gamma_f \gamma_m}$$

where σ_{lim} is the characteristic performance limit of the composite material considered for the application. For design purpose, the structural integrity of a wind turbine can be represented by a set of design situation (normal, extreme operation, shutdown, transport,...) covering the most significant conditions which the turbine may experience [6.11]. For each design situation, the appropriate type of analysis is required: the ultimate and fatigue strength of blades (and other structural members) must be verified.

If a parked wind turbine is considered, as for the previous analysis of the tower, failure analysis with ultimate loads must be considered. If we focus on this condition and by substitution of eq. (6.23) and (6.24) in eq. (6.18), we obtain the functional equation for the blade mass:

$$m_B \propto \left(\rho_a V^2 \gamma_f\right)\left(D^3 \lambda^2 C_{L,R}\right)\frac{\gamma_m}{\frac{\sigma_f}{\rho_s}} \tag{6.25}$$

Loads (γ_f) and material (γ_m) partial safety factor are here related to the blade component.

We can now trace the guideline M_1=200,6 index for a common for GFRP (36%) on the Ashby diagram of Figure 6.4.

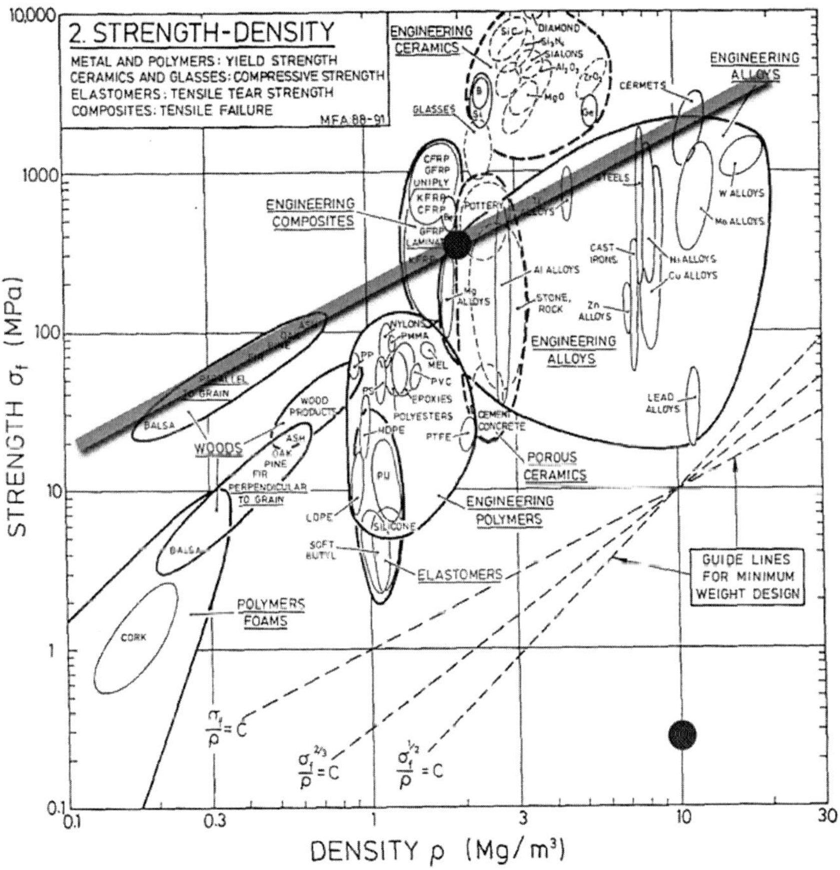

Figure 6.4. Ashby material selection chart and primary constraints (the shaded blue band). [Reproduced by permission of Prof. M. F. Ashby].

Again if the loading and geometrical parameters are not negotiable, high values of the material index M_1 will lead to less mass. Similar considerations developed for the tower can now be replicated for the blade spar, considering buckling and eigenfrequency issues, since the same performance indexes will apply.

A quantitative analysis is difficult to be made for the blade because materials are much less standardized than for other applications. A very wide scatter of data for composite materials are found in the literature. Texture orientation, % of fibres and fabrication methods lead to a large range of available materials. Furthermore, composites exhibit anisotropic behaviour, and the material performance can be tailored on the application. The Young modulus for instance, can easily range from 3 to 40 MPa. It has to be emphasized that the safety factors appearing in eq. (6.25) will vary from material to material and from fibre texture to fibre texture. There is a wide range of data given by advanced materials manufacturers, therefore, as suggested by norms [6.11], better safety factors should be adopted for materials that are not routinely characterized. These large safety factors can offset completely the theoretical gain achievable in mass reduction. Other constraints could be of prior importance in general for offshore situations, such as resistance to corrosion, and the possibility of minimizing the assembly work offshore.

Although nowadays GFRP is the favoured material for internal spars and the external shells, other realistic candidate materials for a lightweight blade, such as Kevlar and Carbon fibre reinforced plastics, can be considered. Kevlar shows conflicting properties: besides beneficial characteristics and strength and being very resistant to impact and abrasion damage, it tends to absorb moisture and its compressive properties are relatively poor. For these reasons, Kevlar fibres cannot substitute for Glass fibres in blade manufacture, unless appropriate protective systems are considered. It may be used as a protective layer on graphite laminates and can be mixed with graphite in hybrid fabrics to provide damage resistance, increased ultimate strains, and to prevent catastrophic failure modes.

Table 6.2 reports the comparative results of the material indexes for epoxy-GFRP (36%) and epoxy-CFRP (50%). At the moment, carbon fibre applications in wind turbine are widely studied: companies have high expectations for the near term, and the intensive and expensive tests lead to a proprietary knowledge, which is not disclosed. Reliable data retrieving is therefore difficult at this stage. As consequence, within the huge possibilities given by composites structure design, only two common consolidated designs are reported here to show the potential capability of new materials. A commercial bidirectional CFRP (50%), due to its excellent ultimate strength properties and elastic modulus, is an attractive solution because it allows a dramatic reduction of the blade mass. For the same loading and geometrical characteristics, the natural frequency is nearly halved compared to GFRP (36%), and this effect shall be analyzed on usual Campbell charts to identify possible dangerous interference with tower natural frequency.

The analysis presented in Table 6.2 is partial because the fatigue loads are not taken into consideration. A comparison on such basis needs a deep knowledge of the blade manufacturing details. Nevertheless, the example shows the potential of the method in investigating candidate materials. New Ashby-like tables can give evidence concerning the combination of properties governing the load case under consideration. Preliminary results show that for large offshore wind turbine blades, the use of carbon-fibre in the load carrying part is becoming a cost effective option.

Table 6.2. Comparison of common material and geometrical indexes for wind turbine blade applications [UTS – ultimate tension strength, UCS – ultimate compression strength].

Material	Density [Mg/m³]	UTS (CTS) [MPa]	E [GPa]	M1	M2	M3	safey factors	mass ratio (M1)	mass ratio (M2)	mass ratio (M3)
GFRP 36%, [±45°,0°] [6.9]	1,8	357,0 (364,0)	17,2	200,6 (202,2)	9,7	3,1	7,4	1	1	1
CFRP 50%, [0°, 90°]	1,6	600,0 (570,0)	70,0	375,0 (356,3)	43,8	6,6	3,7	0,27 (0,28)	0,11	0,47

7. CONCLUSIONS

The complex working environment of offshore wind turbines requires a strong material support programme for properties testing and component characterization. This activity is important to reduce the safety factors currently used, and to free the potential of new materials and fabrication methods.

Finite element approaches could help in transferring laboratory tests on small specimen to high thickness structures.

Fatigue testing should be carried out in real chemical and thermal conditions of the marine environments, and special programs should to be conceived to reduce the test duration without suffering of frequency effect.

8. REFERENCES

[1.1] Ancona, D., McVeigh, J., *"Wind Turbine – Materials and Manufacturing Fact Sheet"*, US Department of Energy, Princeton Energy Resources International LLC, 2001.

[2.1] Sutherland, H.J.*"On the Fatigue Analysis of Wind Turbines"*, Contractor Report SAND99-0089, Sandia National Laboratories, Albuquerque, NM, 1999.

[2.2] Mandell, J.F, Samborsky, D.D., *"DOE/MSU composite material fatigue database, 2003 update"*, Internal Report, Sandia National Laboratories, Albuquerque, NM, 2003.

[2.3] De Smet, B.J., Bach, P.W,. "DATABASE FACT, Fatigue of composites for wind turbines", *3rd IEA Symposium on wind turbine fatigue*, 21–22 April 1994.

[2.3] Manwell, J.F. et al., *"Wind energy explained: theory, design and application"*, John Wiley & Sons LTD, 2002.

[2.4] Veers, P.S. et al., "Trends in the Design, Manufacture and Evaluation of Wind Turbine Blades", *Wind Energy* 2003; 6:245–259.

[2.5] Lilleheden, L. "The quest for higher product reliability and performance by use carbon or smart engineering", *Global Outlook for Carbon Fibre 2004*, October 19-20, Hamburg, Germany.

[2.6] TPI Composites, "Innovative design approaches for large wind turbine blades", SAND 2003-0723, Sandia National Laboratories, Albuquerque, NM, 2003.

[2.7] Griffin, A., "Alternative composite materials for MW scale wind turbine blades: design consideration and recommended testing", *Wind Energy Symposium Proceedings*, pp. 191-201, ASME, 2003.

[2.8] SP Systems, *"Guide of composites"*, www.gurit.com, 1998.

[2.9] DNV/Riso, *"Guidelines for design of wind turbines"*, 2nd edition,2002.

[2.10] Jamieson, P., "Evolution of modern wind turbine rotors design and manufacturing", *Wind Turbine Blade Workshop*, Albuquerque Sheraton Uptown, February 24-25, 2004.

[2.11] Zvanik M., "Composites and Wind Energy: Partners in Performance", *SAMPE Journal*, 38, 2002.

[3.1] Harrison, R. et al., *"Large wind turbines: design and economics"*, John Wiley & Sons LTD, 2000.

[3.2] DNV/Riso, *"Guidelines for design of wind turbines"*, 2nd edition, 2002.

[3.3] Manwell,J.F. et al., *"Wind energy explained: theory, design and application"*, John Wiley & Sons LTD, 2002.

[3.4] NWTC, *"GuidelineDG03, Wind turbine design, yaw & pitch rolling bearing life"*, National Renewable Energy Laboratory, Certification team, 2000.

[3.5] SKF, *"General catalogue"*, Denmark, 1989.

[4.1] Harrison, R. et al., *"Large wind turbines: design and economics"*, John Wiley & Sons LTD, 2000.

[4.2] Tech-wise, *"Offshore wind turbine towers in high strength concrete"*, Report 121020, 2002.

[4.3] DNV/Riso, *"Guidelines for design of wind turbines"*, 2nd edition,2002.

[4.4] Thick,J., *"Offshore Corrosion Protection of Wind Farms"*, October 2004.

[4.5] Petersen, P., Nielsen, K.B., Feld, T., "Design basis for offshore wind structures", *Det Norske Veritas, Copenhagen offshore wind*, Denmark, 2005.

[4.6] Lipman, N.H., Musgrove, P.J., Pontin, G.W., *"Wind energy for the Eighties"*, BWEA, 1982.

[4.7] DNV, *"Design of Offshore Wind Turbine Structures"*, 2004.

[4.8] Germanischer Lloyd, *"Offshore Wind Energy Conversion Systems Guidelines"*, 1999.

[4.9] DEA, *"The Danish Energy Agency's Approval Scheme for Wind Turbines, Recommendation for Technical Approval of Offshore Wind Turbines"*, 2001.

[5.1] IEA-Annex XIX , *"Cold climate turbines operating worldwide"*, www.virtual.vtt.fi, 2008.

[5.2] Manitoba Hydro ,*"Clarification of Wind Turbine Cold Weather Considerations"*, EXHIBIT MH-NCN-1031, 2004.

[5.3] Dutta, P.K. "Structural Fiber Composite Materials for Cold Regions," *Journal of Cold Regions Engineering*, pp. 124 – 134, Vol. 2, No. 3, September 1988.

[5.4] Karbhari, V.M., Rivera, J. and Dutta, P.K. (2000). "Effect of Short-Term Freeze-Thaw Cycling on Composite," *Journal of Composites for Construction*, pp.191-197, Vol. 4, No. 4, November 2000.

[5.5] Lacroix, A., Manwell, J.F., *"Wind energy: cold climates issues"*, 2000.

[5.6] Wigley, D.A., *"Mechanical properties of materials at low temperatures"*, Plenum press New York-London. 1971.

[5.7] Laakso, T., Holttinen, H., Ronsten, G., Horbaty, R., Lacroix, A., Peltola, E., Tammelin, B., *"State-of-the-art of wind energy in cold climates"*, http://arcticwind.vtt.fi, 2003.

[5.8] Campbell, J.E., "Structural alloys at subzero temperatures", in *Metals Handbook*, desk edition, American Society for Metals, 1985, pp.20.24-20.34.

[5.9] Battisti, L., Fedrizzi, R., Dal Savio, S., Giovannelli, A., "Influence of the and Size of Wind Turbines on Anti-icing Thermal Power Requirement ", *Proceedings of EUROMECH 2005 Wind Energy Colloquium*, 4-7 Oct. 2005, Oldenburd – Germany, Springer Verlag 2007.

[5.10] Battisti, L., Fedrizzi, R., Brighenti, A., Laakso, T., "Sea ice and icing risk for offshore wind turbines", *OWEMES*, 20-22 April. Citavecchia, Italy, 2006.

[5.11] Smith, W.F. *"Foundations of Materials Science and Engineering"*, Mc-Graw Hill, p. 247 (1993).

[5.12] Baumeister, T., Editors, *"Marks' Standard Handbook for Mechanical"*, 1978.

[5.13] Diemand, D., "Lubricant at Low Temperatures", *CREEL Technical Digest TD 90-01*, Cold Regions Research & Engineering Laboratory, Hanover, New Hampshire, 1990.

[6.1] Ashby, M F. *"Materials Selection and Process in Mechanical Design"* Butterworth Heinemann, Oxford, 1999.

[6.2] Ashby, M.F., Cebon, D. *"Case studies in Materials Selection"*, Second Edition, Butterworth-Heinemann, Oxford, 1999.

[6.3] Ashby, M.F., Jones, D.R.H., *"Engineering Materials 1"*, Second Edition, Butterworth Heineman, Oxford, 1996.

[6.4] Ashby, M.F., Jones, D.R.H., *"Engineering Materials 2"*, Second Edition, Butterworth Heineman, Oxford, 1998.

[6.5] Nijssen, R.P.L, et al., *"The application of scaling rules in up-scaling and marinisation of a wind turbine"*, EWEC, Copenhagen, Denmark, 2001.

[6.6] Tech-wise A/S, *Summary of report on CRC for off shore wind turbine*, 2002, www.crc-tech.com.

[6.7] Lohaus, L., Anders, S., "High-cycle Fatigue of Ultra-High Performance Concrete and Grouted Joints for Offshore Wind Energy Turbines", *Wind Energy Proceedings of the Euromech Colloquium*, 2007.

[6.8] Baumeister, T., Editors, *"Marks' Standard Handbook for Mechanical"*, 1978.

[6.9] Mandell, J.F, Samborsky, D.D., *"DOE/MSU composite material fatigue database, 2003 update"*, Internal Report, Sandia National Laboratories, Albuquerque, NM, 2003.

[6.10] DNV/Riso, *"Guidelines for design of wind turbines"*, 2nd edition,2002.

[6.11] Standard IEC 61400-1, *Part 1: Design requirements*, 2007.

[6.12] Battisti, L., Soraperra, G., *"Analysis and application of pre-design methods for HAWT rotors"*, EWEC, Milano, Italy, 2007.

[6.13] Battisti, L., Hansen, M.O.L., Soraperra, G., "Aeroelastic simulations of an iced MW-Class wind turbine rotor", *Proceedings of the VII BOREAS Conference*, Saarisalka, Finland, 2005.

[6.14] Milborrow, D.J., "Towards lighter wind turbines", *8th BWEA*, Cambridge, Wind Energy Convertion, 1886.

[6.15] Harrison, R. et al., *"Large wind turbines: design and economics"*, John Wiley & Sons LTD, 2000.

[6.16] Manwell, J.F. et al., *"Wind energy explained: theory, design and application"*, John Wiley & Sons LTD, 2002.

GLOSSARY

CENELEC	European Electrotechnical Commission
DENA	Deutsche Energie-Agentur GmbH (German Energy Agency)
DAF	dynamic amplification factor
DNV	Det Norske Veritas
EC	European Commission
EU	European Union
E&W	England and Wales
FRP	Fibre reinforced plastics
GL	Germanischer Lloyd
GRP	Glass-fibre reinforced polyester
HVAC	high-voltage alternating current
HVDC	high-voltage direct current
IEA	International Energy Agency
IEC	International Electrotechnical Commission
LIDAR	light detection and ranging
NORDEL	(Nordic countries' transmission system operators)
OWEMES	Offshore Wind Energy in Mediterranean and European Seas
RES	renewable energy sources (or supplies)
RESE	renewable energy sources (supplies) [of] electricity
RPM	revolutions per minute
SODAR	sonic detection and ranging
S/N	stress versus number (of cycles)
SASH	(in Swedish) soft boarding to offshore structures
SWATH	small water access twin hull
TLP	tensioned leg platform
TSO	transmission system operator
UCTE	Union for the coordination of transmission of electricity
VARTM	Vacuum assisted resin transfer moulding

INDEX